Printed in the United States of America

Macmillan Publishing Company
866 Third Avenue, New York, New York 10022

Collier Macmillan Canada, Inc.

Library of Congress Cataloging in Publication Data

Helstrom, Carl W.
Probability and stochastic processes for engineers.

Bibliography: p.
Includes index.
1. Engineering mathematics. 2. Probabilities.
3. Stochastic processes. I. Title.
TA340.H425 1984 519.2 83-937
ISBN 0-02-353560-1

Printing: 2 3 4 5 6 7 8 Year: 4 5 6 7 8 9 0 1 2

ISBN 0-02-353560-1

CARL W. HELSTROM

Department of Electrical Engineering
and Computer Sciences
University of California, San Diego

Probability and Stochastic Processes for Engineers

Macmillan Publishing Company
New York
Collier Macmillan Publishers
London

PREFACE

The structure of a book on probability is largely dictated by that of the subject itself, and this book is no exception. Five chapters cover, in order, basic concepts, distributions of a single random variable, bivariate distributions, multivariate distributions, and stochastic processes. An entire chapter is devoted to the properties of a pair of random variables because most of the concepts needed for working with arbitrary numbers of random variables arise with two, and visualization is easier in two dimensions than in n. Examples are drawn from engineering and physics whenever possible, but are of course idealized lest technical details distract from the principles to be illustrated. How a communications receiver acts as a decision-making device is treated in the first chapter not only because of its engineering interest, but also because it provides an instructive application of the concept of conditional probability. Decision problems recur here and there throughout the book.

This is not intended to be a *Handbuch der Wahrscheinlichkeitsrechnung:* amid a plethora of special results, attention to fundamental ideas is liable to be diluted. Some elementary but useful deductions from the theory are featured in certain examples and among the problems at the end of each chapter. Although I am not above tossing coins and dice when necessary, I prefer—*pace* the electronics industry—to let diodes and transistors fail when certain probabilistic concepts need exemplification. I draw no balls from urns, however, and leave for courses on combinatorics the combinatorial problems that occupy so much space in some textbooks on probability.

Because probability is important in modern engineering, engineers should be aware of the philosophical problems that attend it, and the principal interpretations of probability have been described at some length in Sec. 1-4. I am under no illusion that these have been set forth with impeccable logic, and ardent advocates of a particular interpretation may be dissatisfied. The reader should peruse this section lightly at first, but should return to it now and then as his or her experience with probability and its applications grows and matures. A bibliography of works on the foundations of probability appears at the end of Chapter 1.

The customary table of the error-function or normal probability integral is not to be found in this volume. Instructions are instead given for computing it, as well as the binomial and Poisson distributions, on a programmable calculator. Calculators are as common among engineers these days as slide rules were when I went to school, and students should be encouraged to employ them for something more elaborate than simple arithmetic.

I have taught this material, with a few omissions of minor topics, in a two-quarter course for electrical engineering juniors; indeed, the book started as a set of notes for that course. It could perhaps be covered in one semester if Chapter 4 on multivariate distributions were passed over lightly with the remark that most of what was done in Chapter 3 with a pair of random variables is readily extended to any number of variables. Matrix notation is extensively utilized in Chapter 4, particularly in connection with the multivariate Gaussian distribution. The first four chapters require little more mathematics than engineers usually learn during the first two years at a good engineering school. The chapter on stochastic processes also presumes elementary notions of Fourier transforms and linear systems and circuits.

I am immeasurably grateful to Mrs. Pat Norvell for typing the original set of notes and for cheerfully incorporating my numerous corrections and revisions.

CARL W. HELSTROM

La Jolla, California

CONTENTS

CHAPTER 1

Probability

1-1

INTRODUCTION

Tune your radio to a point on the dial where no stations are broadcasting, and turn up the gain control. You will hear a hissing sound called "radio noise." If you replace the speaker with an oscilloscope that records the output of the audio amplifier, it will in the course of time t trace an irregular curve that never repeats itself precisely and cannot be represented by any simple function $f(t)$. Nevertheless, this fluctuating signal seems to possess a certain structure: the amplitude of its excursions stays within vaguely definable bounds and varies widely only during intervals inversely proportional to the bandwidth of the receiver. Observing the outputs of a number of identical receivers tuned to the same frequency, you see that they differ in detail, but exhibit similar characteristics.

These noise fluctuations are caused by the thermally agitated ions and electrons in the components of the receiver and by the excitation of its antenna by random electromagnetic fields arising from the chaotic motions of the constituents of all the surrounding matter. In order to describe and analyze radio noise, we must develop some way of handling the random uncertainty that characterizes it, and that we can do by means of the theory of probability.

Random noise interferes with the reception of weak radio signals. If you are trying to pick up messages from a remote satellite whose radio transmitter broadcasts with only a limited power, the signals may be masked by the noise. Turning up the gain is of no use; the noise is merely amplified along with the signals. Narrowing the passband of the receiver reduces the noise, but by blurring the signals may render them indistinguishable. If the satellite communication system transmits information coded into strings of binary digits, 0's and 1's, the noise causes errors in interpreting the received signal: 0's may be turned into 1's and 1's into 0's. These errors will occur in an erratic fashion and with a frequency depending on the relative strengths of signal and noise. The degree to which the system is prone to error is measured in terms of probability.

In designing the receiver to minimize error, the random behavior of both the signals and the noise is analyzed by means of probability theory. Indeed, the sequence of 0's and 1's that represents the information transmitted by the satellite must also be dealt with in terms of probability. It is an encoding of certain messages to be dispatched to a recipient at a certain rate and with a certain fidelity or reliability. If the sender always said the same thing, there would be no need for the communication system; the less predictable the messages, the more information they carry. Information content and rate of transmission can be quantified through a probabilistic analysis of the messages to be transmitted. They are essential factors in the design of the communication system.

A radar system suffers similar interference by random noise. The echo from a distant airplane or missile may be so weak that it is nearly lost amid the random electromagnetic fluctuations affecting the radar receiver. How should the receiver be designed in order to reduce as far as possible the chance of missing a target? The radar echo contains information about the distance of the target and its rate of approach. The noise introduces random errors into that information, and the designer must minimize them. The radar antenna is to be steered by a control system utilizing the information from the radar in order to track targets moving in azimuth and elevation. The motion of each target contains an element of unpredictability that must be assessed and taken into account in designing the antenna-control mechanism and the electronic system that guides it. Concepts of probability permeate all these design and analysis problems.

Communication and radar systems, like many others, are built of electronic components such as diodes, transistors, and integrated circuits. The components are made by processes that ought to produce identical objects, which when installed would function just as the designer anticipated. That never happens. Erratic variations in raw materials, human fallibility, and uncontrollable deviations in temperatures and flow rates introduce random variations into the manufactured devices, variations that may be so severe that some do not work at all. Others may have physical properties too remote from the norm to behave as the system designer requires. The components must be sampled and tested as they come off the production line. How can that be done economically, and how should the prices charged for them be linked to guarantees of their reliability? Such problems are attacked by means of the theory of probability.

The foreman of this production line likes to play poker after hours. He holds three kings, and someone else has been raising his bet. What should he do? If he is a scientific player, he knows the probability that his opponent holds a better hand than his, and he determines his strategy accordingly. The theory of probability is said to have originated in problems of gambling. Its applications have much ramified since that day in the seventeenth century when the Chevalier de Méré first consulted Blaise Pascal about an obscure betting situation. Probability theory has become indispensable in many an engineering and scientific analysis.

We shall start with basic definitions and introduce one by one the mathematical concepts and techniques needed for coping with probabilistic problems in science and engineering. Often these must be illustrated by and practiced with examples that seem remote from those endeavors, but it is only by beginning with simple situations and working intensively and extensively with the concepts and techniques that you can make them a useful part of your way of reasoning about technical problems.

1-2

THE STUFF OF PROBABILITY: EXPERIMENTS, OUTCOMES, EVENTS

A. Chance Experiments

A theoretical science deals with concepts that are abstractions from the varied situations or phenomena to which it applies—abstractions that preserve the essential and discard the merely accidental. The basic entity treated by the theory of probability is called a *chance experiment*. It is a happening or proceeding that at least conceptually can be repeated arbitrarily often under identical circumstances. It must always have an identifiable outcome belonging to a fixed, known set of possibilities. It is called a "chance" experiment because which of the outcomes actually occurs is fortuitous and unpredictable.

Individual outcomes of the chance experiment will be designated by Greek letters such as ζ, sometimes carrying identifying subscripts, as in ζ_1, ζ_2, and so on. The whole set of possible outcomes is called the *universal set* and is denoted by S. The experiment itself may be designated by a letter such as E. A single instance of performing it is called a *trial* of E. Here are some examples.

E_a: A coin is tossed. The outcomes are ζ_1 = heads (H), ζ_2 = tails (T); $S = (\zeta_1, \zeta_2) = (\text{H, T})$.

E_b: A coin is tossed three times in succession. Outcomes are triplets of heads or tails:

ζ_1	ζ_2	ζ_3	ζ_4	ζ_5	ζ_6	ζ_7	ζ_8
HHH	HHT	HTH	HTT	THH	THT	TTH	TTT

There are eight elements in the universal set S.

E_c: A pentahedral die, whose five faces are labeled with the integers 1 through 5, is cast. ζ_k is the outcome in which the die lands on face k, $k = 1, 2, 3, 4, 5$. The set S has five elements.

E_d: Both a pentahedral and a heptahedral die, whose faces are labeled with integers as in E_c, are cast. The set S has thirty-five elements: (1, 1), (1, 2), . . ., (5, 4), (5, 5), (5, 6), (5, 7).

E_e: A device counts the number of photoelectrons emitted from a photoelectrically emissive surface when a light beam of fixed intensity falls on it for T seconds. An outcome ζ is the number n of emitted electrons counted: ζ_0 = (0 e's counted), ζ_1 = (1 e counted), . . . ζ_n = (n e's counted), and so on. The outcome can just as well be designated by the number n of electrons counted, and only logical nicety distinguishes ζ_n from n. The set S has an infinite number of elements.

E_f: A radio receiver is turned on at a certain time, which we designate as $t = 0$; τ seconds later the voltage v is measured between a certain point in the circuit and ground. Here the value of v itself identifies the outcome ζ of the experiment: $\zeta = v(\tau)$. The range of values of v to be expected in practice may be limited by saturation in electronic or magnetic elements in the circuit or by other conditions. Conceptually we allow the possibility that v may lie in an infinite interval, $-\infty < v < \infty$. Under the assumption that v can be measured with perfect accuracy, the universal set S contains an infinite number of elements, but the infinity is of a higher order than that in E_e.

E_g: The voltage at the same point of the circuit as in E_f is measured at two times τ_1 and τ_2. An outcome is now a pair of numbers, $\zeta = (v_1, v_2)$, with $v_1 = v(\tau_1)$, $v_2 = v(\tau_2)$. We can imagine representing it as a point in a two-dimensional Cartesian space, or plane. In general, S will be the entire plane.

E_h: The voltages at three different points a, b, and c of the circuit of the receiver in E_f are measured at the same time τ. An outcome is now a triplet of numbers $\zeta = (v_a, v_b, v_c)$, which we can represent as a point in a three-dimensional Cartesian space, and S is the entire space.

E_i: A deck of fifty-two cards is shuffled and dealt out among four bridge players, each receiving thirteen cards. Here ζ is a particular distribution of cards among the players; there are $52!/(13!)^4 \doteq 5.364 \times 10^{28}$ such outcomes if we disregard the order in which the players receive and arrange their cards. ($n! = 1 \cdot 2 \cdot 3 \cdot \ldots \cdot n$ = "n factorial.")

E_j: The people arriving at a bank during one hour are observed, and the time each stays in the bank to complete his or her business is recorded. An outcome is a sequence of indefinitely many numbers of minutes. The experiment might be repeated on successive weekdays.

E_k: The sequence of N digits (1's and 0's) emitted by a computer and sent over a communication line to a second computer is recorded during an interval of duration T. There are 2^N outcomes all told.

E_l: Both the sequence of N digits transmitted as in E_k and the sequence as received at the second computer are recorded and compared and the number of errors noted.

B. Events

A collection of outcomes ζ of a chance experiment E forms a subset of the universal set S and is called an *event*. We designate events by capital letters A, B, . . ., possibly with subscripts. An event is defined by the outcomes it contains. The event occurs whenever any outcome in it occurs. If an outcome ζ is in an event A, we write $\zeta \in A$. Here are some examples of events in experiments among those just listed.

E_b: A = "the coin first comes up heads" = $\{\zeta_1, \zeta_2, \zeta_3, \zeta_4\}$
B = "the same face shows in the second and third tosses" = $\{\zeta_1, \zeta_4, \zeta_5, \zeta_8\}$
C = "exactly two tails appear" = $\{\zeta_4, \zeta_6\ \zeta_7\}$
D = "exactly two heads appear" = $\{\zeta_2, \zeta_3, \zeta_5\}$

E_c: A = "the die falls on an odd-numbered face" = $\{1, 3, 5\}$
B = "the die falls on either face 1 or face 4" = $\{1, 4\}$

E_d: A = "both dice fall on even-numbered faces" = $\{(2, 2), (2, 4), (2, 6), (4, 2), (4, 4), (4, 6), (6, 2), (6, 4), (6, 6)\}$
B = "the sum of the numbers on the bottom faces is 10 or more" = $\{(5, 5), (5, 6), (5, 7), (4, 6), (4, 7), (3, 7)\}$

E_e: A = "an odd number of electrons are counted" = $\{1, 3, 5, 7, \ldots\} = \{n : n \text{ odd}\}$*
B = "more than five, but fewer than fifteen electrons are counted" = $\{6, 7, 8, 9, 10, 11, 12, 13, 14\} = \{n : 6 \le n \le 14\}$
C = "fewer than ten electrons are counted" = $\{n : 0 \le n \le 9\}$

E_f: A = "the voltage lies between 5 and 10 volts" = $\{v : 5 < v < 10\}$
$B = \{v : 4 \le v < 8\}$
$C = \{v : 1 < v < 2 \text{ or } 3 < v < 5\}$

E_g: $A = \{\zeta : 1 < v_1 < 2 \text{ and } 3 < v_2 < 5\}, \zeta = (v_1, v_2)$
$B = \{\zeta : 1 < v_1 + v_2 < 3\}$
$C = \{\zeta : 0 < v_1 v_2 < 5 \text{ and } v_1 + v_2 > 2\}$

E_i: A = "player A receives three aces"
B = "players B and C receive the same number of spades"
C = "players A and D receive no aces at all"

E_j: A = "three people out of five take less than 5 minutes in the bank"
B = "five people out of twelve take more than half an hour in the bank"

E_k: A = "the sequence of N digits contains j 0's and $N - j$ 1's"
B = "each 1 is followed by k zeros"
C = "the binary sum of the first three digits equals 1"

E_l: A = "the first three transmitted digits are 101 and are received as 111"
B = "the number of errors in the N digits equals k"
C = "five successive digits are received in error"

The *empty set,* which contains no elements, is denoted by $\varnothing$. It is needed for completeness. In experiment E_f, for instance,

$$\{v: v > a \text{ and } v < b\} = \varnothing$$

when $a > b$.

The simple experiment E_a has only four events, $\varnothing$, {H}, {T}, S, where by {H} we mean the event (subset) with a single outcome "heads"; {T} is the event with the single outcome "tails."

Experiment E_b is richer in events. Each of the eight outcomes can be either included or excluded from the subset defining an event, and there are therefore $2^8 = 256$ events, including $\varnothing$ and S. In a chance experiment with N outcomes, 2^N different events can be defined.

*The colon (:) in such expressions means "such that."

Experiments E_e, E_f, E_g, E_h, and E_j have an infinitude of events, most of which cannot even be defined by any simple enumeration or set of inequalities.

In events E_a through E_e, E_i, E_k, and E_l the outcomes are countable; that is, they can be put into one-to-one correspondence with the integers. The outcomes of such experiments are said to be *discrete*. We can then identify events that contain a single outcome. Thus in E_b, {HHH}, {HTT}, and so on, are such events, of which there are eight. In E_e all events of the form "k e's are counted" contain a single outcome. We call such events *elementary* or *atomic events*. Events with more than one outcome are *composite events*.

In experiments E_f, E_g, and E_h the outcomes are not countable and are said to form a *continuum*. We refrain from identifying atomic events in such experiments.

C. Operations with Sets

New events (subsets of S) can be constructed from old ones by various operations of set theory, which provides us with a useful notation. These operations are the following.

(a) *Complementation:* The elements ζ of S not in a set A constitute its *complement* $A' = \{\zeta : \zeta \notin A\}$. (Read $\zeta \notin A$ as "ζ is not an element of A.") In particular, $\varnothing = S'$, $S = \varnothing'$, $(A')' = A$.

(b) *Intersection:* The outcomes ζ in both event A and event B form their *intersection*

$$A \cap B = \{\zeta : \zeta \in A \text{ and } \zeta \in B\}.$$

("$A \cap B$" is read "A cap B.")

(c) *Union:* The outcomes ζ in either event A or event B or both form their *union*

$$A \cup B = \{\zeta : \zeta \in A \text{ or } \zeta \in B \text{ or both}\}.$$

("$A \cup B$" is read "A cup B.")

If all the outcomes in A are also in B, but B contains some outcomes not in A, we write $A \subset B$; if A and B contain the same outcomes, $A = B$. If either of these situations may obtain, we write $A \subseteq B$. Thus

$$A \cap B \subseteq A, \qquad A \cap B \subseteq B, \tag{1.1}$$

$$A \subseteq A \cup B, \qquad B \subseteq A \cup B. \tag{1.2}$$

The rules

$$(A \cap B)' = A' \cup B' \tag{1.3}$$

$$(A \cup B)' = A' \cap B' \tag{1.4}$$

are sometimes useful, as are the associative laws

$$A \cap (B \cap C) = (A \cap B) \cap C \tag{1.5}$$

$$A \cup (B \cup C) = (A \cup B) \cup C \tag{1.6}$$

and the distributive laws

$$A \cap (B \cup C) = (A \cap B) \cup (A \cap C) \tag{1.7}$$

$$A \cup (B \cap C) = (A \cup B) \cap (A \cup C). \tag{1.8}$$

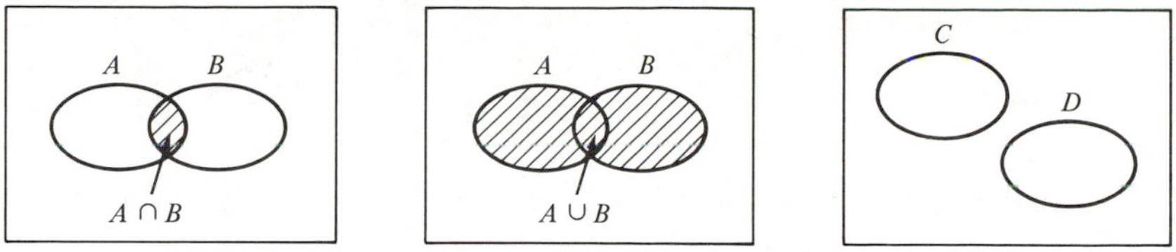

FIGURE 1.1 Venn diagrams.

Relations among sets are often informatively represented by regions in a rectangle, which is called a *Venn diagram* (see Fig. 1.1). The entire rectangle represents the universal set S.

Events with no outcomes in common are called *mutually exclusive*; their sets are disjoint. If C and D are mutually exclusive, $C \cap D = \varnothing$. It is convenient to denote the union of disjoint sets with a "+": $C \cup D = C + D$.

EXAMPLE 1.1. In experiment E_e we defined the events

$$A = \{n : n \text{ is odd}\}$$
$$B = \{n : 6 \leq n \leq 14\}$$
$$C = \{n : 0 \leq n \leq 9\}.$$

Verify the following:

$$A' = \{n : n \text{ is 0 or even}\} = \{0, 2, 4 \ldots\}$$
$$B' = \{n : 0 \leq n \leq 5 \text{ or } n \geq 15\}$$
$$C' = \{n : n \geq 10\}$$
$$A \cap B = \{7, 9, 11, 13\}$$
$$A \cap C = \{1, 3, 5, 7, 9\}$$
$$B \cap C = \{6, 7, 8, 9\}$$
$$A' \cap B = \{6, 8, 10, 12, 14\}$$
$$B' \cap C' = \{n : n \geq 15\}$$
$$A \cup B = \{n : n \text{ is odd or } n = 6, 8, 10, 12, 14\}$$
$$B \cup C = \{n : 0 \leq n \leq 14\}$$
$$B' \cup C' = \{n : 0 \leq n \leq 5 \text{ or } n \geq 10\}.$$ ■

The operations of intersection and union can be extended indefinitely: given n events $A_1, A_2, A_3, \ldots, A_n$, the intersection

$$A_1 \cap A_2 \cap A_3 \cap \cdots \cap A_n = \bigcap_{i=1}^{n} A_i$$

contains all outcomes in A_1 and A_2 and A_3 and $\cdots$ A_n. In Example 1.1, for instance,

$$A \cap B \cap C = \{7, 9\}$$
$$A' \cap B' \cap C' = \{n : n \text{ is even and } n > 15\}.$$

The union

$$A_1 \cup A_2 \cup A_3 \cdots \cup A_n = \bigcup_{i=1}^{n} A_i$$

contains all outcomes in A_1 or A_2 or A_3 or $\cdots$ or A_n. In Example 1.1,

$$A \cup B \cup C = \{n : 0 \leq n \leq 14 \text{ or } n \text{ is odd}\}$$

$$A' \cup B' \cup C' = \{n : 0 \leq n \leq 6, n = 8, n \geq 10\}$$

With an infinite number of outcomes and events we can extend these processes ad infinitum to form

$$\bigcap_{i=1}^{\infty} A_i \quad \text{and} \quad \bigcup_{i=1}^{\infty} A_i.$$

EXAMPLE 1.2. In reference to experiment E_f, consider the mutually exclusive events

$$A_1 = \{v : 0 \leq v < \tfrac{1}{2}\},$$

$$A_2 = \{v : \tfrac{1}{2} \leq v < \tfrac{3}{4}\},$$

$$\vdots$$

$$A_k = \{v : 1 - 2^{-(k-1)} \leq v < 1 - 2^{-k}\},$$

and so on. Then

$$\bigcup_{i=1}^{\infty} A_i = \{v : 0 \leq v < 1\},$$

for any value of v in $0 \leq v < 1$ lies in one of the intervals defining an event A_k for a finite index k. ■

1-3 AXIOMS OF PROBABILITY

Probabilities are numbers assigned to events in a way that is consistent with certain rules. The events are sets of outcomes of a chance experiment E, that is, subsets of the associated universal set S. The rules are expressed in four axioms that form the basis of the mathematical theory of probability. Denote by Pr (A) the probability assigned to event A. The axioms, formulated by the Russian mathematician A. N. Kolmogorov, are these:

Axiom I. $\Pr(A) \geq 0$.

(Probabilities are nonnegative numbers.)

Axiom II. $\Pr(S) = 1$.

(The universal set has probability 1; something must happen.)

Axiom III. If $A \cap B = \varnothing$, then $\Pr(A + B) = \Pr(A) + \Pr(B)$.

(If events A and B are mutually exclusive, the probability that A *or* B occurs is the sum of the probabilities of A and B.)

Axiom IV. If $A_i \cap A_j = \varnothing$ for all i and j, $i \neq j$, then

$$\Pr\left(\bigcup_{i=1}^{\infty} A_i\right) = \sum_{i=1}^{\infty} \Pr(A_i).$$

We return later to an explanation of this technical axiom.

Some simple corollaries of these minimal axioms are deduced immediately.

Corollary 1. With A' the complement of A,

$$\Pr(A') = 1 - \Pr(A).$$

(The probability that event A does not occur equals $1 - \Pr(A)$.)

Proof. $A \cap A' = \varnothing$, $A + A' = S$. Therefore, by Axiom III,

$$\Pr(A) + \Pr(A') = \Pr(S) = 1.$$

Corollary 2. $0 \leq \Pr(A) \leq 1$.

(Probabilities lie between 0 and 1.)

Proof. The left-hand inequality is Axiom I; the right-hand one follows from $\Pr(A') \geq 0$ and Corollary 1.

Corollary 3. If the n events A_i, $1 \leq i \leq n$, are mutually exclusive, that is, $A_i \cap A_j = \varnothing$ for all i and j, $i \neq j$, then

$$\Pr\left(\bigcup_{i=1}^{n} A_i\right) = \sum_{i=1}^{n} \Pr(A_i).$$

Proof. Use mathematical induction on Axiom III.

Axiom III and Corollary 3 state that if two or more events are mutually exclusive, the probability that one or another of them occurs is the sum of their individual probabilities. Axiom IV simply allows us to take $n = \infty$ in Corollary 3 for experiments with an infinite number of outcomes. It is needed for rounding out mathematical proofs of continuity, convergence, and the like, and will be of slight concern hereafter. As long as we assign probabilities only to events with a clear physical meaning, we need not worry about the set-theoretical subleties that entertain readers of purely mathematical treatises on probability.

The axioms endow probability with a kind of substantiality. It is as though we had a unit amount of something distributed over the set S; think of it as mass, if you like. Mass is nonnegative, hence Axiom I. Axiom II gives us a unit quantity of it. The total mass allotted to two disjoint sets is naturally the sum of the masses of the individual sets, hence Axiom III. Corollaries 1 to 3 are just as easily interpreted in terms of this analogy of probability to mass. An assignment of probabilities to events in a way consistent with the four axioms is called a *probability measure*.

From a mathematical point of view, probability theory amounts to calculating new probabilities from some initial probability measure over the events of E. It is a matter of bookkeeping. When you solve a problem in probability, be sure that all the unit quantity of probability you started with is accounted for, that no probability comes out negative, and that by some mistake excess probability has not sprung into existence. In the language of physics, probability is a "conserved" quantity.

When the chance experiment E has a countable set of outcomes ζ, it suffices to attach probabilities initially to only the atomic events, which as we said in Sec. 1-2 contain a single outcome each. Probabilities of composite events can then be calculated by summing those attached to their component atomic events.

EXAMPLE 1.3. Let us illustrate this concept in terms of experiment E_b of Sec. 1-2. Denoting by $\{\zeta_k\}$ the atomic event consisting of the outcome ζ_k alone, we assume an initial assignment of probabilities as in Table 1.1; the numbers sum to 1.

Table 1.1

$\{\zeta_1\}$	$\{\zeta_2\}$	$\{\zeta_3\}$	$\{\zeta_4\}$	$\{\zeta_5\}$	$\{\zeta_6\}$	$\{\zeta_7\}$	$\{\zeta_8\}$
HHH 0.07	HHT 0.31	HTH 0.17	HTT 0.05	THH 0.29	THT 0.01	TTH 0.06	TTT 0.04

Refer to the events defined for experiment E_b on page 5.

A: The probability that the coin first comes up heads is

$$\Pr(A) = 0.07 + 0.31 + 0.17 + 0.05 = 0.60.$$

B: The probability that the same face shows in the second and third tosses is

$$\Pr(B) = 0.07 + 0.05 + 0.29 + 0.04 = 0.45.$$

C: The probability that exactly two tails appear is

$$\Pr(C) = 0.05 + 0.01 + 0.06 = 0.12.$$

D: The probability that exactly two heads appear is

$$\Pr(D) = 0.31 + 0.17 + 0.29 = 0.77.$$

Events *C* and *D* are mutually exclusive. Therefore, the probability that either exactly two heads or exactly two tails appear is

$$\Pr(C + D) = \Pr(C) + \Pr(D) = 0.89.$$

The complementary event $C' \cap D'$ contains outcomes ζ_1 and ζ_8, and its probability is $0.07 + 0.04 = 0.11 = 1 - \Pr(C + D)$. ■

Now we can adjoin a fourth corollary of the axioms to take care of events composed of two overlapping events:

Corollary 4. $\Pr(A \cup B) = \Pr(A) + \Pr(B) - \Pr(A \cap B)$.

Proof. It is apparent from the Venn diagram in Fig. 1.1 that if *A* and *B* overlap and we add their probabilities, we have included the "mass" in their intersection $A \cap B$ twice and must subtract it out once in order to get the mass in the union $A \cup B$. Now we need to know three distinct probabilities in order to calculate Pr $(A \cup B)$.

In Example 1.3, what is the probability that either the first coin shows heads or the same face shows in the second and third tosses? Here we want Pr $(A \cup B)$. The event $A \cap B$ contains the outcomes ζ_1 and ζ_4, and

$$\Pr(A \cap B) = 0.07 + 0.05 = 0.12.$$

Hence by Corollary 4, $\Pr(A \cup B) = 0.60 + 0.45 - 0.12 = 0.93$.

As we have seen, it is possible to set up a probability assignment satisfying the axioms when the outcomes of the chance experiment are discrete, simply by attaching nonnegative numbers to the atomic events and finding the probabilities of composite events by addition. In an experiment such as E_f in which the outcomes form a continuum, this procedure is not feasible, for a single outcome is too slight and insubstantial an object to which to attach a probability. Probabilities must be assigned to events directly and checked for compatibility with the axioms. When we treat random variables we shall see how to do this in general. Here we simply give an example of a possible probability measure on E_f.

EXAMPLE 1.4. We introduce the notation

$$P(a, b) = \Pr\{v : a \leq v < b\}$$

and we postulate

(i) $P(-\infty, 0) = 0$.
(ii) $P(0, x) = x, \quad 0 \leq x \leq 1$.
(iii) $P(1, \infty) = 0$.

All the probability is assigned to the interval $0 \leq v \leq 1$. By Axiom III, for $0 < a < b \leq 1$,

$$\Pr\{v : 0 \leq v < a\} + \Pr\{v : a \leq v < b\} = \Pr\{v : 0 \leq v < b\}$$

because the events on the left-hand side are mutually exclusive. Hence

$$P(a, b) = P(0, b) - P(0, a) = b - a.$$

Probabilities for all events in which the value of the voltage lies in particular intervals can now be calculated by addition on the basis of Axiom III and Corollary 3. Thus with $a < b < c < d$,

$$\Pr\{v : a \leq v < b \text{ or } c \leq v < d\} = P(a, b) + P(c, d) = b - a + d - c$$

since $\{v : a \leq v < b\} \cap \{v : c \leq v < d\} = \emptyset$. And sc on *ad libitum*: the probability of an event is simply equated to the total length of the intervals defining the event. This is a special kind of probability measure.

We can check Axiom IV with respect to the events $A_1, A_2, \ldots, A_k, \ldots$ defined in Example 1.2. The probability of event A_k is

$$\Pr(A_k) = 1 - 2^{-k} - (1 - 2^{-(k-1)}) = 2^{-k},$$

and the events in question are all mutually exclusive. Therefore,

$$\Pr\left(\bigcup_{i=1}^{\infty} A_i\right) = \Pr\{v : 0 \leq v < 1\} = \sum_{k=1}^{\infty} \Pr(A_k) = \tfrac{1}{2} + \tfrac{1}{4} + \cdots + 2^{-k} + \cdots = 1$$

by the rule for summing a geometrical progression. ■

WHAT IS PROBABILITY AFTER ALL?

Most readers are likely to ask how so bizarre a set of numbers as those in Table 1.1 could represent the probabilities of sequences of heads and tails in tossing a coin three times (E_b). Anyone with even a rudimentary knowledge of probability would be apt to assign the probability $\frac{1}{8}$ to all eight atomic events. Whence those numbers in Table 1.1? They were chosen arbitrarily and assigned to the atomic events simply to emphasize that as far as the mathematical theory of probability is concerned, the assignment of probabilities to the events of a chance experiment makes no difference at all, provided that the axioms are satisfied. (This is not to imply that mathematicians do not often find some probability measures more interesting than others.) If probability theory is to be of any use to us, we must certainly pay some attention to which probability measure is assigned to a chance experiment, for all the mathematical theory does is calculate new probabilities on the basis of it. Various interpretations of the concept of probability underlie the attribution of a probability measure, and we shall describe those we believe to be most significant for engineering and science.

The engineer or scientist employs probability in two principal ways: in describing and analyzing randomness in physical systems, and in planning courses of action in situations fraught with uncertainty. Two conceptions of probability are involved, which we call physical probability and hypothetical probability. With *physical probability* the chance experiment is actually being performed on a certain system a great many times, either by the investigator or by nature, and the intent is to understand its apparently erratic outcomes in terms of the physical constitution of the system. With *hypothetical probability* an analyst is asking how a particular course of action would fare in the long run if a certain situation, modeled as a chance experiment, were repeated over and over under reasonable assumptions about the underlying probability measure, but so many trials as that may occur only in the imagination.

Physical probability is concerned in studying the angular distribution of alpha particles scattered by a nucleus, the emission of photoelectrons from the illuminated cathode of a photoelectric cell, the fluctuations in the current passed by an avalanche diode, the noise at the output of a radar receiver, and even the sequence of numbered faces onto which a pentahedral die falls when cast again and again. Hypothetical probability figures in decisions about how many switching circuits are needed in a telephone exchange to handle anticipated frequencies and durations of calls, or about how to select the parameters of a system controlling the antenna of a missile-tracking radar in order to cope with the kinds of targets expected, or about how integrated circuits coming from a production line can be tested to assure a certain degree of reliability. It is involved when the designer of a computer communication system, contemplating the addition of parity-check digits to the digital output of a computer to combat errors introduced by random noise in the channel, assesses different ways of doing so in terms of the rate and reliability with which the system would transport information to the computer at the other end of the line.

In what follows we shall describe various common notions about probability and examine their relevance to its physical and hypothetical modes.*

A. Classical Concept

The simplest, but not necessarily the most reasonable way to assign probabilities in chance experiments with a finite number ν of outcomes is to allocate to each atomic event the same probability $1/\nu$; all atomic events are taken to be equally probable. The probability of a composite event is then equal to the number of outcomes it contains, divided by ν. The probability Pr (A) is often said to equal the number of outcomes "favorable" to A, divided by the total number of outcomes. This manner of assigning probabilities characterizes what is called the *classical concept* of probability.

When appropriate, this assignment is justified by the *principle of indifference* or the *principle of sufficient reason*: one sees no reason to expect one outcome rather than another. One presumes in gaming that the coins and the dice are fair and fairly cast and that the cards are thoroughly and faithfully shuffled and dealt. Strategies are based on that presumption, and if one has doubts about it, one had better not play. Determining the probability of a composite event in situations like these requires counting up favorable outcomes, and that is usually a matter of combinatorics.

EXAMPLE 1.5. What is the probability that a bridge hand of thirteen out of fifty-two cards contains exactly one ace?

There are 52 ways of picking the first card of the hand, 51 ways of picking the second, . . ., 40 ways of picking the last, or $52\cdot 51\cdot \ldots \cdot 40 = 52!/39!$ in all; but rearranging the cards in any one of the 13! possible ways does not change the bridge hand. Therefore, there are $52!/(39!13!)$ possible hands all told. This is the number of outcomes in the universal set S.

Suppose that the one ace is the ace of clubs. Since only one ace is allowed in this "event," the remaining 12 cards in the hand must be selected from the 48 cards that are not aces, and by the same reasoning as before, this can be done in $48!/(36!12!)$ ways. We multiply by 4 to obtain the total number of outcomes in the event because we might have any one of the four aces. Hence the probability sought is

$$\frac{4\cdot 48!}{36!12!} \Big/ \frac{52!}{39!13!} = \frac{39\cdot 38\cdot 37}{51\cdot 50\cdot 49} = 0.439. \qquad \blacksquare$$

Problems like this abound in textbooks on probability, but their probabilistic aspect is trivial, and they are best relegated to courses on combinatorics. We shall not concern ourselves with them.

In terms of the classical concept, the probabilities of the atomic events {H} and {T} in experiment E_a are each $\frac{1}{2}$. Those of the eight atomic events in E_b are $\frac{1}{8}$. To E_c we can hardly apply this concept, for we have no reason to believe that the pentahedral die, which possesses no symmetry, is as likely to land on one face as on another. A reasonable assignment of probabilities in experiment E_c must rest on a study of the dynamics of the pentahedral die in flight, and

*The bibliography at the end of this chapter lists books that discuss these matters at greater length.

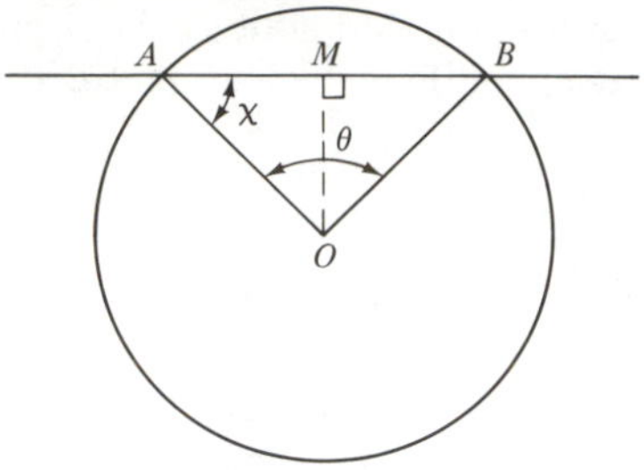

FIGURE 1.2 Bertrand's paradox.

although—as we shall see in part D—the principle of sufficient reason enters, it does so at a deeper level. Again, if the coin in experiment E_a were known to be biased, the classical concept would not satisfy us, and we should seek another approach.

In chance experiments involving a continuum of outcomes, such as E_f, E_g, and E_h, how to apply the classical concept is often unclear, for atomic events cannot be unambiguously defined. The most popular example of its inadequacy is perhaps that known as *Bertrand's paradox*. A straight line is placed "randomly" over a unit circle. What is the probability that the chord AB between its points of intersection with the circle subtends an angle θ less than a given angle ϕ for any value of ϕ in $0 \le \phi < \pi$? (See Fig. 1.2.)

There are various ways in which the principle of sufficient reason might be applied to this problem, and they yield different results. (a) One might believe that the perpendicular distance OM to the chord from the center of the circle is uniformly likely to take any value between 0 and 1. (b) One might assume that the angle $\chi = \angle BAO$ between the chord and the radius vector OA is just as likely to have any value between $-\pi/2$ and $\pi/2$ as any other. (c) The midpoint M of the chord might be assumed just as likely to fall in one part of the circle as in another.

This is an idle problem unless it represents some physical situation. Imagine yourself on a rafter high up in the gymnasium, tossing a long, straight stick down onto a circle drawn on the floor below. If you always aligned the stick, say, north and south before letting it go, assumption (a) might be reasonable. If you twirled the stick as hard as you could before dropping it, assumption (b) might be appropriate. In determining physical probabilities, the physical conditions of the experiment must be carefully taken into account. In the purely geometrical context within which Bertrand's paradox is usually formulated, there is no unique way of making that initial assignment of a probability measure without which the mathematical theory of probability cannot get under way.

The classical concept of probability is directly applicable only to chance experiments having a finite number of outcomes, among which it is reasonable to distribute the probability uniformly. It finds its principal use in the hypothetical mode in which strategies for coping with uncertainty are being planned and evaluated. Gambling with dice or cards comes most readily to mind as furnishing examples. All possible outcomes of dealing the cards or casting the dice are taken into account, and that strategy is preferred whose payoff, averaged over all those outcomes, is maximum. No guarantee can be offered that in future plays of the game the actual distribution of outcomes will approximate the uniform one assumed in the calculation, not even in the long run. In a game like bridge even the most avid player will encounter in a lifetime only an infinitesimal fraction of the possible hands of cards. Adopting the strategy prescribed by an analysis based on the classical concept is hardly more than an act of faith, resting on the belief that one has taken all factors into account in the most judicious possible way.

Comparatively few situations can be formulated as chance experiments with a finite set of atomic events to which the principle of sufficient reason would justifiably assign equal probabilities. We have seen that the classical concept is inadequate when an experiment has a continuum of outcomes. We therefore turn to an alternative concept, that of relative frequency, which may permit an initial attribution of a probability measure on the basis of past experience.

B. Relative Frequency

We said in Sec. 1-2 that a chance experiment E must in principle be repeatable an arbitrary number of times under the same circumstances. The physical experiments to which the theory of probability is ordinarily applied are influenced by chaotic, uncontrollable forces that cause the outcome to vary from one trial to another in an unpredictable manner. For definiteness, think of experiment E_c, in which a pentahedral die is tossed. It is put into a dice box, well shaken, and cast on the table. The face on which it comes to rest, perhaps after a bounce or two, is the outcome of the experiment, labeled by an integer 1, 2, 3, 4, or 5. Let this experiment be repeated over and over many times, or let a great many persons toss identical pentahedral dice from identical dice boxes all at once. Let $n_N(k)$ be the number of times in N trials that the outcome is k, $k = 1, 2, 3, 4, 5$. Then

$$n_N(1) + n_N(2) + n_N(3) + n_N(4) + n_N(5) = N.$$

Define the relative frequency $q_N(k)$ of outcome k in these N trials by

$$q_N(k) = \frac{n_N(k)}{N}$$

$$\sum_{k=1}^{5} q_N(k) = 1.$$

Through a phenomenon called *statistical regularity* these relative frequencies $q_N(k)$ tend to stabilize as the number N of trials increases beyond all bounds. It is believed that as more and more trials are carried out, the relative frequencies $q_N(k)$ converge to numbers $\Pr\{k\}$ that can be taken as the probabilities of the atomic events $\{k\}$:

$$q_N(k) \to \Pr\{k\} \qquad \text{as } N \to \infty.$$

For composite events A in our chance experiment we can likewise define relative frequencies as

$$q_N(A) = \frac{n_N(A)}{N},$$

where $n_N(A)$ is the number of times that event A occurred in N trials. These relative frequencies themselves obey Axioms I to III. The first two are obviously satisfied, and because

$$n_N(A + B) = n_N(A) + n_N(B)$$

when events A and B are mutually exclusive, Axiom III follows upon division by N,

$$q_N(A + B) = q_N(A) + q_N(B), \qquad A \cap B = \varnothing.$$

If these relative frequencies indeed settle down to fixed numbers, $q_N(A) \to \Pr(A)$ as $N \to \infty$, the numbers $\Pr(A)$ will also satisfy the axioms and can be taken as the probabilities of the events A.

In assessments of hypothetical probability the principle of sufficient reason may be inapplicable, and the analyst may look to past experience, if that has furnished what can be taken as a large number of trials of a relevant chance experiment, and adopt as the probability of any event its relative frequency in those trials. An insurance company, for instance, must decide how large a premium to charge for an ordinary life policy to be purchased by a man of, let us say, 43. What is the probability that he will die in the next year and the policy must be paid? The chances that he will and will not die can hardly be taken equal, as the classical concept would entail. The actuary instead looks at the mortality table, which lists the fraction of males who died in their forty-fourth year during some period in the past, and takes this figure as the probability. Premium schedules are thus drawn up in the expectation that the future will nearly enough resemble the past so that the insurance company can anticipate a profit by selling its policies for the prices so calculated.

This use of relative frequencies in a finite, though large number N of trials can be regarded as an empirical approach to the initial assignment of probabilities to atomic events. A gambler intending to play with the pentahedral and heptahedral dice of experiment E_d may privately toss them a great many times beforehand and base his betting strategies on the relative frequencies with which they landed on their several faces during the preliminary trials. He adopts the relative frequencies as the actual probabilities of the atomic events, perhaps believing that he and his friends will tire of these particular dice before so many games with them that any error in his assessment would be of consequence. We must view these relative frequencies, however, as only *estimates* of probabilities that characterize the motions of the two dice in a way that depends on how they are tossed, how they spin and turn in flight, and how they land. If physical probabilities in chance experiments such as this are to be meaningful, they must be related to the nature of the system involved.

The relative-frequency concept of probability has been most extensively propounded in the work of R. von Mises. Imagine a large number N of successive trials of a chance experiment. The digit 1 is written down whenever event A occurs in a trial, and the digit 0 when it does not. In the resulting sequence Σ of N 1's and 0's, according to von Mises, the fraction $n_N(A)/N$ of 1's should approach a limit as $N \to \infty$, and that limit is defined as the probability $\Pr(A)$ of event A. He wished to exclude sequences such as 0 1 0 1 0 1 · · · exhibiting regular patterns of 0's and 1's, however, as objects inappropriate for a theory of probability, which should be confined to sequences exhibiting irregularity or randomness. He therefore imposed on sequences a condition to guarantee that. This condition requires that if a subsequence is selected from Σ by a rule that is specified once for all before the sequence is generated, then in that subsequence as well, the relative frequency of 1's must approach the same limiting value that is to stand as Pr (A). The selection rule might, for instance, direct one to pick every other element in Σ, or to pick the element following each 1, or to pick the elements in places in the sequence indexed by prime numbers. Whether the convergence of the relative frequency to a unique limit in such subsequences can be guaranteed has been questioned, and other methods of defining randomness of sequences have been studied. Kolmogorov and others have been constructing a theory of probability in which a sequence is considered random if a computer program to generate it would require roughly as many digits to write down as there are in the sequence. Regularity of a sequence, on the other hand, would imply that a program generating it could be rather shorter than the sequence itself. If the theory of probability is to be limited to the mathematical properties of sequences, such artifices may be necessary. In physical and engineering applications randomness may be anticipated on other grounds.

C. Subjective Probability

A vague concept denominated "probability" often arises in everyday thought and conversation. You look out at the weather in the morning and think about whether it will rain and whether you should take your umbrella. What you see may determine a degree of inner tension or anxiety ranging from ease if the sky is clear blue to severe mental discomfort if dark, menacing clouds hang overhead. You may say to your spouse something like "It is probably not going to rain" or "It will probably rain before noon." A certain school of thought would assert that you have evaluated the "subjective probability" that it is going to rain.

Subjective probabilities are numbers assigned to possible events in some situation or experiment prone to uncertainty, but it is unnecessary for it to be repeatable under identical conditions; it may represent something that is to happen only once. This assignment is a private matter, and there appears to be no way of checking whether it is correct or not; indeed, the ordinary concept of correctness seems inapplicable. Advocates of this viewpoint usually insist that in evaluating subjective probabilities one must gather and rationally weigh all the available evidence. In a physical problem that would include an analysis of the dynamics of the system concerned.

A serious difficulty with the concept of subjective probability is justifying the additivity of such probabilities, that is, showing that they must satisfy Axiom III. Your subjective probabilities would manifest themselves, it is said, by how much you are willing to bet on the outcome of a chance experiment. They could be measured by your preference for betting either on that experiment or on one of a graded set of alternative experiments, the probabilities of whose outcomes are somehow known. Additivity of subjective probabilities then follows from the additivity of the probabilities for the alternative experiments, but it is unclear how the assignment of the latter can be justified within the subjective theory itself.

For thorough expositions of subjective probability, see the books by Savage and de Finetti. Although one might resort to it, for lack of anything better, in the initial assignment of hypothetical probabilities in a decision problem, it would be difficult to avoid self-deception. The private nature of subjective probability seems to exclude it from scientific and engineering application.

D. Physical Probability

In what follows we shall argue that the probabilities of events in a chance experiment with a physical system are properties of the system itself. They correspond to the propensities of the system to behave in various ways when the experiment is performed. Thus a biased coin may be said to have a propensity to fall on one face rather than on the other, and that propensity is quantified as the probability of heads or tails. What we shall describe may be termed the *propensity interpretation* of probability. Propensities and their associated probabilities are related to the structure of the phase space of the system, which as will be explained governs the motion or behavior of the system during the chance experiment. This structure reveals itself in the sequence of outcomes of the experiment when it is repeated a great many times. Our discussion involves concepts of analytical mechanics, but only in a simple way. For definiteness we shall talk mostly about the pentahedral die of experiment E_c.

Let us scrutinize the physical process involved in casting that pentahedral die. It is well shaken in the dice box and forcefully thrown upward and out. At some time during its flight, its condition or "state" of motion can be specified by the three coordinates x, y, z of its center of gravity; its linear momenta p_x, p_y, and p_z along the x-, y-, and z-directions; its orientation; and its angular momenta. The orientation of the die can be specified by three coordinates such as the Euler angles defined in treatises on rigid-body dynamics. The analytical dynamics of Hamilton associates with each such angular coordinate a conjugate momentum, which can be thought of as an angular momentum. In all, therefore, the state of the moving die at time t is specified by six coordinates and six momenta and can be represented by a point in a 12-dimensional space, which is known as the *phase space*. A point representing the dynamical state in that space is called a *phase point*.

As time passes, the phase point moves along a trajectory in the phase space, but for the moment we are concerned only with a particular time t. The face on which the die finally lands depends uniquely on the state of motion at time t, that is, on the phase point, and we can think of each point in the phase space as labeled with the number of that ultimate outcome of the experiment. Denote the region in which the points are labeled k by R_k, $k = 1, 2, 3, 4, 5$. Each such region is divided into many parts, which are interlaced in a complicated way, for a slight change in a component of angular momentum or orientation at time t will usually suffice to cause the die to land on a different face.

The whole purpose of shaking the dice box is to ensure that the thrower cannot determine where the phase point will be located in the phase space at later times t when the die is in flight, at least not so accurately that it can be put into whichever of the regions R_k the thrower desires. The phase point at time t may be anywhere in a certain part of the phase space, which we can call the *accessible region*. The size of the accessible region is limited because excessive velocities and angular momenta cannot be attained by the tossing mechanism, and the velocities and positions must lie in a bounded region of the phase space if the die is to land on the table at all, and gently enough not to break up on impact. The accessible region contains a great many subregions of R_1, R_2, R_3, R_4, and R_5.

Because the casting of the die cannot be controlled accurately enough to determine in which of the five regions R_k the phase point lies at a time t after the die leaves the box, that point falls here and there in the phase space in successive trials of the experiment, the sequence of faces on which the die lands exhibiting a certain randomness. Let V_k be the volume of region R_k within the accessible region of phase space, whose total volume is V; $k = 1, 2, 3, 4, 5$. As the number N of trials increases, the fraction $n_N(k)/N$ of trials in which the phase point lies in region R_k approaches the fraction V_k/V of the volume V occupied by region R_k. Because the subregions into which each region R_k is divided are so numerous and so intricately mingled, the value of the fraction V_k/V is insensitive to just where the boundaries of the accessible region lie. We identify that ratio V_k/V as the probability of the atomic event $\{k\}$.

Each composite event A, furthermore, is associated with a region R_A of the phase space that is the union of the regions R_k for all outcomes $\zeta_k = k$ in event A. If V_A is the volume of the part of region R_A within the accessible region, we specify

$$\Pr(A) = \frac{V_A}{V}$$

as the probability of event A. Probabilities as defined in this way obviously satisfy Axioms I and II. If events A and B are mutually exclusive, the regions R_A and R_B have no points in common. Volume in the phase space is an additive quantity in the sense that if $A \cap B = \varnothing$, $V_{A+B} = V_A + V_B$, and

$$\Pr(A + B) = \frac{V_{A+B}}{V} = \frac{V_A + V_B}{V} = \Pr(A) + \Pr(B).$$

Axiom III is therefore also satisfied by probabilities defined in this way. Axiom IV follows from the infinite divisibility of the phase space.

In this manner we can give a direct physical meaning to probabilities in a great variety of physical chance experiments, of which the casting of the pentahedral die is only an example. To each event A corresponds a region R_A of the phase space of the physical system, so defined that if at time t the phase point happens to be in R_A, event A occurs. In experiments exhibiting randomness and hence properly treated by the theory of probability, each such region R_A is divided into a large number of subregions interspersed among the finely divided subregions associated with other events exclusive of A. Repeated trials of the experiment yield random outcomes as the phase point takes up unpredictable positions in the phase space, falling now into one subregion, now into another.

Only as a great many trials are carried out do we see the ratio $n_N(A)/N$, the fraction of trials in which event A occurs, approach a characteristic value equal to the relative volume V_A/V of the region R_A of the phase space associated with that event. In a chance experiment with a physical system we can thus identify probabilities with the physical nature of the system itself, and in a way that ensures that the axioms of probability theory are satisfied.

If we had chosen a different time t, would the probabilities specified in this way have taken on different values? The answer is no. A theorem due to Liouville tells us that as time goes on, the volume of the region of phase space occupied by any set of phase points remains unchanged as they move along their trajectories, provided that the system is conservative. We can therefore choose any time t while the die is in free flight and before it has begun to bounce on the table, converting kinetic energy into heat and settling down to its final position. The relative volumes V_A/V remain the same, and the probabilities Pr (A) do not depend on the time t associated with their definition.

Had we used any other set of coordinates to specify the position and orientation of the die at time t, the probabilities, defined as V_A/V, would have turned out to be the same, provided that the conjugate momenta had been chosen as prescribed in Hamiltonian dynamics when we constructed the phase space. A transformation from one such set of coordinates and momenta to another preserves the volumes of the regions R_A associated with events A.

The perfectly symmetrical, homogeneous cubical die of the kind ordinarily used in gambling furnishes one of the commonest examples for elementary probability theory. We can now see why the probabilities of its landing on each of its six faces must be equal to $\frac{1}{6}$. Because of the symmetry and homogeneity of the die all six regions R_1, R_2, R_3, R_4, R_5, and R_6 of its phase space have equal volumes, and hence each occupies one-sixth of the accessible region.

The pentahedral die of experiment E_c is not one of the five regular polyhedra and possesses no such symmetry from which we could draw conclusions about the relative volumes of regions R_1 through R_5. Calculating the probability Pr $\{k\}$ that it lands on face k would be extremely difficult in the framework we have just described. Some simplifying assumptions must be made, and we shall see that invoking the principle of sufficient reason cannot be avoided. Let us sketch how a theory of the "stochastic dynamics" of this die might be constructed.

Suppose that the surface on which the die lands is so soft and yielding that the die does not bounce or even flip over when it first strikes the surface. Imagine a plumb line dropped from the center of gravity of the die. If just before the die lands, that imaginary line intersects face k, it is plausible that the die will come to rest on face k. Assume also that at that time there is no reason for the die to have one orientation rather than another. The probability Pr $\{k\}$ will then equal the fraction of all orientations in which the plumb line intersects face k. If Ω_k is the solid angle subtended by face k from the center of gravity of the die, that fraction will be

$$\Pr\{k\} = \frac{\Omega_k}{4\pi},$$

the total solid angle about the center being 4π steradians.

The plane angle between two straight lines, in radians, is determined by drawing a unit circle centered at their intersection and measuring the arc they intercept. In a similar manner, the solid angle occupied by a cone is determined by constructing a unit sphere centered at the vertex of the cone and measuring the area the cone cuts out on the surface of the sphere. Put a unit sphere at the center of gravity of the die and let a radius vector move along the edges of face k. It will trace out a cone intersecting the surface of the sphere in a spherical polygon whose area equals the solid angle Ω_k in steradians. The area of the entire sphere is 4π.

The principle of sufficient reason entered this analysis when we said that no orientation of the die is preferred over any other. Indeed, in any classical physical theory that determines a probability measure by analyzing the dynamics of a chance experiment, the principle of sufficient reason will be introduced at some stage. That principle is applied not only in probabilistic problems, however, but in other domains of physics as well. When the special theory of relativity postulates that all inertial frames are equivalent, that same principle is being invoked.

It is in statistical mechanics that probability finds its most fundamental application to physics. The phase space of the molecules of a gas or that of the ions and electrons in a solid has an enormous number of dimensions, and the region R_A of any macroscopically observable event A in the dynamics of a gas or a solid must be divided and dispersed throughout that space in an unimaginably complicated way. Indeed, a thorough analysis requires not the phase space of classical mechanics, as in our previous discussion, but the abstract space of quantum-mechanical wave functions. Statistical mechanics cuts through those complexities by judicious application of the principle of sufficient reason, whose ultimate validation lies in the success of the physical predictions that follow from it. The theory of noise in electronic systems originates in statistical mechanics, of which we shall utilize only the results. Before we can appreciate and apply them, however, we must learn to deal with the concepts of probability in an elementary way, and that is the purpose of this book.

E. Summary

We conclude that physical probabilities measure certain propensities of a physical system that manifest themselves in the outcomes of repeated trials of an experiment. Although analogies are perilous, probabilities of events can be likened to properties like specific heat or viscosity. The specific heat of a solid depends in a complicated way on its structure and manifests itself in the amount of heat required to raise the temperature of a solid by 1 degree. The viscosity of a fluid, depending in a complicated way on the molecular structure of the fluid, manifests itself by the force required to move a body through the fluid at constant velocity. The probability of an event in a physical chance experiment also depends in a complicated way on the structure of the system involved. It exists prior to the performance of the experiment, manifesting itself only in the relative frequency of the event in a great many trials.

When planning a course of action in a situation where uncertainty or randomness plays a role, the engineer or scientist applies probability in its hypothetical mode. The situation is modeled as a chance experiment that is conceptually repeated over and over infinitely often. For any physical system involved it is assumed that each identifiable event will occur with a relative frequency equal to its physical probability. The principle of sufficient reason may be invoked to distribute equal probabilities among atomic events when justified by some invariance or symmetry. Past experience may be drawn upon to estimate certain probabilities that cannot be calculated from any well-founded theory, by equating them to the relative frequencies of events in a great many trials of some chance experiment or its equivalent. The behaviors of various strategies or courses of action in all possible sequences of outcomes of the chance experiment, weighted in accordance with their probabilities, are then compared on the basis of some criterion of performance, and the strategy that shows up best is selected. This necessarily, but regrettably abstract description of the use of hypothetical probability will become more meaningful as we encounter elementary examples in the course of this study.

1-5 CONDITIONAL PROBABILITY

A. Definition and Examples

The outcomes common to two events A and M form their intersection $A \cap M$ (Fig. 1.3). The conditional probability of event A, given event M, is defined as

$$\Pr(A \mid M) = \frac{\Pr(A \cap M)}{\Pr(M)} \tag{1.9}$$

in terms of the probability measure assigned to the experiment.

Subjectively, one can think of the conditional probability $\Pr(A \mid M)$ as representing the likelihood of A's having occurred when it is known that event M has occurred. A more satisfactory conception can be based on the notion of relative frequency. Out of a large number N of trials of a chance experiment E, select the subset comprising those trials in which event M occurred; there are $n_N(M)$ of them. Let $n_N(A, M)$ be the number of these in which event A also occurred. Since both A and M occurred,

$$n_N(A, M) = n_N(A \cap M)$$

is the number of trials in which some outcome or other in $A \cap M$ occurred.

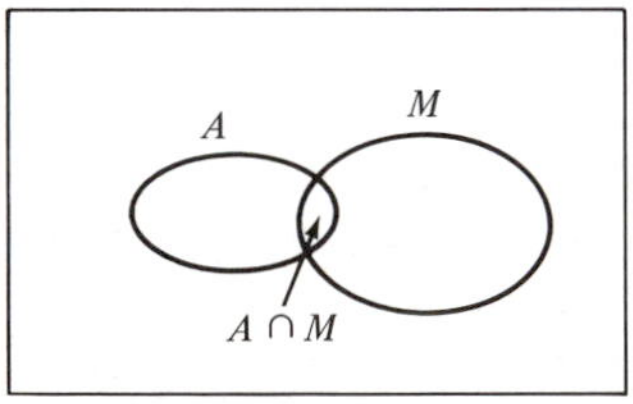

FIGURE 1.3 Venn diagram.

The relative frequency of event A in this subset of trials is

$$\frac{n_N(A, M)}{n_N(M)} = \frac{n_N(A \cap M)/N}{n_N(M)/N},$$

and as the number N of trials grows beyond all bounds, the ratios on the right-hand side approach Pr $(A \cap M)$ and Pr (M), respectively. Hence

$$\frac{n_N(A, M)}{n_N(M)} \to \frac{\Pr(A \cap M)}{\Pr(M)} = \Pr(A \mid M);$$

that is, Pr $(A \mid M)$ is the limit of the relative frequency of event A among all trials in which event M occurred. We also define

$$\Pr(M \mid A) = \frac{\Pr(A \cap M)}{\Pr(A)} \tag{1.10}$$

as the conditional probability of M, given A.

In the propensity interpretation of probability that we presented in Sec. 1-4D, let R_M be the region of the phase space occupied by phase points leading to an outcome in event M, and let $V(M)$ be its volume. Then $R_{A \cap M} = R_A \cap R_M$ and we can define

$$\Pr(A \mid M) = \frac{V(A \cap M)}{V(M)},$$

which also leads to (1.9) because

$$\frac{V(A \cap M)}{V(M)} = \frac{V(A \cap M)/V}{V(M)/V} = \frac{\Pr(A \cap M)}{\Pr(M)}$$

with V again the volume of the accessible region of the phase space. Thus Pr $(A \mid M)$ is the volume of the part of the region R_M containing phase points leading to an outcome in event A, relative to the total volume of R_M.

Conditional probabilities are probabilities in their own right, but are referred to the conditioning event M as their universal set. As in the axioms of Sec. 1-3,

$$\text{(I) } \Pr(A \mid M) \geq 0,$$

$$\text{(II) } \Pr(M \mid M) = 1,$$

$$\text{(III) } \Pr(A \cup B \mid M) = \Pr(A \mid M) + \Pr(B \mid M),$$

if $(A \cap M) \cap (B \cap M) = \varnothing$, that is, if the portions $A \cap M$ and $B \cap M$ of A and B lying in M are disjoint. Indeed, it is evident from the Venn diagram in Fig. 1.4 that the distributive law (1.7) holds,

$$(A \cap M) \cup (B \cap M) = (A \cup B) \cap M.$$

Here $(A \cap M) \cap (B \cap M) = \varnothing$ even though $A \cap B \neq \varnothing$.

EXAMPLE 1.6. Refer to experiment E_b, whose probabilities are listed in Table 1.1. What is the probability that three heads appeared, given that the coin first showed heads? Here the conditioning event is $M = \{\zeta_1, \zeta_2, \zeta_3, \zeta_4\}$ and its probability is

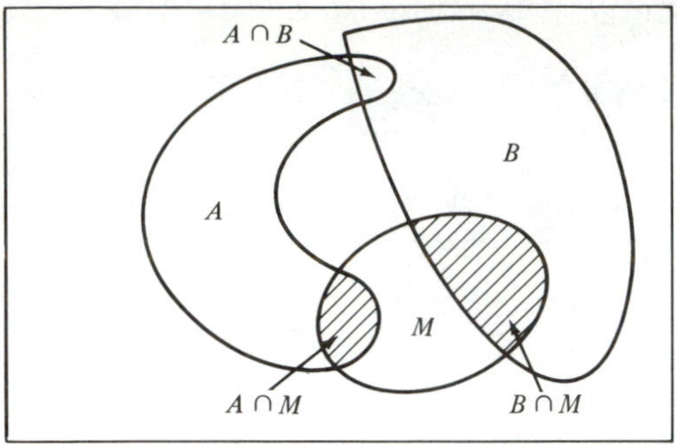

FIGURE 1.4 Venn diagram.

$$\Pr(M) = 0.07 + 0.31 + 0.17 + 0.05 = 0.60.$$

$$\Pr(\text{HHH}) = \Pr\{\zeta_1\} = 0.07.$$

Therefore,

$$\Pr(\text{HHH} \mid M) = \frac{0.07}{0.60} = 0.1167.$$

What is the probability that three heads appeared, given that the coin showed heads on the first two tosses? Now $M = \{\zeta_1, \zeta_2\}$, $\Pr(M) = 0.38$, and $\Pr(\text{HHH} \mid M) = 0.07/0.38 = 0.184$. What is the probability that tails appeared on the second and third tosses (event A) given that the number of tails was odd?

$$M = \{\zeta_2, \zeta_3, \zeta_5, \zeta_8\}, \qquad \Pr(M) = 0.81,$$

$$A = \{\zeta_4, \zeta_8\}, \qquad A \cap M = \{\zeta_8\}, \qquad \Pr(A \cap M) = 0.04.$$

Therefore,

$$\Pr(A \mid M) = \frac{0.04}{0.81} = 0.0494.$$

■

EXAMPLE 1.7. The probability that a certain kind of rectifier lasts at least t hours before it fails is $\exp(-at^2)$. What is the probability that the rectifier will fail between times t_1 and t_2, given that it is still working at time T, $T < t_1 < t_2$?

Here our universal set is the time interval $(0, \infty)$. The conditioning event M is that failure occurs in $T < t < \infty$, and its probability is $\Pr(M) = \exp(-aT^2)$ by hypothesis. Let B be the event that the rectifier fails in the interval $t_1 < t < t_2$. Then

$$\Pr(B = \text{failure in } t_1 < t < t_2) + \Pr(\text{failure in } t_2 < t < \infty)$$

$$= \Pr(\text{failure in } t_1 < t < \infty)$$

because the two events on the left-hand side are mutually exclusive and decompose the event on the right-hand side into two parts. Therefore,

$$\Pr(B) = \exp(-at_1^2) - \exp(-at_2^2).$$

Now $B \cap M = B$, and by definition (1.9),

$$\Pr(B \mid M) = \frac{\exp(-at_1^2) - \exp(-at_2^2)}{\exp(-aT^2)}, \qquad T < t_1 < t_2.$$

■

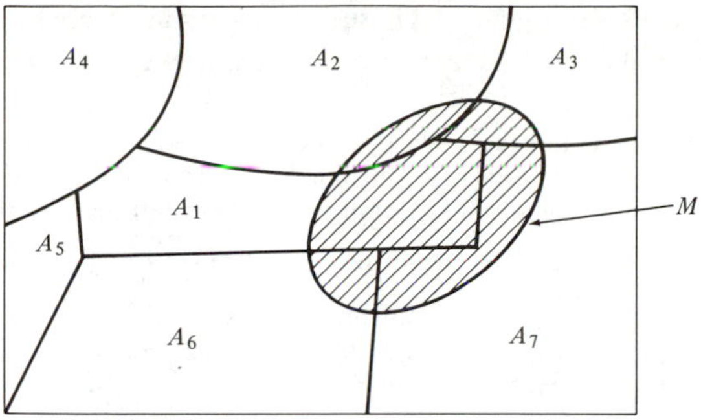

FIGURE 1.5. Venn diagram.

B. Total Probability

As illustrated by the Venn diagram in Fig. 1.5, let $A_1, A_2, \ldots, A_n$ be n mutually exclusive events making up the universal set S,

$$A_1 + A_2 + \cdots + A_n = S, \qquad A_i \cap A_j = 0, \qquad \text{all } i \neq j.$$

Then by Corollary 3 of Sec. 1-3,

$$\Pr(A_1) + \Pr(A_2) + \cdots + \Pr(A_n) = 1.$$

Let M be some other event comprising outcomes in some or all of the events A_1 through A_n. The events $A_1 \cap M, A_2 \cap M, \ldots, A_n \cap M$ are mutually exclusive, as is apparent from Fig. 1.5; if the events A_i have no outcomes in common, the events $A_i \cap M$ can have none in common. Therefore,

$$\Pr(A_1 \cap M) + \Pr(A_2 \cap M) + \cdots + \Pr(A_n \cap M) = \Pr(M), \qquad (1.11)$$

since these events exhaust the event M. Dividing through by $\Pr(M)$ and using the definition of conditional probability, we find that

$$\Pr(A_1 \mid M) + \Pr(A_2 \mid M) + \cdots + \Pr(A_n \mid M) = 1, \qquad (1.12)$$

which we can regard as an expression of Corollary 3 of Sec. 1-3 for conditional probabilities.

By definition (1.10) of conditional probability,

$$\Pr(A_i \cap M) = \Pr(M \mid A_i) \Pr(A_i)$$

for all i. Substituting this into (1.12), we obtain

$$\Pr(M) = \Pr(M \mid A_1)\Pr(A_1) + \Pr(M \mid A_2)\Pr(A_2) + \cdots + \Pr(M \mid A_n)\Pr(A_n), \qquad (1.13)$$

which is known as the *principle of total probability*. It states that the total probability of an event M equals the weighted sum of the conditional probabilities $\Pr(M \mid A_i)$ of its occurring in each event A_i of an exhaustive decomposition of S into mutually exclusive events, each such conditional probability being weighted by the probability $\Pr(A_i)$ that the conditioning event occurs.

EXAMPLE 1.8. Transistors of the same kind are bought from three suppliers, A, B, and C. Let M be the event "the transistor fails before 1000 hours' operation." The probabilities that the transistors from each of the three suppliers fail that soon are

$$\Pr(M \mid A) = 0.15, \qquad \Pr(M \mid B) = 0.40, \qquad \Pr(M \mid C) = 0.25.$$

They are all mixed together in a large bin, from which the assemblers of some equipment pick them at random without noting who supplied them. The probability that a transistor so chosen came from supplier A can be taken as the fraction of transistors A furnished, and so on for B and C. Let those probabilities or fractions be

$$\Pr(A) = 0.45, \qquad \Pr(B) = 0.25, \qquad \Pr(C) = 0.30.$$

These sum to 1. What is the probability that a transistor chosen at random and installed fails before 1000 hours?

By the principle of total probability (1.13), this is

$$\Pr(M) = 0.15 \times 0.45 + 0.40 \times 0.25 + 0.25 \times 0.30 = 0.243.$$

We interpret this by saying that out of a large number of transistors installed by assemblers who ignore from which supplier each one comes, a fraction 0.243 will fail before operating 1000 hours. ■

C. Bayes's Theorem

Let A and M be two events. Then by the definition of conditional probability (1.9),

$$\Pr(A \cap M) = \Pr(A \mid M)\Pr(M) = \Pr(M \mid A)\Pr(A),$$

whence

$$\Pr(A \mid M) = \frac{\Pr(M \mid A)\Pr(A)}{\Pr(M)}, \tag{1.14}$$

which is known as *Bayes's theorem*. More generally, let $A_1, A_2, \ldots A_n$ be n mutually exclusive events exhausting the set S, that is, containing all possible outcomes of the experiment. Then

$$\Pr(A_j \mid M) = \frac{\Pr(M \mid A_j)\Pr(A_j)}{\Pr(M)}, \qquad j = 1, 2, 3, \ldots, n. \tag{1.15}$$

By the principle of total probability (1.13), the denominator of the right-hand side is

$$\Pr(M) = \sum_{k=1}^{n} \Pr(M \mid A_k)\Pr(A_k),$$

which guarantees that the conditional probabilities in (1.15) sum to 1,

$$\sum_{j=1}^{n} \Pr(A_j \mid M) = 1, \tag{1.16}$$

as in (1.12).

EXAMPLE 1.9. In Example 1.8, given that a transistor fails before 1000 hours' operation, what are the probabilities that it came from each of the three suppliers? By Bayes's theorem and the result of Example 1.8 for Pr (M), we calculate

$$\Pr(A \mid M) = \frac{0.15 \times 0.45}{0.243} = 0.2784$$

$$\Pr(B \mid M) = \frac{0.4 \times 0.25}{0.243} = 0.4124$$

$$\Pr(C \mid M) = \frac{0.25 \times 0.30}{0.243} = 0.3092.$$

These sum to 1. Interpret this result in terms of relative frequencies. ■

Further applications of Bayes's theorem will be introduced shortly. Basically, it is a way of calculating conditional probabilities such as Pr $(A_j \mid M)$ from probabilities Pr $(M \mid A_j)$, which in certain problems are more readily available. As we shall see, it lies at the foundation of a branch of statistics known as "hypothesis testing," through which it has become important in communication theory and signal detection.

1-6 INDEPENDENCE

Events A and B are said to be *statistically independent* if

$$\Pr(A \mid B) = \Pr(A); \tag{1.17}$$

that is, the probability of event A is the same whether event B has occurred or not. By the definition of conditional probability (1.9), an equivalent condition for independence is

$$\Pr(A \cap B) = \Pr(A)\Pr(B). \tag{1.18}$$

EXAMPLE 1.10. Let us test the events in experiment E_b for independence. The probabilities of atomic events in that experiment are listed in Table 1.1. We consider only the first two tosses of the coin, and summing over the outcomes of the third toss, we find the following probabilities:

$$\Pr(\mathrm{H_1H_2}) = 0.38, \qquad \Pr(\mathrm{H_1T_2}) = 0.22,$$

$$\Pr(\mathrm{T_1H_2}) = 0.30, \qquad \Pr(\mathrm{T_1T_2}) = 0.10.$$

The subscripts indicate which toss is meant. Summing again, we find that

$$\Pr(\mathrm{H_1}) = 0.60, \qquad \Pr(\mathrm{T_1}) = 0.40,$$

$$\Pr(\mathrm{H_2}) = 0.68, \qquad \Pr(\mathrm{T_2}) = 0.32.$$

Now by (1.9),

$$\Pr(\mathrm{H_2} \mid \mathrm{H_1}) = \frac{\Pr(\mathrm{H_1H_2})}{\Pr(\mathrm{H_1})} = \frac{0.38}{0.6} = 0.633 \neq \Pr(\mathrm{H_2}),$$

$$\Pr(\mathrm{T_2} \mid \mathrm{H_1}) = \frac{\Pr(\mathrm{H_1T_2})}{\Pr(\mathrm{H_1})} = \frac{0.22}{0.6} = 0.367 \neq \Pr(\mathrm{T_2}),$$

and so on. The events referring to the first and second tosses are not independent.

Suppose that instead of the foregoing probabilities we had had

$$\Pr(H_1H_2) = 0.408, \qquad \Pr(H_1T_2) = 0.192,$$

$$\Pr(T_1H_2) = 0.272, \qquad \Pr(T_1T_2) = 0.128,$$

so that $\Pr(H_1) = 0.6$, $\Pr(T_1) = 0.4$, $\Pr(H_2) = 0.68$, and $\Pr(T_2) = 0.32$, as before. You can easily check that now

$$\Pr(H_1H_2) = \Pr(H_1)\Pr(H_2),$$

and so on for the other pairs of heads and tails. The two tosses of the coin yield independent events. Had both coins been fair as well, the probabilities in the table would all have equaled 0.25. ■

EXAMPLE 1.11. Shots are fired at a rectangular target, $a \times b$, by a poor marksman. Any shot that strikes the target can be assumed just as likely to impinge on it at one point as at another. The probability that the shot hits any region B is then proportional to the area of B, which we denote by $A(B)$. Hence with B the event "the shot strikes region B,"

$$\Pr(B) = \frac{A(B)}{ab}.$$

Convince yourself that this probability measure satisfies Axioms I to III.

The probability that the shot strikes at some point (x, y) in the shaded rectangle in Fig. 1.6 is then

$$\Pr(x_1 < x < x_2, y_1 < y < y_2) = \frac{(x_2 - x_1)(y_2 - y_1)}{ab}.$$

The probability that it falls in the vertical strip between x_1 and x_2 is

$$\Pr(x_1 < x < x_2) = \frac{(x_2 - x_1)b}{ab} = \frac{x_2 - x_1}{a},$$

and the probability that it falls in the horizontal strip between y_1 and y_2 is

$$\Pr(y_1 < y < y_2) = \frac{y_2 - y_1}{b}.$$

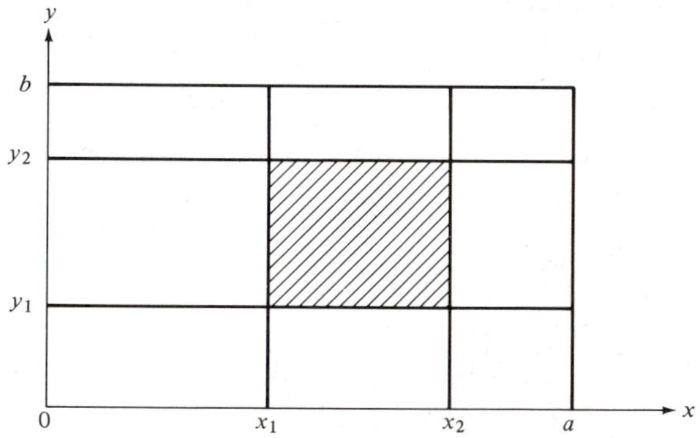

FIGURE 1.6 Rectangular target.

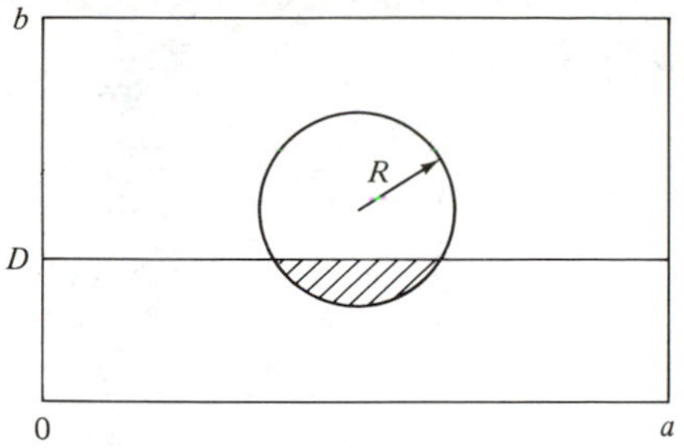

FIGURE 1.7 Rectangular target.

The intersection of these events is

$$\{x : x_1 < x < x_2\} \cap \{y : y_1 < y < y_2\} = \{(x, y): x_1 < x < x_2, y_1 < y < y_2\},$$

and we find that

$$\Pr(x_1 < x < x_2, y_1 < y < y_2) = \Pr(x_1 < x < x_2) \Pr(y_1 < y < y_2),$$

so that the events

$$\{x : x_1 < x < x_2\} \qquad \text{and} \qquad \{y : y_1 < y < y_2\}$$

are statistically independent.

Define the following events as indicated in Fig. 1.7:

(a) C = "the shot falls within radius R of the center of the target." For simplicity take $R < b/2$.
(b) D = "the shot falls within distance D of the bottom of the target," $0 < D < b$.

Now by our postulated probability measure on the events of this experiment,

$$\Pr(C) = \frac{\pi R^2}{ab}, \qquad \Pr(D) = \frac{D}{b}.$$

It is fairly obvious that if $D \neq b/2$, $\Pr(C \cap D) \neq \Pr(C)\Pr(D)$, for $\Pr(C \cap D)$ is the area of the shaded region divided by ab. (You may wish to calculate that area in terms of a, b, D, and R.) The events C and D are therefore independent if and only if $D = b/2$. ■

Several events are said to be independent if the probabilities of all their possible intersections factor as products of the probabilities of the component events. Three events A_1, A_2, and A_3 are independent if and only if

$$\Pr(A_1 \cap A_2) = \Pr(A_1)\Pr(A_2)$$

$$\Pr(A_2 \cap A_3) = \Pr(A_2)\Pr(A_3)$$

$$\Pr(A_3 \cap A_1) = \Pr(A_3)\Pr(A_1)$$

$$\textit{and } \Pr(A_1 \cap A_2 \cap A_3) = \Pr(A_1)\Pr(A_2)\Pr(A_3).$$

The first three relations—termed *pairwise independence*—do not imply the fourth, which must also hold if all three events are to be statistically independent. In order to check the independence of n events, $2^n - n - 1$ such relations need to be tested.

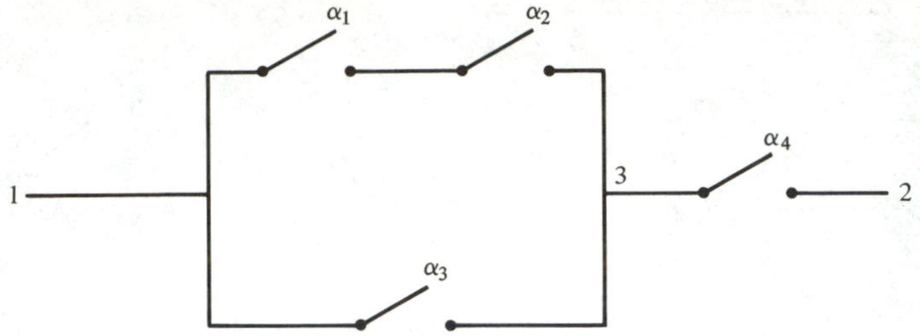

FIGURE 1.8 Switching network.

EXAMPLE 1.12. In the switching circuit in Fig. 1.8, the switches α_1, α_2, α_3, and α_4 are open or closed at random and independently. The probability that any switch is closed at a given time equals p. What is the probability that at any time there is at least one closed path from 1 to 2?

The probability of a closed path from 1 to 2 equals the product of the probabilities that there is a closed path from 1 to 3 and that there is a closed path from 3 to 2, for these events are statistically independent. Furthermore,

$$\begin{aligned}\Pr(\text{closed from 1 to 3}) &= 1 - \Pr(\text{open from 1 to 3})\\ &= 1 - \Pr(\text{top branch open}) \times \Pr(\text{bottom branch open}).\end{aligned}$$

But

$$\begin{aligned}\Pr(\text{top branch open}) &= 1 - \Pr(\text{top branch closed})\\ &= 1 - p^2.\end{aligned}$$

Therefore,

$$\Pr(\text{closed from 1 to 3}) = 1 - (1 - p^2)(1 - p) = p + p^2 - p^3$$

and

$$\Pr(\text{closed from 1 to 2}) = p(p + p^2 - p^3).$$

The rules of set theory introduced in Sec. 1-2 can be conveniently applied to problems such as this. Let A_k be the event that switch α_k is closed, $k = 1, 2, 3, 4$; $\Pr(A_k) = p$. The event that there is a closed path from 1 to 2 can then be written $[(A_1 \cap A_2) \cup A_3] \cap A_4$, in which $\cap$ stands for "and" and $\cup$ stands for "or." By Corollary 4 in Sec. 1-3,

$$\Pr[(A_1 \cap A_2) \cup A_3] = \Pr(A_1 \cap A_2) + \Pr(A_3) - \Pr(A_1 \cap A_2 \cap A_3)$$

and because the events A_k are statistically independent,

$$\Pr[(A_1 \cap A_2) \cup A_3] = p^2 + p - p^3$$

and

$$\Pr\{[(A_1 \cap A_2) \cup A_3] \cap A_4\} = p(p^2 + p - p^3),$$

as before.

Each switch might represent a link in a communication network, the opening of a switch corresponding to failure of that link. By applying the rules of set theory, or "Boolean algebra," given in Sec. 1-2 and assigning probabilities that each link is operative, one could analyze an arbitrarily complicated network and determine the probability that at any given time it is possible to communicate from one point to another. ■

COMPOUND EXPERIMENTS

A. Product Sets

Consider two chance experiments E_a and E_b, which may involve the same or different performances and outcomes. A compound experiment $\mathsf{E}_a \times \mathsf{E}_b$ consists of doing both E_a and E_b, either one before the other, or simultaneously. The outcomes of this compound experiment are pairs of outcomes of the individual experiments. Let E_a have n outcomes in a universal set S_a, $(a_1, a_2, \ldots, a_n)$, and let E_b have m outcomes in a universal set S_b, $(b_1, b_2, \ldots, b_m)$. Then the outcomes of $\mathsf{E}_a \times \mathsf{E}_b$ are the mn pairs

$$\begin{array}{cccc} a_1b_1 & a_2b_1 & \cdots & a_nb_1 \\ a_1b_2 & a_2b_2 & \cdots & a_nb_2 \\ \cdot & \cdot & & \cdot \\ \cdot & \cdot & & \cdot \\ \cdot & \cdot & & \cdot \\ a_1b_m & a_2b_m & \cdots & a_nb_m. \end{array}$$

These form the product set $S = S_a \times S_b$. Events in $\mathsf{E}_a \times \mathsf{E}_b$ are subsets of S, such as $\{a_1b_1, a_2b_3, a_4b_8\}$.

EXAMPLE 1.13. Toss a cubical die and a coin; $n = 2$, $m = 6$. The outcomes are

H1	H2	H3	H4	H5	H6
T1	T2	T3	T4	T5	T6

in an obvious notation. ■

EXAMPLE 1.14. Experiment E_a may be to measure the voltage v_a at point A in a circuit, E_b to measure the voltage v_b at point B. The outcomes of $\mathsf{E}_a \times \mathsf{E}_b$ are pairs of voltages (v_a, v_b). Events in $\mathsf{E}_a \times \mathsf{E}_b$ are such as

$$A = \{(v_a, v_b) : 10 < v_a + v_b < 15\},$$

$$B = \{(v_a, v_b) : v_a < v_b\},$$

$$C = \{(v_a, v_b) : -5 < v_a < 6 \text{ and } -2 < v_b < 8\}.$$ ■

The chance experiments may have countable sets of outcomes, as in our table and in Example 1.13, or continua of outcomes as in Example 1.14. We discuss the former kind, leaving to the reader the easy task of formulating the corresponding concepts for the latter.

To an atomic event such as $\{a_k\}$ in E_a corresponds the event $\{a_kb_1, a_kb_2, \ldots, a_kb_m\}$ in $\mathsf{E}_a \times \mathsf{E}_b$; we can say that what the outcome of E_b was does not matter. This event we denote by $\{a_k\} \times S_b$. Similarly, to any event A of E_a corresponds the event $A \times S_b$ of $\mathsf{E}_a \times \mathsf{E}_b$, which pairs each outcome in A with *all* outcomes in S_b. Furthermore, to any event B of E_b corresponds the event $S_a \times B$ of

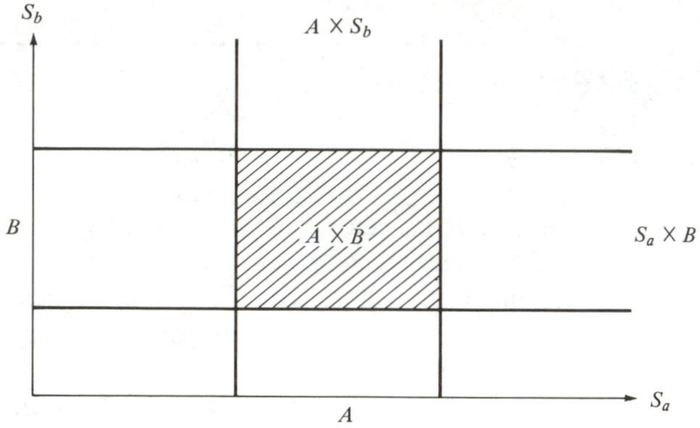

FIGURE 1.9 Product events.

$\mathsf{E}_a \times \mathsf{E}_b$, pairing each outcome of B with all the outcomes in S_a. These can be diagrammed as shown in Fig. 1.9.

The event $A \times B$ of $\mathsf{E}_a \times \mathsf{E}_b$ consists of the pairs formed by all the outcomes of E_a in A with all the outcomes of E_b in B. If, for instance, $A = \{a_1, a_2, a_4\}$ and $B = \{b_1, b_3, b_8\}$, then

$$A \times B = \{a_1b_1, a_1b_3, a_1b_8, a_2b_1, a_2b_3, a_2b_8, a_4b_1, a_4b_3, a_4b_8\}.$$

The diagram shows what is otherwise obvious as well, that

$$A \times B = (A \times S_b) \cap (S_a \times B). \tag{1.19}$$

Not all events of $\mathsf{E}_a \times \mathsf{E}_b$ have this product form, as illustrated by the event $\{a_1b_1, a_2b_3, a_4b_8\}$. The first two events in Example 1.14 are not of product form; the third is.

B. Probability Measures in Compound Experiments

If some probability measure consistent with the axioms has been assigned to all the events of the compound experiment, we can deduce from it the probabilities in the individual experiments. Let $\Pr_a(A)$ be the probability of event A in E_a. Then

$$\Pr_a(A) = \Pr(A \times S_b) \tag{1.20}$$

is known from the assignment to the events of $\mathsf{E}_a \times \mathsf{E}_b$. Similarly,

$$\Pr_b(B) = \Pr(S_a \times B) \tag{1.21}$$

is the probability of an event B in E_b.

We cannot go the other way without additional knowledge or an additional assumption about how the two experiments are related; from $\Pr_a(A)$ and $\Pr_b(B)$ for all $A \subset S_a$ and all $B \subset S_b$ we cannot deduce the probabilities of arbitrary events in $\mathsf{E}_a \times \mathsf{E}_b$.

C. Independent Experiments

The concept of a compound experiment is most useful when the component experiments are statistically independent, whereupon from (1.19)–(1.21),

$$\Pr(A \times B) = \Pr(A \times S_b)\Pr(S_a \times B) = \Pr_a(A)\Pr_b(B) \qquad (1.22)$$

for all events in E_a and E_b. The probability measure on $\mathsf{E}_a \times \mathsf{E}_b$ can then be deduced from those on experiments E_a and E_b individually. Experiments will be statistically independent when carrying out either one does not influence the outcome of the other in any way, that is, when the experiments are independent physically.

The physical independence of two experiments implies that the two systems involved can be described by separate phase spaces. The phase space for the product experiment is formed by adjoining the dimensions of the phase space of the first to those of the phase space of the second; the result is called a *product space*. In this product phase space the volume of the region associated with the event $A \times B$ is found by multiplying the volume associated with the event A in the first space by that associated with event B in the second. The rule (1.22) then follows from the definition in Sec. 1-4D of the probability of an event as the relative volume of the phase space occupied by the phase points leading to outcomes in the event.

For experiments with countable sets of outcomes, the probabilities of atomic events such as $\{a_k b_j\}$ in our table at the beginning of this section are simply

$$\Pr\{a_k b_j\} = \Pr_a\{a_k\}\Pr_b\{b_j\} \qquad (1.23)$$

when the experiments are independent, and by addition the probabilities of composite events in $\mathsf{E}_a \times \mathsf{E}_b$ can be calculated. In Example 1.13, for instance, it is usually plausible that the coin and the die do not interact in any way when cast separately, and we can put Pr (H1) = Pr (H) Pr (1), which equals $\frac{1}{12}$ if coin and die are fair, and so on for the remaining eleven atomic events.

In Example 1.14, whether measurements of the voltages v_a and v_b at points A and B of the circuit are statistically independent experiments depends on the nature of the circuit itself and its inputs. If point A were in one radio receiver and point B in another at some distance, it would ordinarily be reasonable to assume that experiments E_a and E_b are independent, unless perhaps both receivers are tuned to the same radio station.

An event C in $\mathsf{E}_a \times \mathsf{E}_b$ of Example 1.14 can be represented by some region in the (v_a, v_b)-plane, as in Fig. 1.10. As in the calculus, we can break the region C into a great many infinitesimal rectangles $dv_a\, dv_b$. Then

$$\Pr(C) = \int_C \Pr(dv_a\, dv_b),$$

where $\Pr(dv_a\, dv_b)$ is the probability measure assigned to the rectangle $dv_a\, dv_b$ and depends in general on its location in the plane. Statistical independence of E_a and E_b implies that

$$\Pr(dv_a\, dv_b) = \Pr_a(dv_a)\Pr_b(dv_b).$$

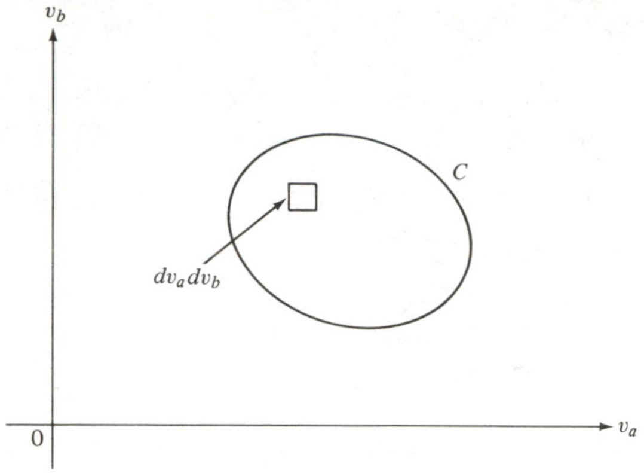

FIGURE 1.10 Events in Example 1.14.

How this is carried through in detail will become clearer when we treat random variables. We merely mention here that even when the outcomes of the experiments E_a and E_b form continua, the statistical independence of the two experiments suffices to determine the probabilities of all events in $\mathsf{E}_a \times \mathsf{E}_b$ from the measures assigned to E_a and E_b individually.

The concept of a compound experiment can be extended to any number of individual experiments. Thus tossing a coin thrice can be considered as a compound experiment $\mathsf{E}_a \times \mathsf{E}_b \times \mathsf{E}_c$, where E_a represents the first cast, E_b the second, and E_c the third. If the three experiments are independent, probabilities to be assigned to the events in the compound experiment can be determined by multiplication from those assigned to the individual experiments; and so on for any number of independent experiments so combined.

EXAMPLE 1.15. Three fair coins and five fair cubical dice are tossed. The compound experiment is eightfold and could be denoted by $\mathsf{E}_1 \times \mathsf{E}_2 \times \mathsf{E}_3 \times \mathsf{E}_4 \times \mathsf{E}_5 \times \mathsf{E}_6 \times \mathsf{E}_7 \times \mathsf{E}_8$; E_1, E_2, and E_3 refer to tossing the coin and E_4 through E_8 to tossing the dice. A typical atomic event in the compound experiment would be a succession of three heads or tails and five integers from the set (1, 2, 3, 4, 5, 6), such as {H T H 1 3 2 6 2}. There are $2^3 \cdot 6^5 = 62{,}208$ such atomic events, and under the reasonable assumption that the casts of the coins and the dice do not influence each other, the probability of each atomic event is $2^{-3} \cdot 6^{-5}$. ■

In some problems it may be necessary to consider experiments that are the product of an infinite number of component experiments, as in the following example.

EXAMPLE 1.16. A coin is tossed repeatedly. What is the probability that heads first appears at the *n*th cast?

Let E_k be the experiment representing the *k*th toss of the coin. Since there is no limit to how many times it may be necessary to toss the coin before heads appears, we must deal with a compound experiment $\mathsf{E}_1 \times \mathsf{E}_2 \times \mathsf{E}_3 \times \cdots \times \mathsf{E}_k \times \cdots$ with the series extended ad infinitum. The event A_n, "heads first appears at the *n*th toss," consists of all outcomes of the form

$$\underbrace{\mathrm{T\ T} \cdots \mathrm{T}}_{(n\,-\,1)\text{ times}}\ \mathrm{H}\quad \text{(queue)}$$

in which (queue) stands for any sequence of H's and T's that might appear thereafter if one went on tossing the coin forever.

Let p be the probability of heads on any toss of the coin, and $q = 1 - p$ the probability of tails, and assume the tosses independent. Then the event in question has probability

$$\Pr(A_n) = q^{n-1}p.$$

As the events A_n for different values of n are mutually exclusive and exhaust all possible events in the compound experiment,

$$A_i \cap A_j = \varnothing, \text{ all } i \neq j, \qquad A_1 + A_2 + \cdots + A_k + \cdots = S,$$

the probabilities must sum to 1, Axiom IV being invoked:

$$\sum_{k=1}^{\infty} \Pr(A_k) = (1 + q + q^2 + \cdots + q^k + \cdots)p = (1 - q)^{-1}p = 1. \qquad \blacksquare$$

1-8

BINOMIAL DISTRIBUTION

A chance experiment E is repeated n times. These n trials form an n-fold compound experiment, which we can designate by

$$\underbrace{\mathsf{E} \times \mathsf{E} \times \cdots \times \mathsf{E}}_{n \text{ times}} = \mathsf{E}^n$$

The outcomes of E^n are n-tuples of outcomes of the component experiments. We focus on a particular event B in experiment E and its complement B', setting

$$\Pr(B) = p, \qquad \Pr(B') = 1 - p = q.$$

An occurrence of B is called a *success* and denoted by a 1; an occurrence of B' is a *failure* and denoted by a 0. Looking thus only at successes and failures, we reduce E^n to an experiment E_n whose outcomes are n-tuples of 0's and 1's, of which there are 2^n all told. We can think of E_n as a compound experiment whose components are experiments E' with only two outcomes, 0 and 1, like the tossing of a biased coin. We seek the probability distribution of the *number* of successes and failures in n trials of E, that is, in a single trial of E_n. Such an experiment is often called a succession of *Bernoulli trials*. We assume the trials statistically independent.

Let A_k be the event that there are k successes and $n - k$ failures in n trials of experiment E. Event A_k is the collection of all n-tuples of k 1's and $n-k$ 0's. We first show that the number of outcomes of E_n in A_k equals

$$\binom{n}{k} = \frac{n!}{k!(n-k)!}, \tag{1.24}$$

which is called the *binomial coefficient*. If $n = 4$, for instance, the sixteen outcomes can be sorted into events A_k as follows:

		$\binom{n}{k}$
A_0	(0000)	1
A_1	(1000), (0100), (0010), (0001)	4
A_2	(1100), (1010), (1001), (0110), (0101), (0011)	6
A_3	(0111), (1011), (1101), (1110)	4
A_4	(1111)	1

We want to write down all sequences of k 1's and $n - k$ 0's. The first 1 can go into any one of n places, the second into any one of the $n - 1$ remaining places, and so on, leaving $n - k + 1$ places for the kth 1. The number of ways this can be done is

$$n(n - 1)(n - 2) \cdots (n - k + 1) = \frac{n!}{(n - k)!}.$$

The remaining places are filled with 0's. All the 1's look alike, however, and can be permuted in $k!$ ways without changing the n-tuple. Hence the net number of distinct n-tuples having k 1's is

$$\frac{n!}{(n - k)!k!} = \binom{n}{k}.$$

This binomial coefficient is the number of outcomes in A_k. Since by the binomial theorem

$$\sum_{k=0}^{n} \binom{n}{k} = (1 + 1)^n = 2^n,$$

we have accounted for all 2^n outcomes of experiment E_n.

An atomic event (single outcome) in E_n consists of k successes and $n - k$ failures in a particular order. Its probability equals $p^k q^{n-k}$ because the trials are independent. The number of such atomic events in A_k being $\binom{n}{k}$, the probability of A_k is, by Sec. 1-3, Corollary 3,

$$\Pr(A_k) = \binom{n}{k} p^k q^{n-k}, \qquad 0 \le k \le n. \tag{1.25}$$

The array of $n + 1$ such probabilities constitutes the *binomial distribution*.

By Axioms II and III these probabilities must sum to 1, for the $n + 1$ mutually exclusive events A_k exhaust the universal set S of E_n, and indeed by the binomial theorem

$$\sum_{k=0}^{n} \Pr(A_k) = \sum_{k=0}^{n} \binom{n}{k} p^k q^{n-k} = (p + q)^n = 1.$$

EXAMPLE 1.17. A pair of fair cubical dice are tossed ten times. What is the probability that the dice total seven points exactly four times?

The event B with which we are concerned consists of the six pairs (1, 6), (2, 5), (3, 4), (4, 3), (5, 2), and (6, 1) and has probability $6 \cdot 6^{-2} = \frac{1}{6} = p$. The probability of failure is $q = \frac{5}{6}$. Hence the probability that event B occurs four times and B' occurs six times is

$$\binom{10}{4}\left(\frac{1}{6}\right)^4\left(\frac{5}{6}\right)^6 = \frac{210 \cdot 5^6}{6^{10}} = 0.0543. \qquad \blacksquare$$

The probability that at most k successes occur in n trials is given by

$$B(k, n; p) = \sum_{r=0}^{k} \binom{n}{r} p^r q^{n-r}, \tag{1.26}$$

and these numbers form what is called the *cumulative* binomial distribution. Extensive tables are available for it*, but are not needed if one has a programmable calculator.

A program for calculating $B(k, n; p)$ is easily written. The terms

$$T_r = \binom{n}{r} p^r q^{n-r}$$

of the sum are calculated iteratively, starting with $T_0 = q^n$, which is stored in two registers, say R_1 and R_2. At stage r, register R_1 holds T_r, from which T_{r+1} is calculated by

$$T_{r+1} = \left(\frac{n - r}{r + 1}\right) \frac{p}{q} T_r.$$

It is stored in R_1 and added into R_2, which is accumulating the sum. The procedure stops when $r = k$, whereupon R_2 holds the cumulative probability $B(k, n; p)$. It is suggested that you write a program to calculate $B(k, n; p)$ on your own calculator. You can check it by means of Table 1.2, which was calculated for $n = 10$, $p = 0.6$, $q = 0.4$.

Table 1.2 Cumulative Binomial Probabilities

k	$B(k, 10; 0.6)$	k	$B(k, 10; 0.6)$
0	0.00010	6	0.61772
1	0.00168	7	0.83271
2	0.01229	8	0.95364
3	0.05476	9	0.99395
4	0.16624	10	1.
5	0.36690		

In Example 1.17 the probability that seven points are scored in at most four casts of the pair of dice is

$$B(4, 10; \tfrac{1}{6}) = 0.9845.$$

*Harvard Computation Laboratory, *Tables of the Cumulative Binomial Probability Distribution*. Cambridge, Mass.: Harvard University Press, 1955.

The probability that seven points are scored five, six, or seven times is

$$B(7, 10; \tfrac{1}{6}) - B(4, 10; \tfrac{1}{6}) = 0.01544.$$

EXAMPLE 1.18 The probability that a certain diode fails before operating 1000 hours is 0.15. The probability that among twenty such diodes, at least five will fail before operating 1000 hours is

$$\sum_{k=5}^{20} \binom{20}{k} 0.15^k 0.85^{20-k} = 1 - B(4, 20; 0.15) = 0.1702. \quad \blacksquare$$

When the numbers n and k are large, even a programmable calculator takes a long time to calculate such binomial probabilities as in (1.26). In Sec. 4-7 we shall obtain a useful formula for approximating them.

1-9 POISSON DISTRIBUTION

If the number n of Bernoulli trials is very large and if the probability p of success in each trial is very small, then the binomial probability is accurately approximated by

$$\begin{aligned} \Pr(A_k) &= \binom{n}{k} p^k q^{n-k} = \frac{n!}{k!(n-k)!} p^k q^{n-k} \\ &\simeq P'_k = \frac{m^k e^{-m}}{k!}, \qquad m = np, \quad 0 \le k < \infty. \end{aligned} \tag{1.27}$$

The probabilities P'_k make up the Poisson distribution; they add up to 1,

$$\sum_{k=0}^{\infty} \frac{m^k e^{-m}}{k!} = e^m e^{-m} = 1$$

by virtue of the power series for e^m. For reasons to appear later, the parameter $m = np$ is called the *mean* or *expected number* of successes.

When a light beam falls on a photoelectrically sensitive surface, photoelectrons are emitted and can be counted by drawing them to an anode by a positive voltage and measuring the current in an external circuit. The number k of electrons emitted in a time interval $(0, T)$ is not predictable in advance, but random. The expected number m of electrons is proportional to the total radiant energy W incident on the surface during $(0, T)$. If the light occupies a narrow band of frequencies about frequency ν, that expected number is

$$m = \frac{\eta W}{h\nu}, \tag{1.28}$$

where h is Planck's constant 6.626×10^{-34} joule-sec and η is a number between 0 and 1 called the quantum efficiency of the device. We can think of $W/h\nu$ as the total number of photons striking the surface and of η as the probability that a single photon ejects an electron that is counted in the external circuit.

The volume to which the light penetrates just beneath the surface contains a huge number n of conduction electrons. The probability p that a single such electron absorbs a photon, makes its way to the surface, and is drawn over to the anode is extremely small. If we term the emission of a photoelectron and its registering in the external circuit a success, the probability that k electrons are counted will be given by a binomial distribution as in (1.25), but because $n >> 1$, $p << 1$, that probability can be approximated by the Poisson formula in (1.27). It is assumed that whether a given electron absorbs a photon and succeeds in being counted is independent of whether any other electron does so, and that the light is not so intense as significantly to deplete the number of available electrons.

EXAMPLE 1.19. Suppose that the expected number of electrons counted in $(0, T)$ is $m = 4.5$. The probability that exactly two electrons are counted is then

$$\Pr(A_2) = \frac{(4.5)^2}{2} e^{-4.5} = 0.1125.$$

The probability that at least six electrons are counted is

$$\Pr(k \geq 6) = 1 - \sum_{k=0}^{5} \frac{(4.5)^k}{k!} e^{-4.5} = 0.2971. \quad \blacksquare$$

A calculator program for the cumulative Poisson distribution

$$P(k; m) = \sum_{r=0}^{k} \frac{m^r}{r!} e^{-m} = \sum_{r=0}^{k} T_r, \qquad T_r = \frac{m^r}{r!} e^{-m}, \tag{1.29}$$

is easily composed. Starting with $T_0 = e^{-m}$, it calculates the terms T_r successively by the recurrent relation

$$T_{r+1} = \left(\frac{m}{r+1}\right) T_r$$

and accumulates their sum until r reaches the upper limit k. You can check your own program by the following table for $m = 2$.

k	0	1	2	3	4	5	6
$P(k; 2)$	0.13534	0.40601	0.67668	0.85712	0.94735	0.98344	0.99547

If m and k are so large that computation of (1.29) consumes too much time, we can approximate that sum by a convenient formula that will be derived in Sec. 4-7.

Here are some other situations in which the Poisson distribution is appropriate. A microprocessor contains, among other features, a large number n of identical circuits on a silicon chip, and the probability p that any one circuit is defective is very small. The number k of defective circuits on the chip has a Poisson distribution (1.27) with mean np. A biologist searches for a particular species of diatom in a bucket of water drawn from the ocean; there are $n >> 1$ diatoms in the sea, but the probability p that a particular one finds itself in the bucket is very small; the Poisson distribution governs the total number of diatoms caught. The number n of cars on the freeway during a certain hour is very large, and the probability p that a particular car would have an accident during that time is very small; the total number of accidents during that hour is described by a Poisson distribution.

The Poisson approximation to the binomial distribution in (1.27) can be derived from Stirling's formula for the factorial,

$$n! \simeq (2\pi)^{1/2} n^{n+1/2} e^{-n}, \qquad n >> 1, \tag{1.30}$$

of which the relative error is of the order of $1/(12n)$ and small even for $n = 10$. Now in (1.27) both n and $n - k$ are huge, and we use Stirling's formula for both $n!$ and $(n - k)!$. Because $p << 1$, $n >> k$, we can put

$$(1 - p)^{n-k} = \left(1 - \frac{m}{n}\right)^{n-k} \simeq \left(1 - \frac{m}{n}\right)^{n} \simeq e^{-m}, \qquad m = np,$$

applying the rule

$$\left(1 + \frac{x}{n}\right)^{n} \simeq e^{x}, \qquad n >> |x|. \tag{1.31}$$

Thus we find that

$$\begin{aligned} \Pr(A_k) &= \frac{n!}{k!(n-k)!} p^k (1 - p)^{n-k} \\ &\simeq \frac{n^{n+1/2} e^{-n}}{k!(n-k)^{n-k+1/2} e^{-(n-k)}} \left(\frac{m}{n}\right)^{k} e^{-m} \\ &= \frac{m^k}{k!} e^{-m} \left(1 - \frac{k}{n}\right)^{-(n-k+1/2)} e^{-k}. \end{aligned}$$

Since $n >> k$, we can apply (1.31) to the next-to-last factor, which becomes e^k and cancels the factor e^{-k}, and we obtain (1.27).

1-10

RANDOMLY SELECTED EXPERIMENTS

Of two possible chance experiments E_1 and E_2 that might be performed, the one that is performed is selected by tossing a coin. Let p be the probability that E_1 is selected, and $1 - p$ the probability that E_2 is selected. This procedure could in principle be repeated arbitrarily often, so that now E_1, now E_2 is performed, with relative frequencies p and $1 - p$, respectively. It corresponds to an experiment E that is called the *sum* of experiments E_1 and E_2 and sometimes designated by $\mathsf{E}_1 + \mathsf{E}_2$, although the notation $p\mathsf{E}_1 + (1 - p)\mathsf{E}_2$ is more informative. The outcomes of the composite experiment E are all the outcomes of E_1 conjoined with all the outcomes of E_2.

EXAMPLE 1.20. Upon entering a gambling den you are asked to toss a coin to determine whether you shall play blackjack or roulette. The probability of winning at blackjack is 0.36, that of winning at roulette is 0.44. If the coin is fair, the probability of your departing a winner is $\frac{1}{2}(0.36) + \frac{1}{2}(0.44) = 0.4$. ■

This concept can be extended to selections among any number of chance experiments. If experiment E_k is chosen with probability p_k, $1 \le k \le n$, the "sum" of these experiments can be designated by $\mathsf{E} = p_1\mathsf{E}_1 + p_2\mathsf{E}_2 + \cdots + p_n\mathsf{E}_n$,

$$\sum_{k=1}^{n} p_k = 1.$$

With respect to this composite experiment E, the probability of event A is, by the principle of total probability,

$$\Pr(A) = \sum_{k=1}^{n} p_k \Pr_k(A), \tag{1.32}$$

where $\Pr_k(A)$ is the probability of event A in experiment E_k and can be considered as a conditional probability with respect to E,

$$\Pr_k(A) = \Pr(A \mid \mathsf{E}_k \text{ chosen}). \tag{1.33}$$

If none of the outcomes of A is among those of E_k, that is, if in E_k $A = \varnothing$, this probability equals zero.

An example of such a composite experiment is Example 1.8, in which picking a transistor from the bin and determining whether it operates for 1000 hours can be viewed as selecting experiment E_a, E_b, or E_c with probability 0.45, 0.25, and 0.30, respectively. Experiment E_a is to test the transistors made by manufacturer A, and so on.

This concept of a composite experiment arises principally in what statisticians call *hypothesis testing*. The outcomes of all n experiments $\mathsf{E}_1, \mathsf{E}_2, \ldots, \mathsf{E}_n$ now lie in the same universal set S, but the probabilities of events in S depend on which experiment is performed. Nature or some other inscrutable agent selects the experiment that is actually performed and picks E_k with relative frequency p_k. The proposition "experiment E_k was selected and performed" is called a *hypothesis* and denoted by H_k, $k = 1, 2, \ldots, n$. It specifies an event in the composite experiment $\mathsf{E} = p_1\mathsf{E}_1 + p_2\mathsf{E}_2 + \cdots + p_n\mathsf{E}_n$.

The outcomes of the composite experiment $\mathsf{E} = p_1\mathsf{E}_1 + p_2\mathsf{E}_2 + \cdots + p_n\mathsf{E}_n$ are pairs (H_k, ζ), of which ζ is an element of S, and $k = 1, 2, \ldots$, or n. The universal set for E is the collection S' of all these pairs. Events in E are pairs (H_k, A), of which A is any event defined in S. The probability of the event (H_k, A) in E is

$$\Pr(H_k, A) = p_k \Pr(A \mid H_k) = p_k \Pr_k(A).$$

The event "hypothesis H_j is true" is (H_j, S), $j = 1, 2, \ldots, n$. Its probability p_j is the probability that experiment E_j is selected and is called the *prior probability*, or the *probability a priori*, of hypothesis H_j.

An observer does not know which of the n experiments was performed, but is informed of the outcome

$$\bar{\zeta} = (H_1, \zeta) + (H_2, \zeta) + \cdots + (H_n, \zeta)$$

or perhaps only of some event

$$\bar{A} = (H_1, A) + (H_2, A) + \cdots + (H_n, A)$$

that occurred, "+" standing for the union of mutually exclusive events. On the basis of that information he is to decide which hypothesis is true. A reasonable procedure in general is to adopt that hypothesis H_k for which the "posterior probability" $\Pr(H_k \mid \bar{A})$ is the largest among all the conditional probabilities $\Pr(H_j \mid \bar{A})$, $j = 1, 2, \ldots, n$, given the event $\bar{A}$ whose occurrence has been observed. These posterior probabilities are calculated by Bayes's theorem (1.14) as in Example 1.9, in which we might decide that a transistor that fails before 1000 hours must have come from supplier B because $\Pr(B \mid M) > \Pr(A \mid M)$ and $\Pr(B \mid M) > \Pr(C \mid M)$. We present two examples of this approach to making decisions among a number of hypotheses.

EXAMPLE 1.21. Let M_t be the event "the transistor fails before t hours." Denote by H_a, H_b, and H_c the hypotheses that a transistor drawn from the bin at random came from suppliers A, B, and C, respectively. As in Example 1.8, the prior probabilities of these hypotheses are

$$p_a = 0.45, \qquad p_b = 0.25, \qquad p_c = 0.3.$$

Let the probabilities that the three different brands of transistors fail before t hours be as follows:

$$\Pr(M_t \mid H_a) = 1 - e^{-1.625 \times 10^{-4} t}$$

$$\Pr(M_t \mid H_b) = 1 - e^{-5.108 \times 10^{-4} t}$$

$$\Pr(M_t \mid H_c) = 1 - e^{-2.877 \times 10^{-4} t}.$$

At time t we inspect the transistor and find that it has failed. From which supplier is it most likely to have come?

The outcomes of the experiment

$$\mathsf{E} = p_a \mathsf{E}_a + p_b \mathsf{E}_b + p_c \mathsf{E}_c$$

are pairs (H_a, τ), (H_b, τ), and (H_c, τ), with τ the time at which the transistor fails. The event (H_a, m_t) that a transistor from supplier A was chosen and failed before time t is the set of outcomes (H_a, τ) such that $0 \le \tau < t$. Its probability is $p_a \Pr(M_t \mid H_a)$. The event observed is the union

$$M_t = (H_a, m_t) + (H_b, m_t) + (H_c, m_t)$$

of three mutually exclusive events.

Bayes's theorem tells us that the probability that hypothesis H_a, "the transistor came from supplier A," is true, given that the transistor was observed to fail before time t, is

$$\Pr(H_a \mid M_t) = \frac{p_a \Pr(M_t \mid H_a)}{\Pr(M_t)}$$

where by the principle of total probability

$$\Pr(M_t) = p_a \Pr(M_t \mid H_a) + p_b \Pr(M_t \mid H_b) + p_c \Pr(M_t \mid H_c).$$

The probabilities that the transistor came from B and C are, similarly,

$$\Pr(H_b \mid M_t) = \frac{p_b \Pr(M_t \mid H_b)}{\Pr(M_t)}$$

$$\Pr(H_c \mid M_t) = \frac{p_c \Pr(M_t \mid H_c)}{\Pr(M_t)}.$$

The most likely supplier is A, B, or C, according to which of these so-called posterior probabilities, or probabilities a posteriori, is largest. The denominator being the same in all three, we need only compare their numerators. If we plot them versus the time t, we find that

$$\Pr(H_b \mid M_t) > \Pr(H_a \mid M_t) \text{ and } \Pr(H_b \mid M_t) > \Pr(H_c \mid M_t), \qquad 0 < t < 4115 \text{ hours},$$

$$\Pr(H_a \mid M_t) > \Pr(H_b \mid M_t) \text{ and } \Pr(H_a \mid M_t) > \Pr(H_c \mid M_t), \qquad t > 4115 \text{ hours}.$$

Hence if we check a transistor before 4115 hours and find that it has failed, we conclude that it came from supplier B. If we check after 4115 hours and find that it has failed, we decide that it came from supplier A. These decisions will sometimes be in error, of course, for some transistors from supplier A will fail before 4115 hours, some from B will fail after 4115 hours, and those from C must fail sometime. (We do not propose this as a sensible method of identifying the supplier of a transistor.) It will be shown in Sec. 1-11D that if we use this decision procedure, the probability of our making an error, averaged over all possible selections of transistors in accordance with our probability model, will be as small as possible. The prescription, "Pick the hypothesis with maximum posterior probability, given the outcome or event observed," is known as "Bayes's rule." ■

EXAMPLE 1.22. A coin is provided that may be biased; the probability p that it comes up heads may be any of the nine multiples of 0.1 from 0.1 to 0.9, all nine possibilities having the same prior probability $\frac{1}{9}$. We can imagine the coin to have been drawn at random from a large heap of coins of nine types, indistinguishable to the eye and present in equal proportions. When a coin of type k is tossed, it comes up heads with probability $\pi_k = 0.1k$, $k = 1, 2, \ldots, 9$. The selected coin is tossed twenty-five times, and seven times it shows heads, eighteen times tails. Which is the most likely of the nine possible values of p for this coin?

Now hypothesis H_k states that a coin of type k was drawn; its prior probability is $\frac{1}{9}$. Bayes's theorem tells us that the posterior probability of hypothesis H_k is

$$\Pr(H_k \mid 7 \text{ H's out of } 25) = \frac{\binom{25}{7}\pi_k^7(1 - \pi_k)^{18}}{\sum_{m=1}^{9}\binom{25}{7}\pi_m^7(1 - \pi_m)^{18}}$$

We could calculate all nine of these probabilities and look for the largest, but as the denominator is the same for all and the binomial factor cancels out, it suffices to calculate the nine numbers $\pi_k^7(1 - \pi_k)^{18}$, $\pi_k = 0.1k$, and pick the largest. Since the function $p^m(1 - p)^{n-m}$ is maximum with respect to p when $p = m/n$, as shown by an easy exercise in calculus, we can expect that the largest of these nine numbers will be the one with $\pi_k = 0.1k$ closest to $m/n = \frac{7}{25} = 0.28$, and it can be checked directly that hypothesis H_3 indeed has the largest posterior probability. We decide, therefore, that the coin we selected is of type 3. We might, of course, be wrong; it might have been a coin of one of the other eight types that we were tossing. ■

CONDITIONAL PROBABILITIES IN DIGITAL COMMUNICATIONS

A. Digital Communication System

A digital communication system consists of a source of digits, a transmitter, a channel, and a receiver, as shown in Fig. 1.11. The source produces a digit every T seconds, selected from a fixed alphabet of symbols, such as $\{0, 1\}$ or $\{A, B, C\}$. The symbols express messages of some sort. To each symbol of the alphabet is assigned a different signal, which the transmitter sends along the channel whenever that symbol is emitted by the source. The signals last no longer than T seconds. The receiver sees each signal attenuated, possibly distorted by the channel, and corrupted by random noise. We assume for simplicity that the signals, even if distorted, do not overlap when they reach the receiver. On the basis of what the receiver perceives, it makes a decision about which of the possible signals was transmitted, and it issues the associated digit. How it makes these decisions will be described later. The decisions may be wrong, and the stream of digits emerging from the receiver may contain errors.

The digits emitted by the source contain an element of unpredictability or randomness, by virtue of which they convey information. If the receiver knew exactly what digits were forthcoming, it would not need to observe the incoming signals, nor indeed would the transmitter need to send them. It is appropriate, therefore, to describe the output of the source in terms of probability. That output may consist of the letters and spaces in English text, or of numbers spewed out by a computer; or the English text or the computer output may have been transformed into a new sequence of symbols by some encoding process. The mechanism of the source will not concern us, and for simplicity we shall ignore relationships or "correlations" that may exist among the digits emerging from the source. Each instance of the process of digit emission, signal transmission, and receiver decision will be treated separately. We need therefore only to specify the probabilities with which the several digits are emitted by the source.

For definiteness let us consider binary messages in which the digits are 0's and 1's. Denote their probabilities by P_0 and P_1, $P_0 + P_1 = 1$. Roughly speaking, in a long stream of digits emitted by the source we should find a fraction P_0 of 0's and a fraction P_1 of 1's. They are to be transmitted by an optical communication system.

Whenever a 1 appears, the transmitter emits a pulse of light, which lasts no longer than T seconds and travels along an optical fiber to the receiver. Whenever the source produces a 0, no light pulse is sent. (This is sometimes called an

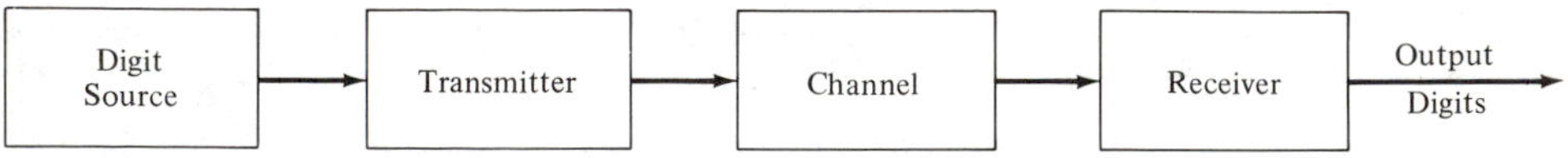

FIGURE 1.11 Digital communication system.

"on–off" system.) At the far end of the fiber is a detector, which for simplicity we take to be a photoelectric cell. During the interval of T seconds' duration when the light pulse, if sent, would arrive, the receiver counts the number of electrons that have passed through the cell into the external circuitry. Denote this number by the integer k. It tends to be larger when a light pulse emerges from the fiber, the message digit having been a 1, than when no pulse has been transmitted, the message digit having been a 0. The electrons observed to pass through the photoelectric cell when no light is incident are known as "dark current." In either case the number k is not definite, but random; its value depends on whether electrons in the cathode, agitated by light and heat, happen to burrow to its surface, escape, and impinge on the anode. The random variability of the number k of electrons counted is a kind of noise. Again a probabilistic description is necessary.

The required probabilities must be specified in terms of the event structure of a certain chance experiment E. From the standpoint of an external observer it consists of the transmission or lack of transmission of a light pulse and the counting of so many electrons at the output of the photoelectric cell. It can be considered as the weighted sum

$$\mathsf{E} = P_0\mathsf{E}_0 + P_1\mathsf{E}_1$$

of two subexperiments E_0 and E_1. In subexperiment E_0 the number k of electrons emitted by the detector during T seconds is counted in the absence of any incident light. In subexperiment E_1 the number k of electrons is counted during the arrival of a T-second pulse of light from the transmitter. Which of these subexperiments is actually carried out in any interval is determined by the transmitter: E_0 when digit 0 is issued by the message source, E_1 when digit 1 is issued. The receiver does not know whether E_0 or E_1 was performed, but must decide on the basis of the number k of electrons it counts during the interval. It chooses between two hypotheses,

H_0: The message digit was 0; E_0 was performed.

H_1: The message digit was 1; E_1 was performed.

Subexperiment E_0 is selected by the transmitter with probability P_0, which is the prior probability of hypothesis H_0; subexperiment E_1 is selected with probability P_1, the prior probability of H_1.

The physical nature of the optical fiber and the photoelectric cell and the manner in which light emerging from the fiber impinges on the cell determine the probabilities $\Pr(k \mid H_0)$ and $\Pr(k \mid H_1)$ of counting k electrons in subexperiments E_0 and E_1, respectively. With respect to the composite experiment E these are conditional probabilities. In typical circumstances they will have the Poisson form

$$\Pr(k \mid H_1) = \frac{\lambda_1^k}{k!}e^{-\lambda_1}, \qquad \Pr(k \mid H_0) = \frac{\lambda_0^k}{k!}e^{-\lambda_0}, \tag{1.34}$$

with $\lambda_0 < \lambda_1$. They are sketched as bar graphs in Fig. 1.12. That the peak of the distribution $\Pr(k \mid H_1)$ lies to the right of that of $\Pr(k \mid H_0)$ corresponds to the statement that when a light pulse falls on the detector, more electrons are

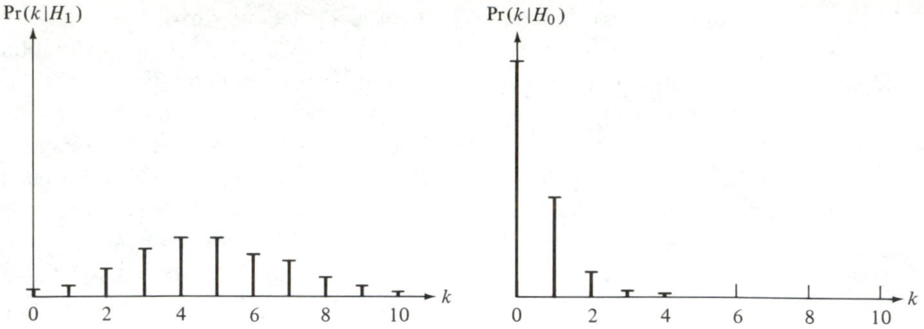

FIGURE 1.12 The conditional probabilities.

usually emitted than when one does not. By using the power series for e^x you can easily verify that, as for (1.27),

$$\sum_{k=0}^{\infty} \Pr(k \mid H_0) = \sum_{k=0}^{\infty} \Pr(k \mid H_1) = 1. \tag{1.35}$$

An outcome of the composite experiment E can be specified by a pair (H_i, k), where i is 1 or 0 depending on whether a pulse is or is not transmitted, and k is the number of electrons counted, $k = 0, 1, 2, \ldots$. Since these outcomes form a countable set, we can take each outcome (H_i, k) as an elementary event in E. By the definition of conditional probability the basic probability measure in the composite experiment E consists of probabilities of the form

$$\Pr(H_0, k) = P_0 \Pr(k \mid H_0) = \frac{P_0 \lambda_0^k e^{-\lambda_0}}{k!} \tag{1.36}$$

$$\Pr(H_1, k) = P_1 \Pr(k \mid H_1) = \frac{P_1 \lambda_1^k e^{-\lambda_1}}{k!} \tag{1.37}$$

for all k in $0 \le k < \infty$. All these probabilities add up to 1,

$$\sum_{i=0}^{1} \sum_{k=0}^{\infty} \Pr(H_i, k) = 1. \tag{1.38}$$

B. The Receiver as a Decision Mechanism

The receiver must decide every T seconds whether a pulse was transmitted or not, that is, whether the current message digit is a 0 or a 1. In the terminology of hypothesis testing, the receiver chooses between hypotheses H_0 and H_1. It must base its decision on the number k of electrons that it counts. Once and for all it will have divided the range $(0, \infty)$ of possible values of k into two parts, which we denote by R_0 and R_1. If k turns out to lie in R_0—we denote this by $k \in R_0$—the receiver decides that no pulse arrived and that the message digit is 0; if $k \in R_1$, it decides that a pulse arrived and that the message digit is 1. In

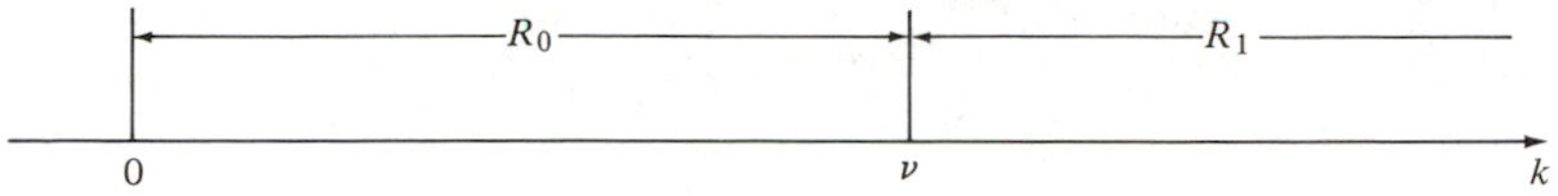

FIGURE 1.13 Decision regions.

the present system those regions of the scale of integers will be as indicated in Fig. 1.13, R_0 consisting of integers k in $0 \le k \le \nu$, R_1 of integers k in $k > \nu$, where ν is some integral point of dichotomy. How can ν best be determined?

Once the value of ν—that is, the classification of the integers into R_0 and R_1—has been specified, events known as errors and correct decisions can be defined. A *correct decision* is made if a pulse was sent and $k \in R_1$, or if no pulse was sent and $k \in R_0$; this event consists of all outcomes (H_0, k) with $0 \le k \le \nu$ and all outcomes (H_1, k) with $k > \nu$. Its probability is

$$\Pr(\text{correct decision}) = \sum_{k=0}^{\nu} \Pr(H_0, k) + \sum_{k=\nu+1}^{\infty} \Pr(H_1, k). \tag{1.39}$$

The complementary event is an *error*; its probability is

$$\Pr(\text{error}) = \sum_{k=\nu+1}^{\infty} \Pr(H_0, k) + \sum_{k=0}^{\nu} \Pr(H_1, k). \tag{1.40}$$

The designer wants this system to attain the smallest possible value of Pr (error), or the largest possible value of Pr (correct decision), and will select the point ν of dichotomy on that basis.

It is shown in part D that this goal of minimum error probability is attained if the receiver in effect bases its decisions on the conditional probabilities $\Pr(H_1 \mid k)$ and $\Pr(H_0 \mid k)$. These are the conditional probabilities that a pulse was or was not sent, given that k electrons were counted. According to Bayes's theorem,

$$\Pr(H_1 \mid k) = \frac{\Pr(H_1, k)}{\Pr(k)} = \frac{P_1 \Pr(k \mid H_1)}{\Pr(k)}, \tag{1.41}$$

where

$$\begin{aligned} \Pr(k) &= \Pr(H_0, k) + \Pr(H_1, k) \\ &= P_0 \Pr(k \mid H_0) + P_1 \Pr(k \mid H_1) \end{aligned} \tag{1.42}$$

is the probability that k electrons are counted without regard to whether a pulse was or was not sent. [In a long series of trials of the composite experiment E, that is, for a long sequence of transmitted symbols, 0's and 1's, exactly k electrons are counted in a fraction $\Pr(k)$ of the trials.] Similarly,

$$\Pr(H_0 \mid k) = \frac{\Pr(H_0, k)}{\Pr(k)} = \frac{P_0 \Pr(k \mid H_0)}{\Pr(k)}. \tag{1.43}$$

$\Pr(H_0 \mid k)$ and $\Pr(H_1 \mid k)$ are the posterior probabilities of the hypotheses H_0 and H_1, the number k of counts being given. They sum to 1,

$$\Pr(H_0 \mid k) + \Pr(H_1 \mid k) = 1.$$

The receiver emits a 1 if

$$\Pr(H_1 \mid k) > \Pr(H_0 \mid k); \tag{1.44}$$

otherwise it emits a 0. To choose the hypothesis with the greater posterior probability is to apply Bayes's rule.

From (1.41) and (1.43) you can easily show that the receiver decides for H_1 if

$$\frac{\Pr(k \mid H_1)}{\Pr(k \mid H_0)} > \frac{P_0}{P_1}. \tag{1.45}$$

The quantity on the left is called the *likelihood ratio*. From (1.34) it is given in this problem by

$$\frac{\Pr(k \mid H_1)}{\Pr(k \mid H_0)} = \left(\frac{\lambda_1}{\lambda_0}\right)^k \exp[-(\lambda_1 - \lambda_0)].$$

Hence the receiver decides for H_1 and issues the digit 1 if

$$k > \frac{\lambda_1 - \lambda_0 + \ln(P_0/P_1)}{\ln(\lambda_1/\lambda_0)}; \tag{1.46}$$

otherwise it decides for H_0 and issues a 0. The boundary integer ν in Fig. 1.13 is the greatest integer in the quantity on the right-hand side of (1.46). Having determined ν from (1.46) or otherwise, one can calculate the error probability from (1.40).

C. Ternary Communication System

The same principles apply to more complex communication systems. Suppose, for instance, that messages are encoded into an alphabet of three symbols $\{A, B, C\}$. The relative frequencies with which they occur are P_A, P_B, and P_C,

$$P_A + P_B + P_C = 1.$$

The transmitter has emitters of red, yellow, and blue light. Each A in the output of the digit source causes it to send a pulse of red light, each B a pulse of yellow light, and each C a pulse of blue light. At the far end of the optical fiber a prism imperfectly divides the light, passing most of the red light to detector 1, most of the yellow to detector 2, and most of the blue to detector 3. These three detectors are photoelectric cells, and during each pulse interval of duration T the receiver counts the number of electrons passing from each. Denote these three numbers by k_1, k_2, and k_3.

Now three subexperiments E_A, E_B, and E_C are at hand, and in the composite experiment

$$\mathsf{E} = P_A\mathsf{E}_A + P_B\mathsf{E}_B + P_C\mathsf{E}_C$$

the outcomes are denoted by (H_i, k_1, k_2, k_3), where $i = A, B,$ or C, and

$$0 \le k_1 < \infty, \qquad 0 \le k_2 < \infty, \qquad 0 \le k_3 < \infty.$$

The basic elements of the probability measure on E are the probabilities

$$\Pr(H_i, k_1, k_2, k_3) = P_i \Pr(k_1, k_2, k_3 \mid H_i), \qquad i = A, B, C;$$

$\Pr(k_1, k_2, k_3 \mid H_A)$ is the probability that detectors 1, 2, and 3 emit k_1, k_2, and k_3 photoelectrons, respectively, when a red pulse arrives along the fiber. It might have the form

$$\Pr(k_1, k_2, k_3 \mid H_A) = \frac{\lambda_{1A}^{k_1} \exp(-\lambda_{1A})}{k_1!} \frac{\lambda_{2A}^{k_2} \exp(-\lambda_{2A})}{k_2!} \frac{\lambda_{3A}^{k_3} \exp(-\lambda_{3A})}{k_3!} \tag{1.47}$$

for certain known constants $\lambda_{1A} > \lambda_{2A} > \lambda_{3A}$. The conditional probabilities $\Pr(k_1, k_2, k_3 \mid H_B)$ and $\Pr(k_1, k_2, k_3 \mid H_C)$ would be given by similar expressions.

The receiver must now choose among the three hypotheses labeled H_A, H_B, and H_C; H_A is the hypothesis "a red pulse was transmitted; subexperiment E_A is being performed," and so on. The receiver is equipped with a decision rule that divides the set of all triads (k_1, k_2, k_3) of nonnegative integers into three subsets R_A, R_B, and R_C. If $(k_1, k_2, k_3) \in R_A$, the receiver decides for hypothesis H_A and issues the digit A; if $(k_1, k_2, k_3) \in R_B$, it issues a B; if $(k_1, k_2, k_3) \in R_C$, a C. Again the events "correct decision" and "error" can be identified and their probabilities calculated:

$$\begin{aligned}\Pr(\text{correct decision}) = & \sum_{(k_1,k_2,k_3)\in R_A} \Pr(H_A, k_1, k_2, k_3) \\ & + \sum_{(k_1,k_2,k_3)\in R_B} \Pr(H_B, k_1, k_2, k_3) \\ & + \sum_{(k_1,k_2,k_3)\in R_C} \Pr(H_C, k_1, k_2, k_3)\end{aligned} \tag{1.48}$$

and $\Pr(\text{error}) = 1 - \Pr(\text{correct decision})$.

If the probability of error is to be minimum, or the probability of correct decision maximum, the receiver will apply Bayes's rule. That is, for the triad of numbers k_1, k_2, and k_3 of electrons counted at the three detectors, it will calculate the three posterior probabilities

$$\Pr(H_A \mid k_1, k_2, k_3), \quad \Pr(H_B \mid k_1, k_2, k_3), \quad \Pr(H_C \mid k_1, k_2, k_3).$$

It will decide for hypothesis H_A and issue the digit A if the first of these conditional probabilities is the largest of the three; if the second is largest, it issues digit B; and if the third is largest, digit C. This rule can be translated into a division of the set of all triads (k_1, k_2, k_3) into decision regions R_A, R_B, and R_C by using formulas like those in (1.47). The first posterior probability, for instance, is

$$\begin{aligned}\Pr(H_A \mid k_1, k_2, k_3) &= \frac{\Pr(H_A, k_1, k_2, k_3)}{\Pr(k_1, k_2, k_3)} \\ &= \frac{P_A \Pr(k_1, k_2, k_3 \mid H_A)}{\Pr(k_1, k_2, k_3)}\end{aligned} \tag{1.49}$$

with

$$\Pr(k_1, k_2, k_3) = P_A \Pr(k_1, k_2, k_3 \mid H_A) + P_B \Pr(k_1, k_2, k_3 \mid H_B) + P_C \Pr(k_1, k_2, k_3 \mid H_C)$$

by the principle of total probability. The remaining posterior probabilities are given by similar expressions. The region R_A consists of all triads for which

$$\Pr(H_A \mid k_1, k_2, k_3) > \Pr(H_B \mid k_1, k_2, k_3)$$

and

$$\Pr(H_A \mid k_1, k_2, k_3) > \Pr(H_C \mid k_1, k_2, k_3),$$

and R_B and R_C are similarly specified.

D. Derivation of Bayes's Rule

We shall prove that Bayes's rule maximizes the probability of correct decision for a ternary communication system. It will be apparent that the result can be generalized to a system transmitting any number of signals, or indeed to any problem in testing hypotheses.

From (1.48) and (1.49) we put

$$\begin{aligned}\Pr(\text{correct decision}) = &\sum_{\mathbf{k}\in R_A} \Pr(\mathbf{k}) \Pr(H_A \mid \mathbf{k}) \\ &+ \sum_{\mathbf{k}\in R_B} \Pr(\mathbf{k}) \Pr(H_B \mid \mathbf{k}) \\ &+ \sum_{\mathbf{k}\in R_C} \Pr(\mathbf{k}) \Pr(H_C \mid \mathbf{k}),\end{aligned} \tag{1.50}$$

with $\mathbf{k}$ denoting the triad (k_1, k_2, k_3) for short. The regions R_A, R_B, and R_C in the three-dimensional lattice of all triads have yet to be determined. Each triad $\mathbf{k}$ must lie in one of them. Imagine the process of choosing the region to which a given triad should be assigned, building up the regions one triad at a time. For that triad $\mathbf{k}$ we calculate the three conditional probabilities $\Pr(H_A \mid \mathbf{k})$, $\Pr(H_B \mid \mathbf{k})$, and $\Pr(H_C \mid \mathbf{k})$. The way to make the quantity on the right-hand side of (1.50) as large as possible is to put $\mathbf{k}$ into R_A if $\Pr(H_A \mid \mathbf{k})$ is the largest of these three, to put $\mathbf{k}$ into R_B if $\Pr(H_B \mid \mathbf{k})$ is the largest, and to put $\mathbf{k}$ into R_C if $\Pr(H_C \mid \mathbf{k})$ is the largest. (When there is a tie among two or three of these conditional probabilities for first place, it does not matter to which of the corresponding regions the triad $\mathbf{k}$ is assigned.) If we do this for all triads in the lattice, Pr (correct decision) will be as large as it can possibly be. The resulting division of the set of all triads into R_A, R_B, and R_C is equivalent to applying Bayes's rule, "Choose the hypothesis with the greatest posterior probability."

In setting up a receiver to make decisions by Bayes's rule, probability is being applied in the hypothetical mode defined at the beginning of Sec. 1-4. Just what sequence of digits will be issued by the message source, and how many photoelectrons each received pulse will induce in the detector, cannot be predicted.

Any proposed strategy by which the receiver might make its decisions must be tested with respect to the entire ensemble of possible combinations of digits and photoelectron counts. In effect the number of errors that result is determined for each such combination and weighted in accordance with the probability that that combination occurs. The probabilities are drawn from a model of the communication system that takes into account the nature of message source, channel, and detector. Prior probabilities such as P_0 and P_1, or P_A, P_B, and P_C, may be estimated from a long sequence of message digits observed during some past interval, or from studies of the language expressing the messages and the manner of their encoding. Conditional probabilities such as $\Pr(k \mid 0)$ and $\Pr(k \mid 1)$ for the numbers k of photoelectrons counted are derived from the physics of channel and detector. The model is justified by the past experience that validated the scientific theories on which it is founded. The probability Pr (error) calculated from this model assesses the effectiveness of a given decision strategy, for among all combinations of very long sequences of $N >> 1$ transmitted digits and the ensuing numbers of counts, by far the greatest probability will attach to those in which the number of errors is close to N Pr (error)—in a sense whose precise formulation must await further development of this subject. The designer builds into the receiver that strategy for which the probability Pr (error) is minimum. We have shown how it can be derived.

PROBLEMS

1-1 For the events A, B, and C defined for experiment E_f on page 5, determine the events A', B', C', $A \cap B$, $A \cap C$, $B \cap C$, $A' \cap B$, $B' \cap C'$, $A \cup B$, $B \cup C$, and $B' \cup C'$. Be careful to distinguish "$<$" from "$\leq$," "$>$" from "$\geq$."

1-2 Illustrate the rules (1.1) through (1.8) by means of Venn diagrams. Use colored pencils if helpful.

1-3 Referring to experiment E_g, represent on a diagram the events A, B, and C defined on page 5, and indicate on similar diagrams the events C', $A' \cap B$, $A \cap C$, $(A \cap B) \cup C$.

1-4 Let S be the set of positive integers $n = 1, 2, 3, \ldots$. Given the following three subsets of S,

$$A = \{n : n \text{ is divisible by } 3\} = \{3, 6, 9, 12, \ldots\},$$
$$B = \{n : n \text{ is divisible by } 4\} = \{4, 8, 12, 16, \ldots\},$$
$$C = \{n : n \text{ is divisible by } 5\} = \{5, 10, 15, 20, \ldots\},$$

find the sets A', B', C', $A \cap B$, $A \cup B$, $A \cap C$, $A \cup C$, and $A \cap B \cap C$.

1-5 Let S be the upper-right quadrant of the XY-plane, that is

$$S = \{(x, y) : x \geq 0, y \geq 0\}.$$

Given the following three subsets of S,

$$A = \{(x, y) : x \geq 2y\},$$
$$B = \{(x, y) : x + y < 5\},$$
$$C = \{(x, y) : xy > 1\},$$

sketch in the XY-plane the sets A', B', C', $A \cap B$, $A \cup B$, $A \cap C$, $A \cup C$, and $A \cap B \cap C$.

1-6 Write a detailed proof of Corollary 4 by using Axioms I–III and the rules of set theory.

1-7 Derive the counterpart of Corollary 4 for three overlapping events A, B, and C, that is, three events that may have some outcomes in common.

1-8 Show that for any n events

$$\Pr (A_1 \cup A_2 \cup A_3 \cup \cdots \cup A_n) = \Pr \left(\bigcup_{i=1}^{n} A_i \right) \le \sum_{i=1}^{n} \Pr (A_i).$$

Communication theorists call this the "union bound" and find it useful when the event on the left is too complicated for its probability to be calculated exactly.

1-9 Suppose that the intervals in Example 1.4 overlap: $a < c < b < d$. Determine $\Pr \{v : a \le v < b \text{ or } c \le v < d\}$. Verify Corollary 4 for this example, taking $A = \{v : a \le v < b\}$, $B = \{v : c \le v < d\}$ and determining $\Pr (A)$, $\Pr (B)$, and $\Pr (A \cap B)$ as well.

1-10 A child's game consists of spinning two arrows, each pivoted about its center of gravity. An arrow is just as likely to stop at one azimuth as at another, and where the one stops is independent of where the other stops. The board under the first arrow is divided into angular sectors numbered 1 through 5 and encompassing the following angles:

$$1{:}\ 0° < \theta < 30°, \quad 2{:}\ 30° < \theta < 80°, \quad 3{:}\ 80° < \theta < 150°,$$
$$4{:}\ 150° < \theta < 230°, \quad 5{:}\ 230° < \theta < 360°.$$

The board under the second arrow is similarly divided:

$$1{:}\ 0° < \theta < 70°, \quad 2{:}\ 70° < \theta < 100°, \quad 3{:}\ 100° < \theta < 160°,$$
$$4{:}\ 160° < \theta < 280°, \quad 5{:}\ 280° < \theta < 360°.$$

One's score is the sum of the numbers of the sectors in which the two arrows stop. Calculate the probability $\Pr (k)$ that one achieves a score of k points for all integers k from 2 through 10. Be sure that your answers add up to 1.

1-11 Under each of the assumptions (a), (b), and (c) in Sec. 1-4A, calculate the probability $\Pr (0 \le \theta < \phi)$ that the angle θ subtended by the chord AB in Fig. 1.2 is less than ϕ, $0 \le \phi < \pi$, and sketch these probabilities as functions of the angle ϕ. (Bertrand's paradox is usually discussed for $\phi = 2\pi/3$, whereupon the chord AB is required to be shorter than the side of an equilateral triangle inscribed in the circle, but geometrical elegance is the only virtue of that special case.)

1-12 A rectangular aquarium is $a \times b \times c$. A certain fish is just as likely to be in one place in the aquarium as in another; that is, the probability of finding it in a region B of volume $\text{Vol}(B)$ is $\Pr (B) = \text{Vol}(B)/abc$. Show that the three events

$$A_1 = \{x : x_1 < x < x_2\}, \qquad A_2 = \{y : y_1 < y < y_2\}, \qquad A_3 = \{z : z_1 < z < z_2\},$$

where (x, y, z) are the rectangular coordinates of the fish, are statistically independent.

1-13 Show that two mutually exclusive events A and B are statistically independent if and only if one or the other or both have probability 0.

1-14 In the circuit in Fig. 1P-14 the switches $\alpha_1, \alpha_2, \ldots, \alpha_5$ open and close randomly and independently. The probability that switch α_k is closed at any given time equals p_k. Calculate the probability that at any time there is at least one closed path from point 1 to point 2. (*Hint*: One method is to use Boolean algebra, as in Example 1.12, and the result of Problem 1-7. Another is to write down all possible combinations of open and closed switches and note for which combinations there exists at least one closed path from 1 to 2.)

1-15 **(a)** Two persons, A and B, alternately cast dice according to a rule that results in a win on any cast with probability p. A starts, and they continue one after the other until one of them makes a winning throw. Calculate the probability that A eventually wins. (*Hint*: Write down the possible sequences of failures and successes that lead to A's winning, and add up their probabilities. The throws are, of course, statistically independent.)

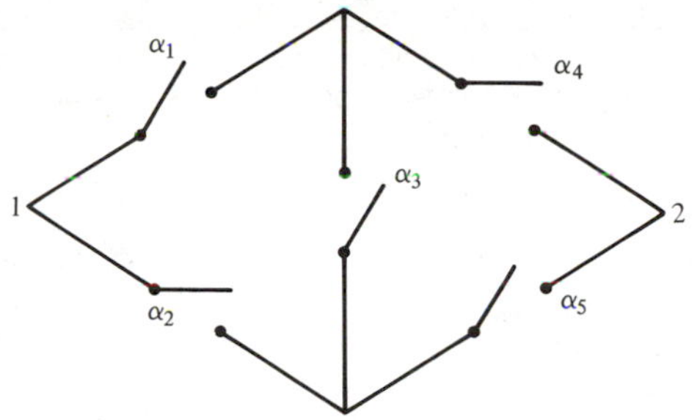

FIGURE 1P-14 Switching network.

(b) Three persons, A, B, and C, cast the same dice according to the same rule for winning, one after another in that order, A starting, until someone wins. Calculate each player's probability of winning this game.

1-16 Five cars start out on a cross-country race. The probability that a car breaks down and drops out of the race is 0.2. What is the probability that at least three cars finish the race?

1-17 A biased coin has a known probability p of coming up heads when tossed. It is tossed six times by someone else, who tells you that heads turned up in more than half the tosses. What is the probability, given that information, that heads appeared in all six tosses?

1-18 Three mutually exclusive events A_1, A_2, and A_3 exhaust all possible outcomes of chance experiment E,

$$A_i \cap A_j = \varnothing, \qquad \text{all } i \neq j, \qquad A_1 + A_2 + A_3 = S.$$

Their probabilities are $\Pr(A_i) = p_i$, $p_1 + p_2 + p_3 = 1$. If n independent trials of experiment E are conducted, what is the probability that event A_1 occurs k_1 times, event A_2 k_2 times, and event A_3 k_3 times, $k_1 + k_2 + k_3 = n$? Generalize your result to m such mutually exclusive events, occurring respectively $k_1, k_2, \ldots k_m$ times in n trials, $k_1 + k_2 + \cdots + k_m = n$.

1-19 Calculate the binomial probability in (1.25) for $k = 2$ and
(a) $n = 10, p = 0.1$,
(b) $n = 20, p = 0.05$,
(c) $n = 40, p = 0.025$.
Compare the results with the Poisson probability in (1.27) with $m = np$, $k = 2$.

1-20 Light falls on a photoelectric emitter with fixed intensity. The probability that n electrons are emitted in any interval of duration T seconds is

$$\Pr(n\ e\text{'s in } T \text{ sec}) = \frac{(\lambda T)^n e^{-\lambda T}}{n!}$$

where λ is a constant proportional to the intensity. The numbers n_1 and n_2 of photoelectrons emitted in disjoint intervals of time are statistically independent. Thus the probability that n_1 are emitted in $(0, t_1)$ and n_2 in (t_1, t_2) is

$$\Pr(n_1\ e\text{'s in } (0, t_1) \text{ and } n_2\ e\text{'s in } (t_1, t_2)) = \frac{(\lambda t_1)^{n_1} \exp(-\lambda t_1)}{n_1!} \frac{[\lambda(t_2 - t_1)]^{n_2} \exp[-\lambda(t_2 - t_1)]}{n_2!}$$

Calculate the conditional probability that n_1 electrons are emitted in $(0, t_1)$, given that a total of $N = n_1 + n_2$ are emitted in the combined interval $(0, t_2)$. Show that it is independent of λ.

1-21 Light beams of different intensities are incident on two photoelectric detectors. Our experiment consists of counting the numbers n_1 and n_2 of photoelectrons emitted in each detector during a fixed interval of time. The atomic events ζ are thus pairs of nonnegative integers (n_1, n_2), and the probability of atomic event (n_1, n_2) is

$$\Pr(n_1, n_2) = (1 - v_1)(1 - v_2)\, v_1^{n1} v_2^{n2}, \qquad n_1 \geq 0, \quad n_2 \geq 0.$$

Find the conditional probability of the event $n_1 \geq n_2$, given that the second detector emits exactly k electrons; that is, find

$$\Pr(n_1 \geq n_2 \mid n_2 = k).$$

Now calculate the total probability $\Pr(n_1 \geq n_2)$ that $n_1 \geq n_2$, with no condition on n_2. [*Hint*: Draw the (n_1, n_2)-plane and note that the atomic events are points on a lattice in this plane. Now identify the composite events in question.]

1-22 In the context of Example 1.21, suppose that we check at time t and find that a transistor is still working. Which was the most probable supplier, A, B, or C? Calculate the probability of error in a decision as to the supplier based on Bayes's rule.

1-23 **(a)** Calculate the probability of error in Example 1.21 as a function of time t of observation.

(b) If at time t when we check the transistor, we find it working, we decide as to its supplier as in Problem 1-22; if we find that it has failed, we decide as in Example 1.21. Calculate the average probability of error of this procedure as a function of the time t of observation. For what observation time t is the probability of error minimum?

1-24 In Example 1.22, utilize Bayes's rule to divide the range of values of the number m of heads, $0 \leq m \leq 25$, into nine intervals R_k, such that if $m \in R_k$, hypothesis H_k, "A coin of the type k was tossed; the probability of heads is $\pi_k = 0.1k$," is chosen. That is, for each value of m from 0 to 25, do the same as was done in Example 1.22 for $m = 7$. The technique of Sec. 1-11 will simplify this determination. Calculate the overall probability of error incurred when this strategy is used for deciding which of the nine types of coins one had on the basis of how many heads occur in $n = 25$ tosses. Again assume that each hypothesis is equally probable.

1-25 A communication system transmits signals labeled 1, 2, and 3. The probability that symbol j is sent *and* symbol k is received is listed in the table for each pair (j, k) of sent and received symbols. For example, the probability is 0.12 that 1 is sent and, owing to noise in the channel, 3 is received.

Received

			k	
	j	**1**	**2**	**3**
	1	0.10	0.06	0.12
Sent	**2**	0.07	0.15	0.05
	3	0.10	0.15	0.20

Calculate the probability that symbol k was sent, given that symbol k is received, for $k = 1, 2, 3$, and calculate the probability of error incurred in using this system. An error is defined as the reception of any symbol other than the one transmitted.

1-26 In the binary communication system of Sec. 1-11, an "error of the first kind" is made when a 0 is sent, but the receiver decides for a 1. It is sometimes called a "false alarm." An "error of the second kind," or "false dismissal," is made when a 1 is sent, but the receiver decides for a 0. In terms of the probabilities introduced in (1.34) to (1.37), write down the conditional probabilities of errors of the first and second kinds. Express the probability Pr (error) in terms of these and P_0 and P_1.

1-27 In the binary communication system of Sec. 1-11, take $P_0 = 0.4$, $P_1 = 0.6$, $\lambda_0 = 0.5$, and $\lambda_1 = 7.0$. Determine the point ν of dichotomy in Fig. 1.13 so that the probability of error, Pr (error), is minimum; use (1.46). Calculate the resulting probabilities of errors of the first and second kinds as defined in Problem 1-26, and calculate Pr (error) and Pr (correct decision) as defined in (1.39) and (1.40).

1-28 **(a)** A cereal manufacturer puts a picture of one of three famous engineers into each box of cereal, distributing them at random, so that the probability that a box contains a picture of a given engineer equals $\frac{1}{3}$. You buy the cereal until you have a complete collection of all three pictures. What is the probability that you reach your goal upon buying your nth box of cereal, for $n = 3, 4, 5$, and 6? [*Hint*: Your collection is in state k when it contains k different pictures, no matter how many of each kind you have. It starts in state 1 with the first box purchased. When it reaches state 3, you can stop buying cereal. Let $P_k(n)$ be the probability that the collection is in state k after n boxes have been bought, $k = 1, 2, 3$. Show by the principle of total probability that these probabilities obey the equations

$$P_1(n + 1) = \tfrac{1}{3}P_1(n),$$

$$P_2(n + 1) = \tfrac{2}{3}P_1(n) + \tfrac{2}{3}P_2(n),$$

$$P_3(n + 1) = \tfrac{1}{3}P_2(n) + P_3(n).$$

Solve these successively, starting with the initial values $P_1(1) = 1$, $P_2(1) = P_3(1) = 0$. If you are familiar with z-transforms, you should be able to find the general solution for all values of n.]

(b) Suppose that there are M different pictures in the collection instead of three. The probability that a box contains a particular picture equals $1/M$. Write down the equations, corresponding to those in the hint in part (a), for the M probabilities $P_k(n)$, $k = 1, 2, \ldots, M$. Do not bother to solve them.

1-29 In a binary optical communication system the receiver counts the number of photoelectrons ejected by the light incident on a photocell during an interval $(0, T)$. When no light signal has been transmitted toward the photocell (hypothesis H_0), the probability that k electrons are counted is

$$\Pr(k \mid H_0) = (1 - v_0)v_0^k, \qquad k = 0, 1, 2, \ldots .$$

When a signal has been transmitted, on the other hand (hypothesis H_1),

$$\Pr(k \mid H_1) = (1 - v_1)v_1^k, \qquad k = 0, 1, 2, \ldots$$

with $0 < v_0 < v_1 < 1$. A signal is just as likely as not to have been transmitted; that is, its prior probability equals $\frac{1}{2}$.

(a) Determine the conditional or posterior probability that a signal was sent, given that exactly m photoelectrons were counted.

(b) When $k \geq n_0$ the receiver decides that a signal was indeed sent; when $0 \leq k < n_0$, it decides that no signal was sent, where n_0 is some positive integer. Calculate the probability P_e of error incurred by the receiver in terms of n_0, v_0, and v_1. For what value of n_0 is the probability P_e of error minimum?

BIBLIOGRAPHY

FOUNDATIONS OF PROBABILITY

AYER, A. J., *Probability and Evidence*. New York: Columbia University Press, 1972.

CARNAP, R., *Logical Foundations of Probability*. Chicago: University of Chicago Press, 1962.

DAVID, F. N., *Games, Gods, and Gambling*. New York: Hafner, 1962.

DAVID, F. N. and BARTON, D. E., *Combinatorial Chance*. London: Griffin, 1962.

DE FINETTI, B., *Theory of Probability*, 2 vols. New York: Wiley, 1974.

FINE, T., *Theories of Probability*. New York: Academic Press, 1973.

GOOD, I. J., *Probability and the Weighing of Evidence*. London: Griffin, 1950.

JEFFREYS, H., *Scientific Inference*, 2nd ed. Cambridge: Cambridge University Press, 1957.

KEYNES, J. M., *A Treatise on Probability*. London: Macmillan, 1963.

KNEALE, W. C., *Probability and Induction*. Oxford: Oxford University Press, 1952.

KOLMOGOROV, A. N., *Foundations of the Theory of Probability*. New York: Chelsea, 1956.

KOLMOGOROV, A. N., "Logical basis for information theory and probability theory," *IEEE Transactions on Information Theory, IT-14,* 662–664, Sept. 1968.

KÖRNER, S., ed. *Observation and Interpretation*. London: Butterworths, 1957.

MELLOR, D. H. *The Matter of Chance*. Cambridge: Cambridge University Press, 1971.

POINCARÉ, H., *Science et Méthode*. Paris: Flammarion, n.d. See Chap. IV, "Le hasard," pp. 67–98.

REICHENBACH, H., *The Theory of Probability*. Berkeley: University of California Press, 1949.

SAVAGE, L. J., *The Foundations of Statistics*. New York: Wiley, 1954 (Reprinted by Dover, New York, 1972).

TODHUNTER, I., *A History of the Mathematical Theory of Probability*. New York: Chelsea, 1949.

VON MISES, R., *Probability, Statistics, and Truth,* 2nd ed. New York: Macmillan, 1957.

TEXTBOOKS AND SOURCES OF EXAMPLES AND PROBLEMS

BECKMANN, P., *Probability in Communication Engineering*. New York: Harcourt Brace & World, 1967.

COOPER, G. R. and MCGILLEM, C. D., *Probabilistic Methods of Signal and System Analysis*. New York: Holt, Rinehart and Winston, 1971.

DAVENPORT, W. F., *Probability and Random Processes: An Introduction for Applied Scientists and Engineers*. New York: McGraw-Hill, 1970.

FELLER, W., *An Introduction to Probability Theory and Its Applications,* 2 vols. New York: Wiley, 1950, 1966.

PAPOULIS, A., *Probability, Random Variables, and Stochastic Processes*. New York: McGraw-Hill, 1965.

PARZEN, E., *Modern Probability Theory and its Applications*. New York: Wiley, 1960.

PEEBLES, P. Z., *Probability, Random Variables, and Random Signal Principles*. New York: McGraw-Hill, 1980.

SPIEGEL, M. R., *Schaum's Outline of Theory and Problems of Probability and Statistics*. New York: McGraw-Hill, 1975.

THOMAS, J. B., *An Introduction to Statistical Communication Theory*. New York: Wiley, 1969.

THOMAS, J. B., *An Introduction to Applied Probability and Random Processes*. New York: Wiley, 1971.

CHAPTER 2

Random Variables

2-1 PROBABILITY DISTRIBUTIONS

A. Random Variables Defined

Nearly all applications of probability to science and engineering are concerned with chance experiments whose outcomes are labeled by numbers. Voltages and currents are measured in circuits; electrons, protons, or other particles are counted; arrival times and Doppler shifts of radar signals are estimated; intensities of acoustic or electromagnetic fields are determined; velocities of turbulent fluids are measured. We continue to refer to chance experiments in the sense of Sec. 1-2; events are sets of outcomes of a chance experiment. Numbers labeling outcomes of experiments to whose events a probability measure has been assigned are called *random variables*.

The outcome ζ of such an experiment may be labeled by a single number, by a pair of numbers, by an n-tuple of numbers, or even—as we shall learn in Chapter 5—by a whole function of a parameter such as the time. We begin by considering chance experiments whose outcomes are labeled by single numbers. To avoid prolixity we shall often say that the outcome of the experiment *is* a number, which is termed a random variable if a probability measure has been assigned to events. A random variable is represented by a boldface letter, such as **x, y, v, n.** A random variable takes on numerical values in some universal set *S,* such as the nonnegative integers or the natural numbers. Subsets of *S* are events, and the theory concerns itself only with those subsets to which probabilities can be assigned in a manner consistent with the four axioms of Sec. 1-3. These sets suffice in practice.

Refer to the chance experiments exemplified in Sec. 1-2. Once a probability measure has been assigned to the events of experiment E_e, the number of photoelectrons counted becomes a random variable and can be represented, say, by the letter **n.** Now the atomic events of E_e are designated by $(\mathbf{n} = 0)$, $(\mathbf{n} = 1)$, $(\mathbf{n} = 2)$, and so on. The probabilities of these might, for instance, be given by the Poisson distribution of Sec. 1-9. The probability that the random variable **n** takes on the particular integral value k is then

$$\Pr(\mathbf{n} = k) = \frac{m^k e^{-m}}{k!}$$

for some positive parameter m. Observe that an italic letter like k stands for a particular numerical value that the random variable might take on in a trial of the experiment. Other, composite events could be designated by inequalities and their probabilities calculated by applying Corollary 3 of Sec. 1-3 and adding the probabilities of the component atomic events. For instance,

$$\Pr(5 \le \mathbf{n} \le 10) = \sum_{k=5}^{10} \Pr(\mathbf{n} = k).$$

In Sec. 1-3 we assigned a simple probability measure to the events of experiment E_f, in which the voltage is measured at a certain point of a circuit τ seconds after the circuit is turned on. That assignment made the voltage into a random variable, which we now represent by the letter **v.** An event in E_f might be that the outcome of the measurement is labeled by a number lying between a and b volts; we designate this event by $(a < \mathbf{v} < b)$. Our probability measure was such that

$$\Pr(a < \mathbf{v} < b) = b - a$$

for $0 \le a \le b \le 1$, zero probability having been assigned to the events $(\mathbf{v} < 0)$ and $(\mathbf{v} > 1)$.

B. Cumulative Distribution Functions

A random variable is the numerical label of an outcome of a chance experiment to whose events a probability measure has been assigned. That assignment is usually made in one of two standard forms—through either the cumulative distribution function or the probability density function. We introduce the former in this part, the latter in the next.

Let **x** be our random variable. The cumulative distribution function $F_{\mathbf{x}}(x)$ is defined by

$$F_{\mathbf{x}}(x) = \Pr(\mathbf{x} \le x) \tag{2.1}$$

for all values x in $-\infty < x < \infty$, that is, the cumulative distribution function $F_{\mathbf{x}}(x)$ is the probability of the event, "The random variable $\mathbf{x}$ takes on a value equal to or less than x in a trial of the chance experiment E."

From the cumulative distribution function, the probability of an event such as $(x_1 < \mathbf{x} \le x_2)$ can be deduced. The events $(\mathbf{x} \le x_1)$ and $(x_1 < \mathbf{x} \le x_2)$ are mutually exclusive, and

$$(\mathbf{x} \le x_1) + (x_1 < \mathbf{x} \le x_2) = (\mathbf{x} \le x_2).$$

Therefore by Axiom III and (2.1),

$$F_{\mathbf{x}}(x_1) + \Pr(x_1 < \mathbf{x} \le x_2) = F_{\mathbf{x}}(x_2)$$

and hence

$$\Pr(x_1 < \mathbf{x} \le x_2) = F_{\mathbf{x}}(x_2) - F_{\mathbf{x}}(x_1). \tag{2.2}$$

Because probabilities must always be nonnegative, (2.2) shows us that the function $F_{\mathbf{x}}(x)$ can never decrease as x increases. It is customary and sensible to disallow $-\infty$ and $+\infty$ as possible outcomes of an experiment, and therefore

$$F_{\mathbf{x}}(-\infty) = 0, \qquad F_{\mathbf{x}}(\infty) = 1$$

because the event $(\mathbf{x} \le -\infty)$ is empty and the event $(\mathbf{x} < \infty)$ must include the entire set S of possible outcomes of the experiment. The cumulative distribution function $F_{\mathbf{x}}(x)$ generally rises from 0 to 1 as x goes from $-\infty$ to ∞, but it may remain constant over certain ranges of values of x, as our examples will show.

EXAMPLE 2.1 The Poisson Distribution on E_e

The cumulative distribution function is

$$F_{\mathbf{n}}(x) = \Pr(\mathbf{n} \le x), \qquad -\infty < x < \infty.$$

Because the outcomes of E_e are the nonnegative integers,

$$\Pr(\mathbf{n} \le x) = F_{\mathbf{n}}(x) \equiv 0, \qquad x < 0.$$

When $0 \le x < 1$,

$$\Pr(\mathbf{n} \le x) = \Pr(\mathbf{n} = 0) = e^{-m},$$

for the only outcome in the event $(\mathbf{n} \le x)$ is now the number 0. For $1 \le x < 2$, there are two outcomes, 0 and 1, in the event $(\mathbf{n} \le x)$, and hence

$$\Pr(\mathbf{n} \le x) = \Pr(\mathbf{n} = 0) + \Pr(\mathbf{n} = 1) = e^{-m} + me^{-m}.$$

Continuing thus, we see that with $[x]$ the greatest integer in x, the cumulative distribution function of the random variable $\mathbf{n}$ can be written in the following equivalent ways:

$$\begin{aligned} F_{\mathbf{n}}(x) &= \sum_{k=0}^{[x]} \Pr(\mathbf{n} = k) = \sum_{k=0}^{[x]} \frac{m^k}{k!} e^{-m} \\ &= \sum_{k=0}^{\infty} \frac{m^k}{k!} e^{-m} U(x - k), \qquad x \ge 0, \\ &\equiv 0, \qquad x \le 0. \end{aligned} \tag{2.3}$$

Here

$$U(x) = \begin{cases} 1, & x \ge 0, \\ 0, & x < 0, \end{cases}$$

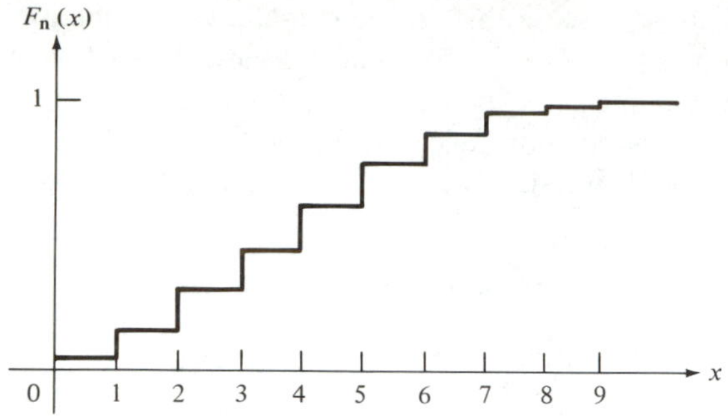

FIGURE 2.1 Cumulative Poisson distribution.

is the unit step function. Thus the cumulative distribution function has the staircase form shown in Fig. 2.1. The steps occur at the integers, and the height of the step at integer k equals the Poisson probability Pr $(\mathbf{n} = k)$. ■

EXAMPLE 2.2 The Uniform Distribution on E_f

The probability measure in Example 1.4, to which we alluded in part A, implies the cumulative distribution function

$$F_{\mathbf{v}}(x) = \begin{cases} 0, & x < 0, \\ x, & 0 \le x < 1, \\ 1, & x > 1, \end{cases} \tag{2.4}$$

and this so-called uniform distribution function looks as shown in Fig. 2.2. By (2.2) we find as before that if $0 \le a \le b \le 1$,

$$\Pr (a < \mathbf{v} < b) = b - a. \tag{2.5}$$

■

EXAMPLE 2.3 Diodes of a certain kind are tested to see how long they last before burning out under typical stressful conditions. An experiment is the test of an individual diode starting at time 0, and the outcome is the time $\mathbf{t}$ at which it fails. The cumulative distribution function of this random variable might have the form

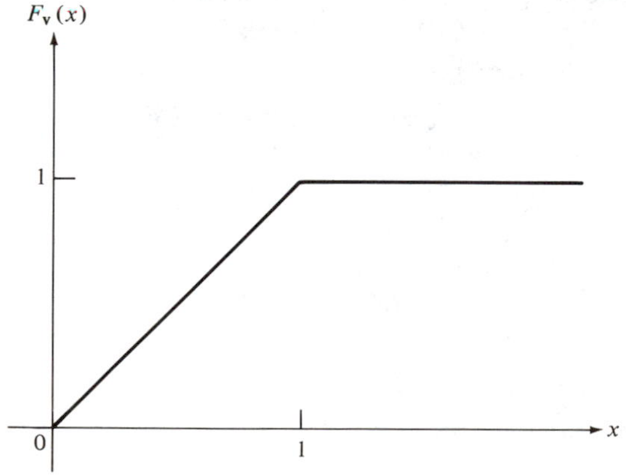

FIGURE 2.2 Uniform distribution.

$$F_t(t) = \Pr(\mathbf{t} \le t) = \begin{cases} 0, & t < 0, \\ 1 - e^{-\mu t}, & t \ge 0, \end{cases} \tag{2.6}$$

for some constant μ. It is shown in Fig. 2.3. The probability that the diode fails between times a and b, $0 \le a \le b$, is then, by (2.2),

$$\Pr(a < \mathbf{t} < b) = e^{-\mu a} - e^{-\mu b}.$$

The probability that its lifetime is longer than c is

$$\Pr(\mathbf{t} > c) = e^{-\mu c}, \qquad c \ge 0.$$

■

EXAMPLE 2.4 The life test of the diodes of Example 2.3 was modified to include applying to those diodes still operating a brief, strong surge of voltage exactly T hours after the test begins. The diode may either burn out then or survive with altered properties. Let p be the probability that it burns out under the surge, given that it has survived to time T. Then the cumulative distribution function of the random variable $\mathbf{t}$, the time of failure, might have the form

$$F_t(t) = \begin{cases} 0, & t < 0, \\ 1 - e^{-\mu t}, & 0 \le t < T, \\ 1 - (1 - p)e^{-\mu T - \nu(t - T)}, & t \ge T, \end{cases} \tag{2.7}$$

which is depicted in Fig. 2.4.

Now the probability that the diode fails between times $T - \varepsilon$ and $T + \varepsilon$ equals

$$\begin{aligned} \Pr(T - \varepsilon < \mathbf{t} \le T + \varepsilon) &= F_t(T + \varepsilon) - F_t(T - \varepsilon) \\ &= 1 - (1 - p)e^{-\mu T - \nu\varepsilon} - (1 - e^{-\mu(T - \varepsilon)}) \\ &= e^{-\mu(T - \varepsilon)} - (1 - p)e^{-\mu T - \nu\varepsilon}, \qquad \varepsilon > 0, \end{aligned}$$

and as ε goes to zero,

$$\lim_{\varepsilon \to 0} \Pr(T - \varepsilon < \mathbf{t} \le T + \varepsilon) = pe^{-\mu T}$$

is the probability that the diode fails at time T; it equals the height of the jump in the cumulative distribution function $F_t(t)$ occurring at $t = T$. We can write this probability as

$$\Pr(\mathbf{t} = T) = pe^{-\mu T}.$$

For any other value of t, $\Pr(\mathbf{t} = t) = 0$, for elsewhere the cumulative distribution function is continuous. ■

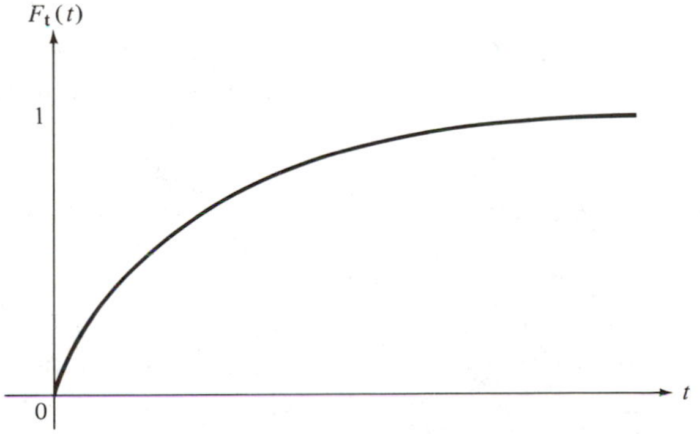

FIGURE 2.3 Cumulative distribution of diode lifetimes.

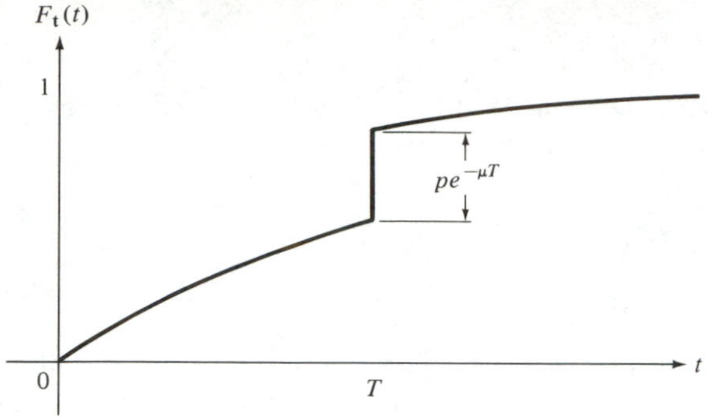

FIGURE 2.4 Cumulative distribution of lifetimes with surge test.

From these examples we see that the cumulative distribution function $F_{\mathbf{x}}(x)$ may be continuous at all values of x, or it may have jumps at certain values of x. If $F_{\mathbf{x}}(x)$ jumps by Δ at, say, $x = a$, then for $\varepsilon \geq 0$,

$$\lim_{\varepsilon \to 0} [\Pr(\mathbf{x} \leq a + \varepsilon) - \Pr(\mathbf{x} \leq a - \varepsilon)] = \lim_{\varepsilon \to 0} \Pr(a - \varepsilon < \mathbf{x} \leq a + \varepsilon) = \Delta,$$

and we find that

$$\Pr(\mathbf{x} = a) = \Delta.$$

The height of a jump therefore equals the probability that the random variable takes on exactly the value at which the jump occurs.

If the probability $\Pr(\mathbf{x} = a)$ is nonzero, the jump in $F_{\mathbf{x}}(x)$ at point a occurs infinitesimally to the left of that point because the value a is included in the defining event $(\mathbf{x} \leq a)$. Thus for $\varepsilon > 0$,

$$\lim_{\varepsilon \to 0} [F_{\mathbf{x}}(a) - F_{\mathbf{x}}(a - \varepsilon)] = \lim_{\varepsilon \to 0} \Pr(a - \varepsilon < \mathbf{x} \leq a) = \Pr(\mathbf{x} = a).$$

On the other hand,

$$\lim_{\varepsilon \to 0} [F_{\mathbf{x}}(a + \varepsilon) - F_{\mathbf{x}}(a)] = \lim_{\varepsilon \to 0} \Pr(a < \mathbf{x} \leq a + \varepsilon) = 0,$$

for outcome $\mathbf{x} = a$ is not included in the event $a < \mathbf{x} \leq a + \varepsilon$. One says, therefore, that the cumulative distribution function $F_{\mathbf{x}}(x)$ is "continuous from the right."

When the cumulative distribution function $F_{\mathbf{x}}(x)$ is continuous everywhere, as in Figs. 2.2 and 2.3, the random variable $\mathbf{x}$ is said to be "of the continuous type"; for short, one calls $\mathbf{x}$ a *continuous* random variable. When $F_{\mathbf{x}}(x)$ is made up entirely of steps, as in Fig. 2.1, with $F_{\mathbf{x}}(x)$ constant between each of its jumps, $\mathbf{x}$ is "of the discrete type" and is called a *discrete* random variable. If the cumulative distribution function $F_{\mathbf{x}}(x)$ has one or more jumps, but rises continuously over at least one stretch of values of x, the random variable $\mathbf{x}$ is said to be of *mixed* type.

A discrete random variable $\mathbf{x}$ takes on one of a countable set of possible values $x_1, x_2, x_3, \ldots$. The array of probabilities $\Pr(\mathbf{x} = x_k)$, $k = 1, 2, 3, \ldots$, of the events $\mathbf{x} = x_1$, $\mathbf{x} = x_2$, and so on, is called the *probability distribution* of **x**. The cumulative distribution function of this random variable is then

$$F_{\mathbf{x}}(x) = \sum_{k=1}^{\infty} \Pr(\mathbf{x} = x_k) U(x - x_k)$$

in terms of the unit step function.

C. Probability Density Functions

For the time being we shall deal only with continuous random variables, returning to those of discrete and mixed types at the end of this part. Let

$$F_{\mathbf{x}}(x) = \Pr(\mathbf{x} \le x)$$

be the cumulative distribution function of the continuous random variable **x**. Then its probability density function is defined as the first derivative of the cumulative distribution function,

$$f_{\mathbf{x}}(x) = \frac{dF_{\mathbf{x}}(x)}{dx}. \tag{2.8}$$

Thus if the cumulative distribution function looks like the function shown in Fig. 2.5(a), the probability density function will look like the curve in Fig. 2.5(b). The cumulative distribution function can be recovered from the probability density function by integration,

$$F_{\mathbf{x}}(x) = \int_{-\infty}^{x} f_{\mathbf{x}}(y)\, dy. \tag{2.9}$$

Furthermore, by (2.2), the probability that the random variable **x** lies between a and b equals

$$\begin{aligned} \Pr(a < \mathbf{x} < b) &= F_{\mathbf{x}}(b) - F_{\mathbf{x}}(a) \\ &= \int_{-\infty}^{b} f_{\mathbf{x}}(x)\, dx - \int_{-\infty}^{a} f_{\mathbf{x}}(x)\, dx = \int_{a}^{b} f_{\mathbf{x}}(x)\, dx, \end{aligned} \tag{2.10}$$

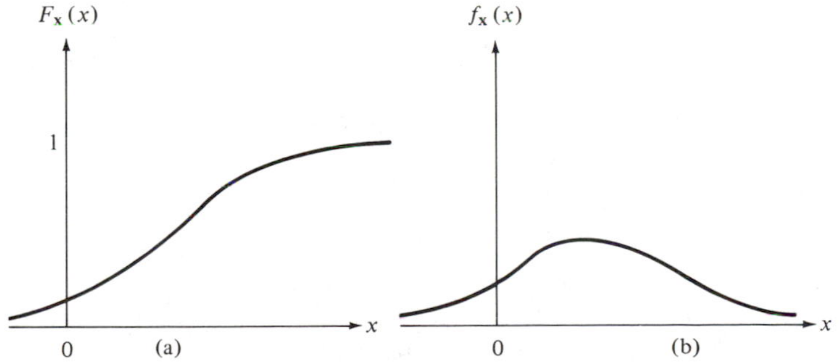

FIGURE 2.5 Cumulative distribution and probability density functions.

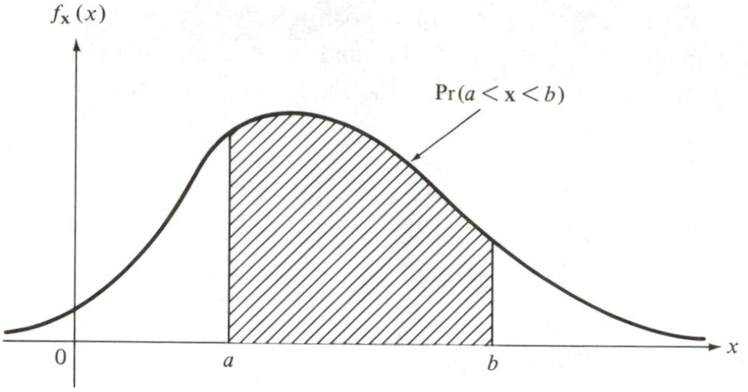

FIGURE 2.6 Probability of an event as the area under the density function.

that is, it equals the area under the probability density function between $x = a$ and $x = b$, as shown in Fig. 2.6. Because this probability must be nonnegative, however short the interval (a, b), the probability density function $f_{\mathbf{x}}(x)$ must be nonnegative everywhere,

$$f_{\mathbf{x}}(x) \geq 0, \qquad -\infty < x < \infty. \tag{2.11}$$

We often consider infinitesimal intervals $(x, x + dx)$ and say that the probability that the random variable $\mathbf{x}$ takes on a value between x and $x + dx$ is

$$\Pr(x < \mathbf{x} < x + dx) = f_{\mathbf{x}}(x)\, dx. \tag{2.12}$$

By virtue of Axiom II, furthermore,

$$\int_{-\infty}^{\infty} f_{\mathbf{x}}(x)\, dx = 1. \tag{2.13}$$

This is called the *normalization integral* for the density function $f_{\mathbf{x}}(x)$. Equations (2.11) to (2.13) express the basic characteristics of the probability density function.

EXAMPLE 2.2 Uniform Distribution (continued)

Continuing this example, we find that the probability density function of the random variable $\mathbf{v}$ is

$$f_{\mathbf{v}}(x) = \begin{cases} 1, & 0 < x < 1, \\ 0, & x < 0, \quad x > 1. \end{cases}$$

It is sketched in Fig. 2.7. One says that $\mathbf{v}$ is *uniformly distributed* over the interval (0, 1). ■

EXAMPLE 2.3 (continued) The probability density function of the lifetime $\mathbf{t}$ of the diodes is

$$f_{\mathbf{t}}(t) = \begin{cases} 0, & t < 0, \\ \mu e^{-\mu t}, & t \geq 0. \end{cases}$$

It is sketched in Fig. 2.8. As before, for $0 < a < b$,

$$\Pr(a < \mathbf{t} \leq b) = \int_a^b \mu e^{-\mu t}\, dt = e^{-\mu a} - e^{-\mu b}$$

from (2.10). ■

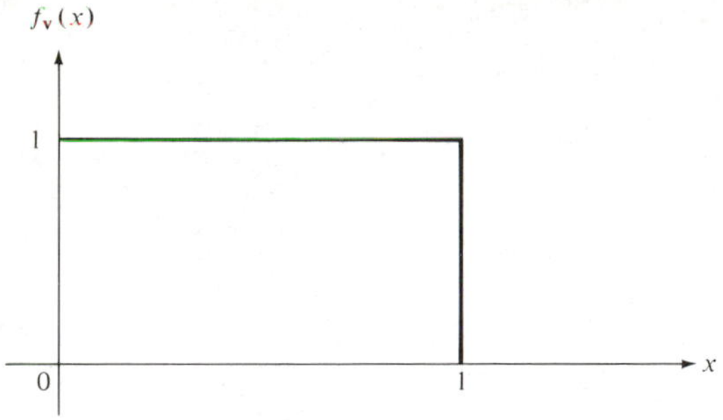

FIGURE 2.7 Uniform density function.

To random variables of discrete and mixed types, probability density functions can be ascribed if we utilize the Dirac delta function. Remember the defining formula for the delta function,

$$\int_b^c g(x)\delta(x-a)\,dx = \begin{cases} g(a), & b < a < c, \\ 0, & a < b, \quad a > c. \end{cases}$$

Think of $\delta(x - a)$ as an infinitely tall spike of infinitesimal width, standing at the point $x = a$; the area under the spike equals 1. The delta function is the derivative of the unit step function,

$$\frac{d}{dx} U(x - a) = \delta(x - a).$$

Let a discrete random variable $\mathbf{x}$ take on only the values $x_1, x_2, x_3, \ldots$ with probabilities $\Pr(\mathbf{x} = x_k)$, $k = 1, 2, 3, \ldots$, respectively,

$$\sum_{k=1}^{\infty} \Pr(\mathbf{x} = x_k) = 1.$$

Its cumulative distribution function will have steps of heights $\Pr(\mathbf{x} = x_k)$ at all points x_k, and it can be written as

$$F_{\mathbf{x}}(x) = \sum_{k=1}^{\infty} \Pr(\mathbf{x} = x_k) U(x - x_k). \tag{2.14}$$

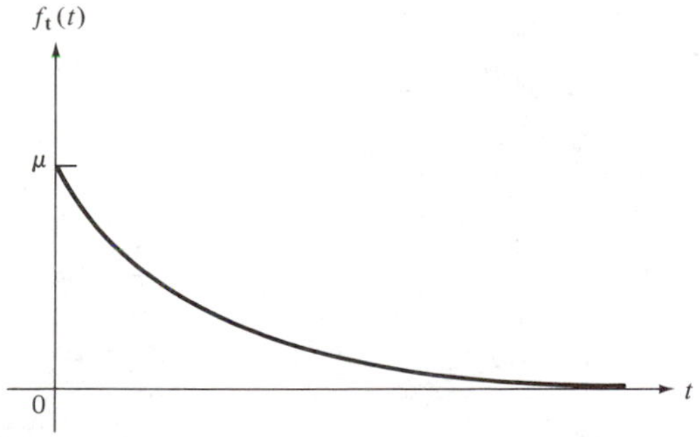

FIGURE 2.8 Density function of diode lifetimes.

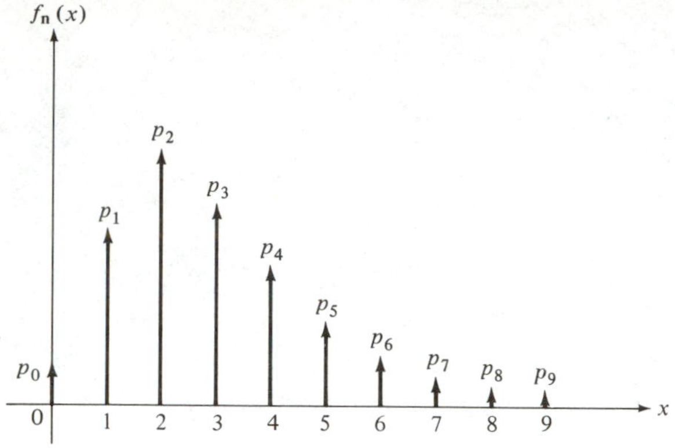

FIGURE 2.9 Density function: discrete random variable.

Applying the definition in (2.8) for the probability density function, we find that

$$f_{\mathbf{x}}(x) = \sum_{k=1}^{\infty} \Pr(\mathbf{x} = x_k)\delta(x - x_k). \tag{2.15}$$

Thus in Example 2.1,

$$f_{\mathbf{n}}(x) = \sum_{k=0}^{\infty} \frac{m^k e^{-m}}{k!} \delta(x - k). \tag{2.16}$$

Depicting the delta function by a vertical arrow, we could represent this probability density function as shown in Fig. 2.9, with $p_k = \Pr(\mathbf{n} = k)$ its multiplying factors.

The probability density function of a mixed random variable will have a delta function at each of its jumps. Thus in Example 2.4, the probability density function of the lifetime $\mathbf{t}$ of the diodes can be taken as

$$f_{\mathbf{t}}(t) = \begin{cases} 0, & t < 0, \\ \mu e^{-\mu t} + pe^{-\mu T}\delta(t - T), & 0 < t \leq T, \\ \nu(1 - p)e^{-\mu T - \nu(t-T)}, & t > T. \end{cases} \tag{2.17}$$

This probability density function looks like the sketch shown in Fig. 2.10.

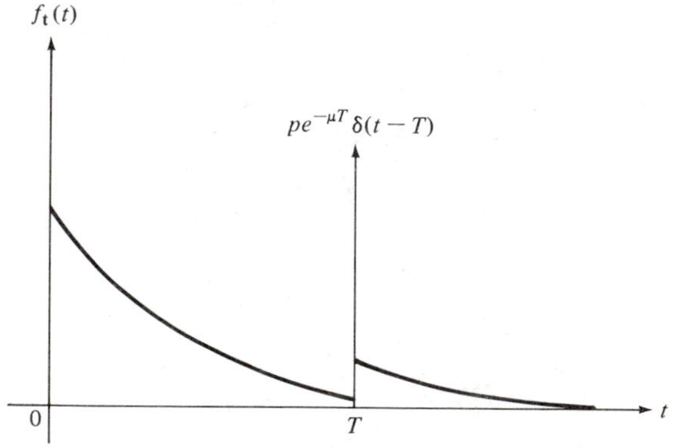

FIGURE 2.10 Density function: mixed random variable.

Because the cumulative distribution function is defined with a "≤" sign in (2.1), its jumps occur just to the left of a value of the random variable to which positive probability is assigned. In the probability density function of a discrete or mixed random variable, therefore, the delta function stands just to the left of such a value. If the event $\mathbf{x} = a$ has positive probability Δ, the probability density function $f_{\mathbf{x}}(x)$ contains a term $\Delta\delta(x - a^-)$, where a^- lies infinitesimally to the left of a. Thus in the preceding example the delta function occurs just before time T and was included in the interval $0 < t \le T$, but not in $t > T$. We usually need not distinguish a^- from a.

D. Random Variables as Functions of Outcomes

A random variable may arise through the assignment of numerical values to the outcomes ζ of a chance experiment. Suppose, for instance, that upon tossing a fair coin, you gain one dollar each time heads appears and lose one dollar each time tails appears. Your gain $\mathbf{x}$ is a function of the outcome of the experiment, $\mathbf{x} = \mathbf{x}(\zeta)$, according to the table

$$\mathbf{x}(\mathrm{H}) = +1$$

$$\mathbf{x}(\mathrm{T}) = -1.$$

Since $\Pr(\mathrm{H}) = \Pr(\mathrm{T}) = \frac{1}{2}$, the probability density function of $\mathbf{x}$ will be

$$f_{\mathbf{x}}(x) = \tfrac{1}{2}\delta(x+1) + \tfrac{1}{2}\delta(x-1),$$

and its cumulative distribution function will be

$$F_{\mathbf{x}}(x) = \Pr(\mathbf{x} \le x) = \begin{cases} 0, & x < -1, \\ \frac{1}{2} & -1 \le x < 1, \\ 1, & x \ge 1, \end{cases} \tag{2.18}$$

as depicted in Fig. 2.11.

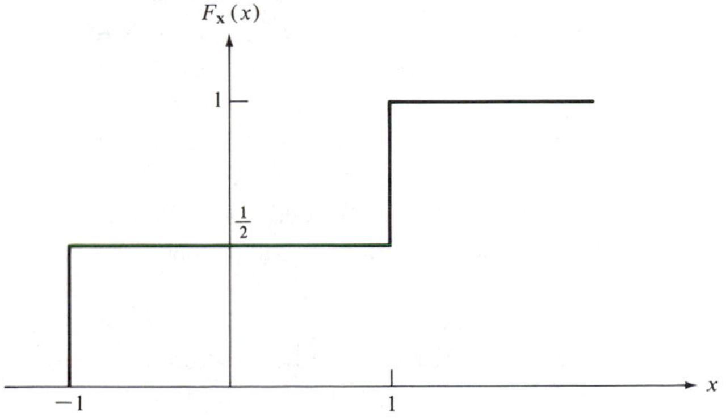

FIGURE 2.11 Cumulative distribution of gain **x**.

EXAMPLE 2.5 Binomial Distribution

The outcomes of the Bernoulli-trials experiment E_n discussed in Sec. 1-8 are n-tuples of 0's and 1's, representing failures and successes of an event A in n trials of an experiment E. The number $\mathbf{k}$ of successes in n trials is a function $\mathbf{k}(\zeta)$ of outcome ζ of E_n. Thus for $n = 4$,

$$\mathbf{k}(0000) = 0, \quad \mathbf{k}(0001) = 1, \quad \mathbf{k}(0110) = 2,$$

$$\mathbf{k}(1010) = 2, \quad \mathbf{k}(1011) = 3, \quad \mathbf{k}(1111) = 4,$$

and so on for the rest of the sixteen outcomes of E_4. The random variable $\mathbf{k}$ thus takes on the values 0, 1, 2, 3, or 4 in trials of this experiment. The probability distribution of the random variable $\mathbf{k}$ arises from the probability measure on the events of the experiment E_n. An atomic event of E_n, represented by an n-tuple of k 1's and $n - k$ 0's in any order, has probability $p^k q^{n-k}$, $q = 1 - p$. On the basis of that measure we deduce as in Sec. 1-8 that

$$\Pr(\mathbf{k} = k) = \binom{n}{k} p^k q^{n-k},$$

so that the probability density function of the random variable $\mathbf{k}$ is

$$f_{\mathbf{k}}(x) = \sum_{r=0}^{n} \binom{n}{r} p^r q^{n-r} \delta(x - r).$$

Its cumulative distribution function will have the staircase form characteristic of a discrete random variable and can be written

$$F_{\mathbf{k}}(x) = \sum_{r=0}^{n} \binom{n}{r} p^r q^{n-r} U(x - r)$$

in terms of the unit step function. ■

EXAMPLE 2.6 To each outcome ζ of an experiment E the same number $\mathbf{x}(\zeta) \equiv a$ is assigned. The cumulative distribution function of this "random" variable is simply

$$F_{\mathbf{x}}(x) = U(x - a)$$

and its probability density function is

$$f_{\mathbf{x}}(x) = \delta(x - a).$$ ■

Some writers on probability take the view that any random variable $\mathbf{x}$ must be considered a function $\mathbf{x}(\zeta)$ of the outcome of some experiment. They aver that there exist an experiment E with outcomes ζ and an actual or imaginary table that assigns a unique number $\mathbf{x}(\zeta)$ to each outcome of E. This function $\mathbf{x}(\zeta)$ is said to "map" the set S of outcomes ζ of E "onto the real line." The probability measure on the events of E thereby induces the probability distribution of the random variable $\mathbf{x}$. The event $(\mathbf{x} \le x)$, for any number x, consists of all outcomes ζ of E such that $\mathbf{x}(\zeta) \le x$, and its probability $\Pr(\mathbf{x} \le x) = F_{\mathbf{x}}(x)$ is determined by the initial assignment of probabilities to all possible events of E.

In problems involving continuous random variables, however, and in many involving discrete random variables as well, one has no choice but to label the outcomes ζ of the underlying experiment with numbers, in effect putting $\mathbf{x}(\zeta) = \zeta$, and the distinction between the outcome and the random variable can hardly be maintained. Random variables are specified by their cumulative distribution function or their probability density function, and the probability measure on that underlying experiment E is seldom invoked. For the number **n** of photoelectrons counted in a certain interval, as in experiment E_e, page 4, for instance, one specifies Pr $(\mathbf{n} = k)$ as the probability distribution of the random variable **n** and need not draw distinctions among the outcome "k e's were counted," the atomic event consisting of that one outcome, and the event $(\mathbf{n} = k)$. We mention this alternative viewpoint for the sake of those who may encounter it in treatises on probability.

E. Important Probability Distributions

Besides the binomial and the Poisson distributions, introduced in Chapter 1, the following probability distributions often arise.

1. NORMAL (GAUSSIAN) DISTRIBUTION

The normal or Gaussian probability density function is

$$f_{\mathbf{x}}(x) = (2\pi\sigma^2)^{-1/2} \exp\left[-\frac{(x-m)^2}{2\sigma^2}\right]. \tag{2.19}$$

For reasons to appear in Sec. 2-4, m is called the *mean* and σ^2 the *variance* of the *normal* or *Gaussian* random variable **x.** The name "Gaussian" is more common in the literature of electrical engineering. Figure 2.12(a) exhibits $f_{\mathbf{x}}(x)$ for $m = 0$, $\sigma = 1$.

The cumulative normal distribution function is expressed in terms of the error-function integral,

$$\begin{aligned} \operatorname{erfc} x &= (2\pi)^{-1/2} \int_x^{\infty} e^{-t^2/2}\, dt, \\ \operatorname{erfc}(-\infty) &= 1, \qquad \operatorname{erfc} 0 = \tfrac{1}{2}, \quad \operatorname{erfc} \infty = 0, \\ \operatorname{erfc}(-x) &= 1 - \operatorname{erfc} x. \end{aligned} \tag{2.20}$$

This cumulative distribution function is

$$F_{\mathbf{x}}(x) = 1 - \operatorname{erfc}\left(\frac{x-m}{\sigma}\right). \tag{2.21}$$

The function erfc x cannot be expressed in closed form. Figure 2.12(b) shows $F_{\mathbf{x}}(x)$ for $m = 0$, $\sigma = 1$.

COMPUTATIONAL DIGRESSION

Extensive tables of the error-function integral exist, but if one has a programmable calculator, one can do without them. Because of the importance of this function in probability and statistics, we shall explain a few ways of calculating it and urge the reader to try them on a calculator.

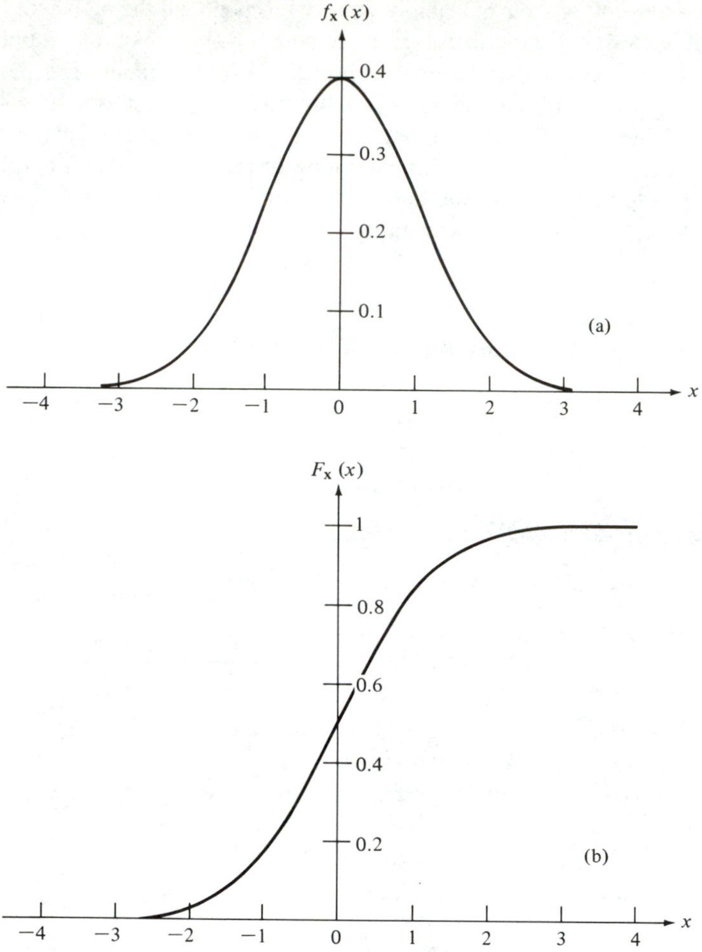

FIGURE 2.12 Normal (Gaussian) distribution.

(a) For small values of x, positive or negative, one can use the power series for $\operatorname{erf} x = \frac{1}{2} - \operatorname{erfc} x$,

$$\begin{aligned}
\operatorname{erfc} x &= \tfrac{1}{2} - (2\pi)^{-1/2} \int_0^x e^{-t^2/2}\, dt \\
&= \tfrac{1}{2} - (2\pi)^{-1/2} \sum_{r=0}^{\infty} \int_0^x \frac{(-t^2/2)^r}{r!}\, dt \\
&= \tfrac{1}{2} - (2\pi)^{-1/2}\, x \sum_{r=0}^{\infty} \frac{(-x^2/2)^r}{(2r+1)r!} \\
&= \tfrac{1}{2} - \sum_{r=0}^{\infty} \frac{T_r}{2r+1},
\end{aligned}$$

where

$$T_0 = x(2\pi)^{-1/2}, \qquad T_{r+1} = \frac{-x^2/2}{r+1}\, T_r.$$

Thus the series can be calculated iteratively; the factors T_r are stored in one register and the sum of terms $T_r/(2r + 1)$ is accumulated in another. The summation is stopped when the absolute value of the additional term is less than a preset bound. This method takes a long time when $|x|$ is larger than 3 or so.

(b) The following polynomial approximation* gives an error less than 7.5×10^{-8} over $0 < x < \infty$:

$$\operatorname{erfc} x = F(t) \exp\left(-\frac{x^2}{2}\right),$$

$$t = (1 + px)^{-1}, \qquad p = 0.2316419,$$

$$F(t) = a_1 t + a_2 t^2 + a_3 t^3 + a_4 t^4 + a_5 t^5$$

$$= \left\{\{[(a_5 t + a_4)t + a_3]t + a_2\}t + a_1\right\}t,$$

for which the coefficients are given in Table 2.1. The second form represents the more rapid way of evaluating the polynomial.

In some problems one is given the value $Q = \operatorname{erfc} x$ and needs to find the value of $x = \operatorname{erfc}^{-1} Q$. For $0 < Q < \frac{1}{2}$ an easily programmed iterative method starts with a trial value x_0 and calculates a new trial value by

$$x_1 = \left\{2 \ln\left[\frac{F(t_0)}{Q}\right]\right\}^{1/2}, \qquad t_0 = (1 + px_0)^{-1},$$

with $F(\cdot)$ the foregoing polynomial. This procedure is repeated until the value of x ceases to change significantly. The choice of the initial trial value is not critical. Both the forward calculation of $Q = \operatorname{erfc} x$ and the inverse calculation of $x = \operatorname{erfc}^{-1} Q$ can be incorporated in the same program if a subroutine is used for the polynomial $F(t)$.

(c) A method that is rapid for $x > 2$ and does not require remembering all those numerical coefficients and storing them in the calculator memory is based on the continued fraction†

$$\operatorname{erfc} x = (2\pi)^{-1/2} e^{-x^2/2} G(x),$$

$$G(x) = \frac{1}{x+}\,\frac{1}{x+}\,\frac{2}{x+}\,\frac{3}{x+}\cdots\frac{k}{x+}\,\frac{k+1}{x+}\cdots$$

Table 2.1 Polynomial Coefficients for erfc x

a_1	=	0.127414796
a_2	=	−0.142248368
a_3	=	0.710706871
a_4	=	−0.726576013
a_5	=	0.530702714

*M. Abramowitz and I. Stegun, *Handbook of Mathematical Functions,* National Bureau of Standards, Applied Mathematics Series No. 55. Washington, D.C.: U.S. Government Printing Office, 1964. See p. 932.

†Ibid.

Table 2.2 Starting Values of Continued Fraction

x	1.5	2	3	4	5	6	8	11	15	20
n	25	14	8	7	6	5	4	3	2	1

Cutting off the continued fraction at a sufficiently high starting value $k = n$, the calculator successively forms

$$\frac{n}{x}, \quad x + \frac{n}{x} = p_n, \quad x + \frac{n-1}{p_n} = p_{n-1}, \quad x + \frac{n-2}{p_{n-1}} = p_{n-2},$$

$$\ldots, \quad x + \frac{2}{p_3} = p_2, \quad x + \frac{1}{p_2} = p_1, \quad G(x) = \frac{1}{p_1},$$

then multiplies by $(2\pi)^{-1/2} \exp(-x^2/2)$. Table 2.2 lists starting values n for the continued fraction that lead to an accuracy of about six significant figures in $G(x)$. The inverse function $x = \operatorname{erfc}^{-1} Q$ can be calculated by the same kind of iteration as before, to wit,

$$x_1 = \left\{2 \ln \left[\frac{G(x_0)}{Q\sqrt{2\pi}}\right]\right\}^{1/2}$$

where x_0 is one trial value, x_1 the next. Tables 2.3 and 2.4 furnish values of the error-function integral and its inverse, by which you can check your program.

Different writers often use different definitions of erfc x and erf x, and it is important to ascertain which has been adopted. Be careful to determine the definition of computer functions such as ERF(·) when programming. It is wise to check the accuracy with which a computer evaluates ERF(·) for large values of its argument, especially if you want accurate values of the "tail probability" erfc $x = \frac{1}{2} - \operatorname{erf} x$. It may be necessary to write your own subroutine for erfc x using one of the methods described above.

2. UNIFORM DISTRIBUTION

If the random variable **x** is uniformly distributed over the interval (a, b), its probability density function is

$$f_{\mathbf{x}}(x) = \begin{cases} 0, & x < a, \\ (b-a)^{-1}, & a < x < b, \\ 0, & x > b, \end{cases} \tag{2.22}$$

Table 2.3

x	erfc x
1	0.15865525
3	0.001349898
5	2.866516×10^{-7}
7	1.279813×10^{-12}
9	1.128588×10^{-19}

Table 2.4

Q	$\text{erfc}^{-1} Q$
10^{-1}	1.28155
10^{-3}	3.09023
10^{-5}	4.26489
10^{-7}	5.19934
10^{-9}	5.99781

[Fig. 2.13(a)], and its cumulative distribution function is

$$F_{\mathbf{x}}(x) = \begin{cases} 0, & x \le a, \\ \dfrac{x - a}{b - a}, & a < x \le b, \\ 1, & x > b. \end{cases} \tag{2.23}$$

[Fig. 2.13(b)].

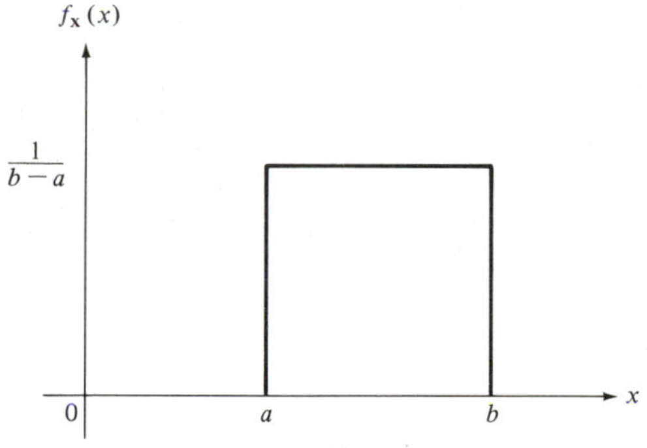

(a)

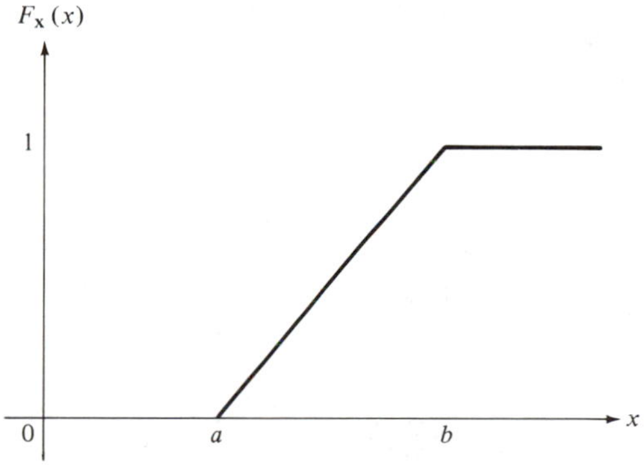

(b)

FIGURE 2.13 Uniform distribution.

3. RAYLEIGH DISTRIBUTION

The Rayleigh probability density function is

$$f_{\mathbf{x}}(x) = \frac{x}{b^2} e^{-x^2/2b^2} U(x), \tag{2.24}$$

and the associated cumulative distribution function is

$$F_{\mathbf{x}}(x) = (1 - e^{-x^2/2b^2}) U(x). \tag{2.25}$$

These are plotted in Fig. 2.14 for $b = 1$.

4. GAMMA DISTRIBUTION*

The gamma probability density function is

$$f_{\mathbf{x}}(x) = [\Gamma(c)]^{-1} x^{c-1} e^{-x} U(x), \qquad c > 0, \tag{2.26}$$

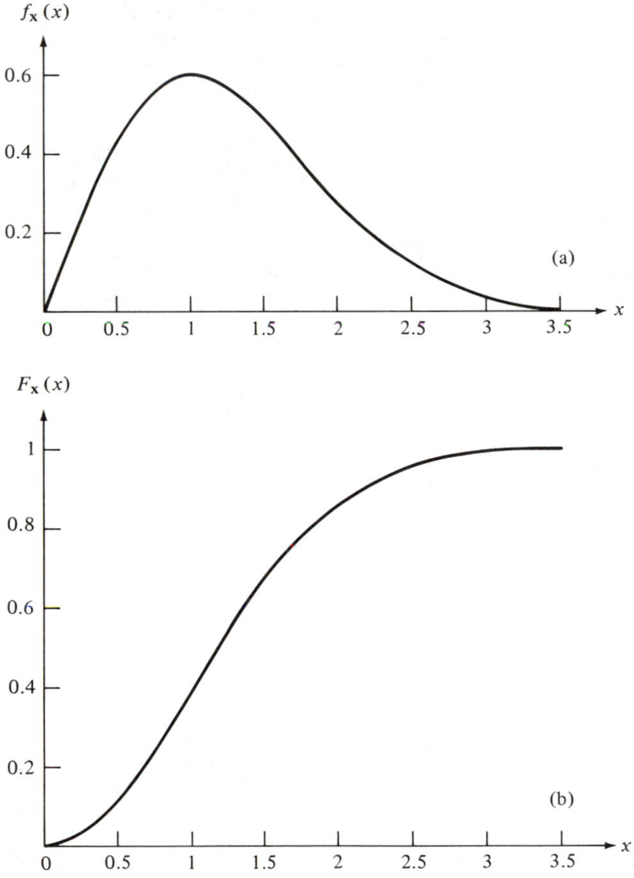

FIGURE 2.14 Rayleigh distribution.

*These formulas need not be memorized.

[Fig. 2-15(a)], where $\Gamma(c)$ is the gamma function,

$$\Gamma(c) = \int_0^\infty x^{c-1} e^{-x}\, dx. \tag{2.27}$$

Important properties of the gamma function are

$$\Gamma(c + 1) = c\Gamma(c), \tag{2.28}$$

$$\Gamma(n + 1) = n! \qquad \text{for } n \text{ a positive integer,} \tag{2.29}$$

$$\Gamma(\tfrac{1}{2}) = \pi^{1/2}, \quad \Gamma(\tfrac{3}{2}) = \tfrac{1}{2}\,\pi^{1/2}, \quad \text{etc.} \tag{2.30}$$

The argument c of the gamma function must not equal zero or a negative integer.

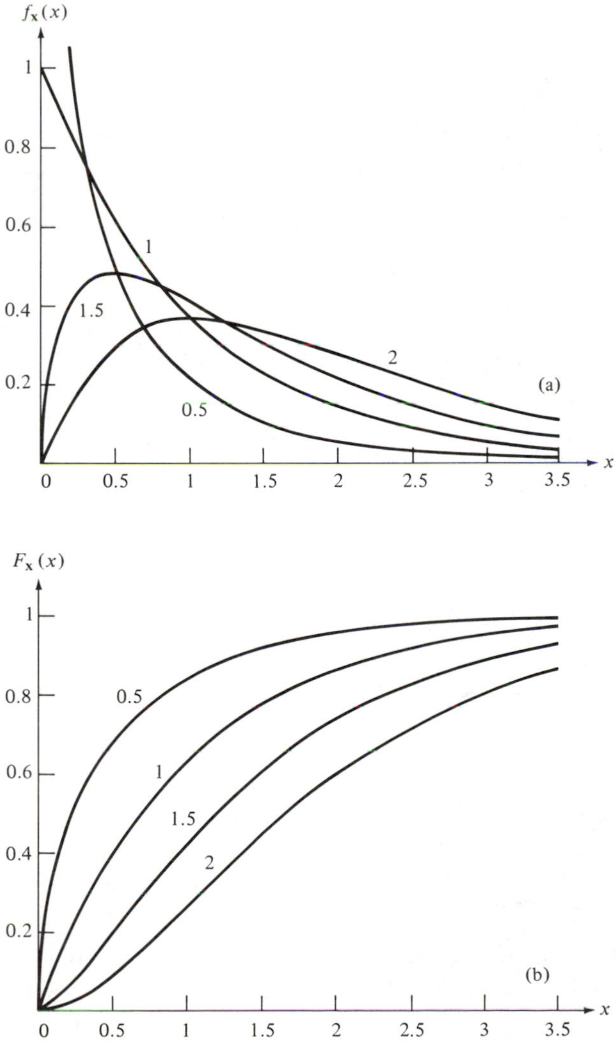

FIGURE 2.15 Gamma distribution.

The cumulative gamma distribution function [Fig. 2.15(b)]

$$F_{\mathbf{x}}(x) = U(x) \int_0^x y^{c-1} e^{-y} \frac{dy}{\Gamma(c)} \tag{2.31}$$

is expressed in terms of the incomplete gamma function

$$I(u, p) = \int_0^{u\sqrt{p+1}} v^p e^{-v} \frac{dv}{\Gamma(p+1)} \tag{2.32}$$

as

$$F_{\mathbf{x}}(x) = I(c^{-1/2}x, c-1)U(x).$$

The incomplete gamma function has been tabulated by Pearson.* For c a positive integer the cumulative distribution function can be written in closed form,

$$F_{\mathbf{x}}(x) = \left[1 - \sum_{k=0}^{c-1} \frac{x^k}{k!} e^{-x}\right] U(x), \qquad c \text{ integral}, \quad c > 0, \tag{2.33}$$

as can be verified by differentiation. This cumulative distribution function can be calculated by the same program as for the cumulative Poisson distribution described in Sec. 1-9.

For $c = \frac{1}{2}$,

$$F_{\mathbf{x}}(x) = [1 - 2 \operatorname{erfc}(\sqrt{2x})]U(x), \qquad c = \tfrac{1}{2} \tag{2.34}$$

For c equal to half of any odd positive integer, the cumulative distribution function can be reduced to this by successive integrations by parts.

5. CAUCHY DISTRIBUTION

The Cauchy random variable has the probability density function

$$f_{\mathbf{x}}(x) = \frac{a}{\pi} \frac{1}{x^2 + a^2}, \tag{2.35}$$

and its cumulative distribution function is

$$F_{\mathbf{x}}(x) = \frac{1}{2} + \pi^{-1} \tan^{-1}\left(\frac{x}{a}\right), \tag{2.36}$$

taking the function $\tan^{-1} x$ as lying between $-\pi/2$ and 0 for $-\infty < x < 0$. These are plotted in Fig. 2.16 for $a = 1$.

6. ARCSINE DISTRIBUTION

The "arcsine law" refers to a random variable whose probability density function is

*K. Pearson, *Tables of the Incomplete Gamma-Function*. Cambridge: Cambridge University Press, 1934.

$$
f_{\mathbf{x}}(x) = \begin{cases} (a\pi)^{-1}\left[1 - \left(\dfrac{x}{a}\right)^2\right]^{-1/2}, & |x| < a, \\ \\ 0, & |x| > a, \end{cases} \tag{2.37}
$$

Its cumulative distribution function is

$$
F_{\mathbf{x}}(x) = \begin{cases} 0, & x < -a, \\ \dfrac{1}{2} + \pi^{-1} \sin^{-1}\left(\dfrac{x}{a}\right), & -a < x < a, \\ 1, & x > a, \end{cases} \tag{2.38}
$$

taking the function $\sin^{-1} x$ as lying between $-\pi/2$ and 0 for $-1 < x < 0$. These are plotted in Fig. 2.17 for $a = 1$.

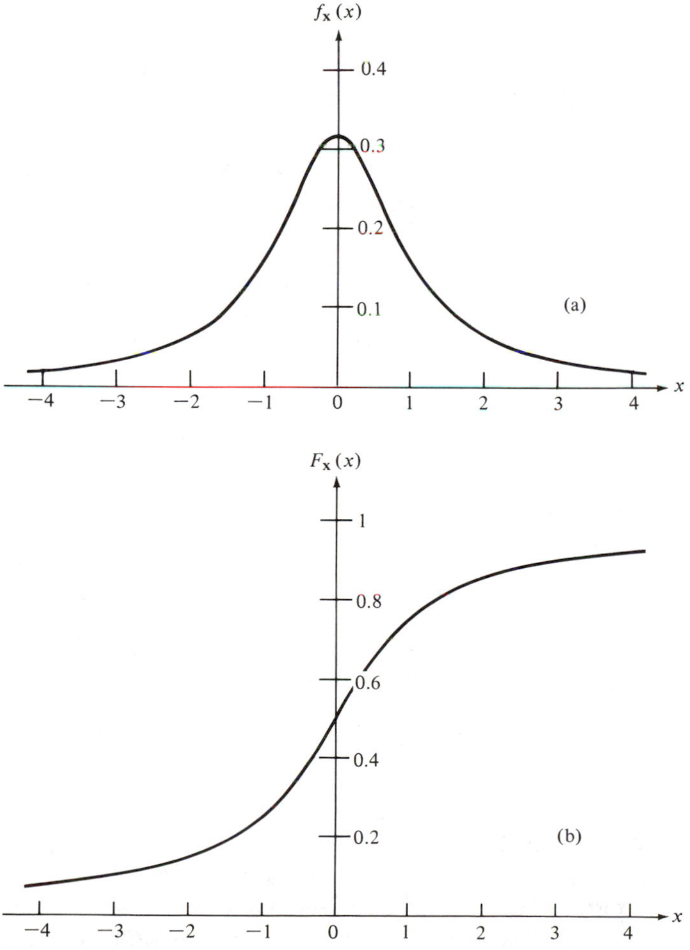

FIGURE 2.16 Cauchy distribution.

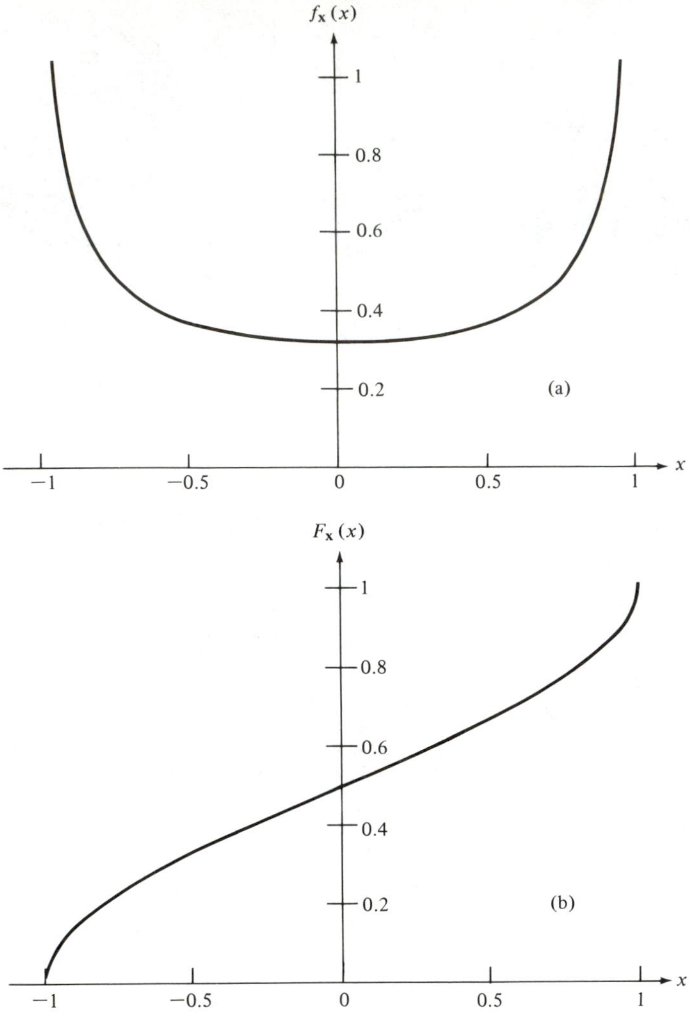

FIGURE 2.17 Arcsine distribution.

2-2

CONDITIONAL DISTRIBUTIONS

A. Nonnumerical Attributes

The outcome of a chance experiment may have both numerical and nonnumerical attributes. In Examples 1.8 and 1.21, for instance, transistors are picked out of a bin in which transistors of the same type from three suppliers, A, B, and C, were mixed together. Each transistor is set into a circuit, and the time until it fails is recorded. The outcome of the experiment is labeled by both the time of failure and the source (A, B, or C) of the transistor; the latter is a nonnumerical attribute.

The time of failure is a random variable and denoted by $\mathbf{t}$, and we are given the probabilities $\Pr(\mathbf{t} \le t, A)$, $\Pr(\mathbf{t} \le t, B)$, and $\Pr(\mathbf{t} \le t, C)$ for all values of t in $(0, \infty)$, where $(\mathbf{t} \le t, A)$ is the event, "The transistor failed before time t and came from supplier A," and so on. By the definition of conditional probability in (1.9) the conditional probability that the transistor failed before time t, given that it came from supplier A, is

$$\Pr(\mathbf{t} \le t \mid A) = \frac{\Pr(\mathbf{t} \le t, A)}{\Pr(A)}. \tag{2.39}$$

In Example 1.21 this probability was denoted by $\Pr(M_t \mid A)$; it took the form

$$\Pr(\mathbf{t} \le t \mid A) = (1 - e^{-k_a t})U(t). \tag{2.40}$$

This conditional probability defines the conditional cumulative distribution function

$$F_{\mathbf{t}}(t \mid A) = \Pr(\mathbf{t} \le t \mid A)$$

of the random variable $\mathbf{t}$, given event A. Similarly,

$$\begin{aligned} \Pr(\mathbf{t} \le t \mid B) &= F_{\mathbf{t}}(t \mid B) = (1 - e^{-k_b t})U(t), \\ \Pr(\mathbf{t} \le t \mid C) &= F_{\mathbf{t}}(t \mid C) = (1 - e^{-k_c t})U(t). \end{aligned} \tag{2.41}$$

The constants k_a, k_b, and k_c were given on page 42. The unit step function $U(t)$ indicates that these conditional cumulative distribution functions are zero for $t < 0$. Our example shows how conditional distribution functions can be defined for conditioning events involving nonnumerical attributes.

By differentiation we can define the conditional probability density functions of the random variable $\mathbf{t}$,

$$\begin{aligned} f_{\mathbf{t}}(t \mid A) &= \frac{dF_{\mathbf{t}}(t \mid A)}{dt} = k_a e^{-k_a t}U(t), \\ f_{\mathbf{t}}(t \mid B) &= k_b e^{-k_b t}U(t), \\ f_{\mathbf{t}}(t \mid C) &= k_c e^{-k_c t}U(t). \end{aligned} \tag{2.42}$$

We can say, for instance, that the probability that a transistor fails between times t and $t + dt$, given that it came from supplier A, is, for $t > 0$,

$$\Pr(t < \mathbf{t} < t + dt \mid A) = f_{\mathbf{t}}(t \mid A)\, dt = k_a e^{-k_a t} dt. \tag{2.43}$$

Conditional cumulative distribution functions and probability density functions have all the properties of ordinary cumulative distribution functions and probability density functions given in parts B and C of Sec. 2-1. For instance,

$$\int_{-\infty}^{\infty} f_{\mathbf{t}}(t \mid A)\, dt = 1.$$

The "overall" cumulative distribution function of the random variable $\mathbf{t}$ is, by the principle of total probability,

$$\begin{aligned} F_{\mathbf{t}}(t) &= \Pr(\mathbf{t} \le t) = \Pr(\mathbf{t} \le t, A) + \Pr(\mathbf{t} \le t, B) + \Pr(\mathbf{t} \le t, C) \\ &= F_{\mathbf{t}}(t \mid A)\Pr(A) + F_{\mathbf{t}}(t \mid B)\Pr(B) + F_{\mathbf{t}}(t \mid C)\Pr(C); \end{aligned} \tag{2.44}$$

A, B, and C exhaust the set of nonnumerical attributes in this example. The cumulative distribution function $F_{\mathbf{t}}(t)$ is the probability that a transistor picked out of the bin at random will fail before time t. Differentiating with respect to t, we find the overall probability density function of $\mathbf{t}$ to be

$$f_{\mathbf{t}}(t) = f_{\mathbf{t}}(t \mid A)\Pr(A) + f_{\mathbf{t}}(t \mid B)\Pr(B) + f_{\mathbf{t}}(t \mid C)\Pr(C). \tag{2.45}$$

The probability that a transistor picked at random fails between t and $t + dt$ is $f_{\mathbf{t}}(t)\,dt$.

Suppose that a transistor fails between time a and time b, $0 < a < b$. What is the probability that it came from supplier A? We invoke Bayes's theorem,

$$\Pr(A \mid a < \mathbf{t} \le b) = \frac{\Pr(a < \mathbf{t} \le b \mid A)\Pr(A)}{\Pr(a < \mathbf{t} \le b)}, \tag{2.46}$$

where by (2.2),

$$\Pr(a < \mathbf{t} \le b \mid A) = F_{\mathbf{t}}(b \mid A) - F_{\mathbf{t}}(a \mid A), \tag{2.47}$$

$$\begin{aligned} \Pr(a < \mathbf{t} \le b) &= F_{\mathbf{t}}(b) - F_{\mathbf{t}}(a) \\ &= \Pr(A)[F_{\mathbf{t}}(b \mid A) - F_{\mathbf{t}}(a \mid A)] \\ &\quad + \Pr(B)[F_{\mathbf{t}}(b \mid B) - F_{\mathbf{t}}(a \mid B)] \\ &\quad + \Pr(C)[F_{\mathbf{t}}(b \mid C) - F_{\mathbf{t}}(a \mid C)]. \end{aligned} \tag{2.48}$$

Hence

$$\Pr(A \mid a < \mathbf{t} \le b) = \frac{[F_{\mathbf{t}}(b \mid A) - F_{\mathbf{t}}(a \mid A)]\Pr(A)}{\Pr(a < \mathbf{t} \le b)}. \tag{2.49}$$

We can calculate the three posterior probabilities $\Pr(A \mid a < \mathbf{t} \le b)$, $\Pr(B \mid a < \mathbf{t} \le b)$, and $\Pr(C \mid a < \mathbf{t} \le b)$ and utilize Bayes's rule to decide from which supplier a transistor failing between times a and b most likely came.

Now let the interval (a, b) be very short, say $a = t - \frac{1}{2}\varepsilon$, $b = t + \frac{1}{2}\varepsilon$; ε could be considered as the accuracy with which the time of failure can be measured. We suppose that $k_a\varepsilon << 1$, $k_b\varepsilon << 1$, $k_c\varepsilon << 1$. Then because the conditional distribution functions are continuous, we can put

$$\Pr(t - \tfrac{1}{2}\varepsilon < \mathbf{t} < t + \tfrac{1}{2}\varepsilon \mid A) \doteq \varepsilon f_{\mathbf{t}}(t \mid A),$$

$$\Pr(t - \tfrac{1}{2}\varepsilon < \mathbf{t} < t + \tfrac{1}{2}\varepsilon) \doteq \varepsilon f_{\mathbf{t}}(t),$$

in terms of the conditional and overall probability density functions of the failure time $\mathbf{t}$. Putting these into (2.49), we see that ε cancels from numerator and denominator. Assuming ε negligible, we can represent the conditioning event $t - \frac{1}{2}\varepsilon < \mathbf{t} < t + \frac{1}{2}\varepsilon$ by $\mathbf{t} = t$ and write (2.46) for this situation as

$$\Pr(A \mid \mathbf{t} = t) = \frac{f_{\mathbf{t}}(t \mid A)\Pr(A)}{f_{\mathbf{t}}(t)}, \tag{2.50}$$

$$f_{\mathbf{t}}(t) = f_{\mathbf{t}}(t \mid A)\Pr(A) + f_{\mathbf{t}}(t \mid B)\Pr(B) + f_{\mathbf{t}}(t \mid C)\Pr(C),$$

which is another form of Bayes's theorem. We call $\Pr(A \mid \mathbf{t} = t)$ the probability of event A, given that the random variable $\mathbf{t}$ took on the value t. It is often written simply as $\Pr(A \mid t)$.

EXAMPLE 2.7 A transistor is picked at random from the bin as described in Example 1.21, and it is observed to fail at time t. Using the values of the constants k_a, k_b, and k_c as given there, apply Bayes's rule to determine the most likely supplier of that transistor.

It is necessary to calculate $\Pr(A \mid t)$, $\Pr(B \mid t)$, and $\Pr(C \mid t)$, given as in (2.50), and find which is the largest. From (2.50) we see that it suffices to compare their numerators, and these are

$$\begin{aligned} f_{\mathbf{t}}(t \mid A)\Pr(A) &= 0.45k_a e^{-k_a t}, \\ f_{\mathbf{t}}(t \mid B)\Pr(B) &= 0.25k_b e^{-k_b t}, \\ f_{\mathbf{t}}(t \mid C)\Pr(C) &= 0.30k_c e^{-k_c t}. \end{aligned}$$

The prior probabilities of the events A, B, and C were taken from Example 1.8. We can plot these three functions and determine whether they intersect, and if so, where. We find that $\Pr(B \mid t)$ is the largest of the three conditional probabilities for $0 < t < 1600.67$ hours, and that $\Pr(A \mid t)$ is largest for $t > 1600.67$ hours. All transistors failing before 1600.67 hours will therefore be attributed to supplier B, all those failing afterward to supplier A.

These decisions will sometimes, of course, be wrong: some of the transistors supplied by A will fail before 1600 hours, some supplied by B will fail after 1600 hours, and all supplied by C will fail sometime or other. The probability P_e of error with respect to the entire collection of transistors from all three suppliers is therefore

$$\begin{aligned} P_e &= \Pr(A)\Pr(\mathbf{t} < 1600.67 \mid A) + \Pr(B)\Pr(\mathbf{t} > 1600.67 \mid B) + \Pr(C) \\ &= 0.45(0.229) + 0.25(0.441) + 0.30 = 0.513. \end{aligned}$$

This is the minimum attainable probability of error among all possible ways of deciding the supplier on the basis of an observation of when the transistor fails. It turns out to be unnecessary to watch over the transistor until it fails. We merely check it at time $t = 1600.67$ hours. If it is working, we decide that it came from supplier A; if it has failed, we decide that it came from supplier B. ■

B. Numerically Defined Conditioning Events

A conditioning event may be described by an interval in which a random variable is observed to lie, or by a particular value that it is observed to take on. Consider a chance experiment E whose outcome is a continuous random variable $\mathbf{x}$ with probability density function $f_{\mathbf{x}}(x)$ and cumulative distribution function $F_{\mathbf{x}}(x) =$

$\Pr(\mathbf{x} \leq x)$. The conditional cumulative distribution function of $\mathbf{x}$, given $a < \mathbf{x} \leq b$, is defined as

$$F_{\mathbf{x}}(x \mid a < \mathbf{x} \leq b) = \Pr(\mathbf{x} \leq x \mid a < \mathbf{x} \leq b) \tag{2.51}$$
$$= \frac{\Pr[(\mathbf{x} \leq x) \cap (a < \mathbf{x} \leq b)]}{\Pr(a < \mathbf{x} \leq b)}$$

by the definition of conditional probability in (1.9). Now

$$(\mathbf{x} \leq x) \cap (a < \mathbf{x} \leq b) = \begin{cases} \varnothing, & x \leq a, \\ (a < \mathbf{x} \leq x), & a < x \leq b, \\ (a < \mathbf{x} \leq b), & x > b. \end{cases}$$

Hence

$$F_{\mathbf{x}}(x \mid a < \mathbf{x} \leq b) = \begin{cases} 0, & x \leq a, \\ \dfrac{F_{\mathbf{x}}(x) - F_{\mathbf{x}}(a)}{F_{\mathbf{x}}(b) - F_{\mathbf{x}}(a)}, & a < x \leq b, \\ 1, & x > b. \end{cases} \tag{2.52}$$

It will look somewhat like Fig. 2.18.

The conditional probability density function of $\mathbf{x}$, given $a < \mathbf{x} \leq b$, is obtained by differentiation,

$$f_{\mathbf{x}}(x \mid a < \mathbf{x} \leq b) = \begin{cases} 0, & x \leq a, \\ \dfrac{f_{\mathbf{x}}(x)}{F_{\mathbf{x}}(b) - F_{\mathbf{x}}(a)}, & a < x \leq b, \\ 0, & x > b, \end{cases} \tag{2.53}$$

and is illustrated in Fig. 2.19. Since we are told that the outcome $\mathbf{x}$ of the experiment lies between a and b, the conditional probability density function of $\mathbf{x}$ must be zero outside that interval. Inside the interval it is proportional to $f_{\mathbf{x}}(x)$, the constant of proportionality being just what is required to satisfy the condition

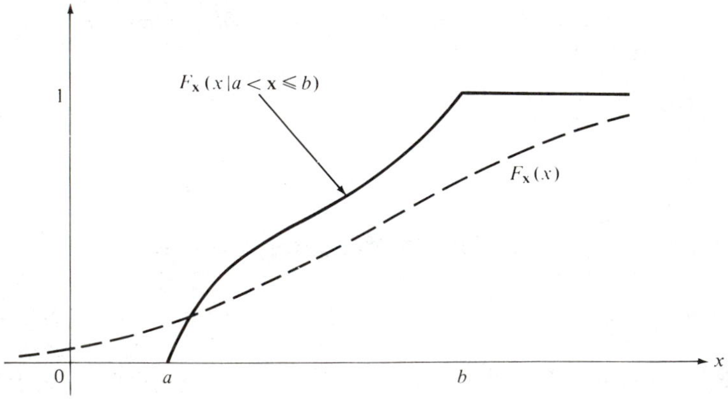

FIGURE 2.18 Conditional cumulative distribution.

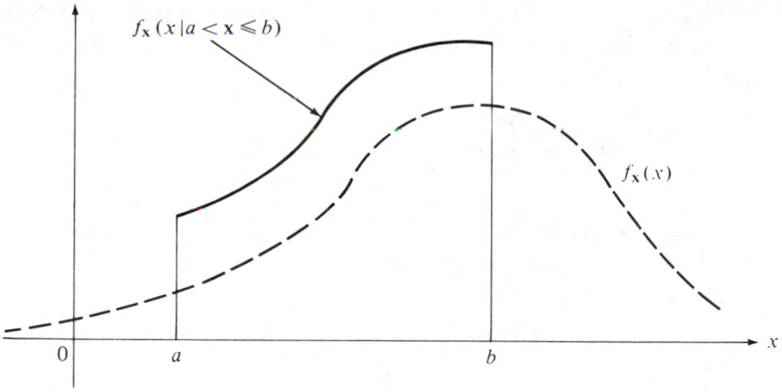

FIGURE 2.19 Conditional probability density function.

$$\int_{-\infty}^{\infty} f_{\mathbf{x}}(x \mid a < \mathbf{x} \leq b)\, dx = 1. \tag{2.54}$$

Letting the interval (a, b) shrink down to some point $\xi \in (a, b)$, we see from (2.52) and (2.53) that

$$F_{\mathbf{x}}(x \mid \mathbf{x} = \xi) = U(x - \xi), \tag{2.55}$$

$$f_{\mathbf{x}}(x \mid \mathbf{x} = \xi) = \delta(x - \xi), \tag{2.56}$$

which should be obvious on general considerations.

C. Failure Probabilities

An instructive application of these conditional distributions is to be found in analyzing the failures of equipment. The conditional failure rate $\beta(t)$ of a device is defined so that $\beta(t)\, dt$ is the conditional probability that it fails between time t and time $t + dt$, given that it has survived until time t; that is, with $\mathbf{t}$ the failure time,

$$\Pr(t < \mathbf{t} < t + dt \mid \mathbf{t} > t) = \beta(t)\, dt. \tag{2.57}$$

Observe that $\beta(t)$ is not a probability density function.

Let $f_{\mathbf{t}}(t)$ be the probability density function, $F_{\mathbf{t}}(t)$ the cumulative distribution function of the failure time $\mathbf{t}$. Then as in (2.52), with $a = t$, $b = \infty$,

$$\Pr(t < \mathbf{t} < t + dt \mid \mathbf{t} > t) = F_{\mathbf{t}}(t + dt \mid \mathbf{t} > t) - F_{\mathbf{t}}(t \mid \mathbf{t} > t)$$

$$= \frac{F_{\mathbf{t}}(t + dt) - F_{\mathbf{t}}(t)}{1 - F_{\mathbf{t}}(t)} \doteq \frac{f_{\mathbf{t}}(t)\, dt}{1 - F_{\mathbf{t}}(t)} = \beta(t)\, dt,$$

whence the conditional failure rate is

$$\beta(t) = \frac{f_{\mathbf{t}}(t)}{1 - F_{\mathbf{t}}(t)}. \tag{2.58}$$

If one has plausible grounds on which to hypothesize the functional form of $\beta(t)$, one can calculate the probability density function $f_{\mathbf{t}}(t)$ of the failure time **t** by integration. From (2.8),

$$\beta(t) = \frac{dF_{\mathbf{t}}(t)/dt}{1 - F_{\mathbf{t}}(t)} = -\frac{d}{dt}\{\ln[1 - F_{\mathbf{t}}(t)]\}.$$

Integrating, and using the fact that $F_{\mathbf{t}}(0) = 0$, the device being supposed working when turned on at $t = 0$, we find that

$$\ln[1 - F_{\mathbf{t}}(t)] = -\int_0^t \beta(s)\, ds,$$

and hence

$$F_{\mathbf{t}}(t) = 1 - \exp\left[-\int_0^t \beta(s)\, ds\right] \tag{2.59}$$

is the cumulative distribution function of the failure time. Under the assumption that any device will fail sometime, the conditional failure rate must be such that

$$\int_0^\infty \beta(s)\, ds = \infty,$$

so that $F_t(\infty) = 1$. The probability density function of the failure time **t** is found by differentiation,

$$f_{\mathbf{t}}(t) = \beta(t) \exp\left[-\int_0^t \beta(s)\, ds\right]. \tag{2.60}$$

The model we adopted for the failure of the diodes in Example 2.3 has $\beta(t) \equiv \mu$, a constant: the diode is just as likely to burn out in the next Δt seconds, no matter how long it has already operated. This model may be unrealistic for diodes. It is correct for radioactive atoms if the radioactive decay

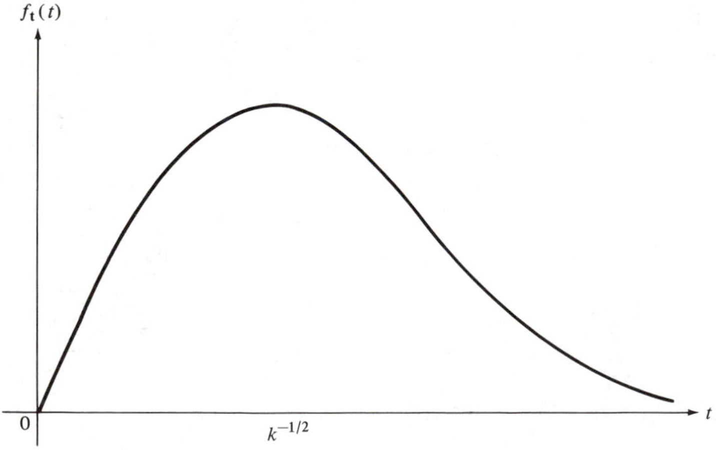

FIGURE 2.20 Rayleigh failure-time density function.

of an atom is considered its failure. To a device consisting of a multitude of small parts that wear out continually, the linear conditional failure rate

$$\beta(t) = kt, \qquad t > 0,$$

might be attributed. Then (2.60) gives the Rayleigh probability density function

$$f_t(t) = kte^{-kt^2/2}U(t)$$

for the failure time. This probability density function is plotted in Fig. 2.20. It peaks at $t = k^{-1/2}$, by which time a fraction $1 - e^{-1/2} = 0.3935$ of the devices have broken down.

The conditional failure rates $\beta(t)$ of the male (M) and female (F) human body are graphed in Fig. 2.21, and their cumulative failure-time distribution

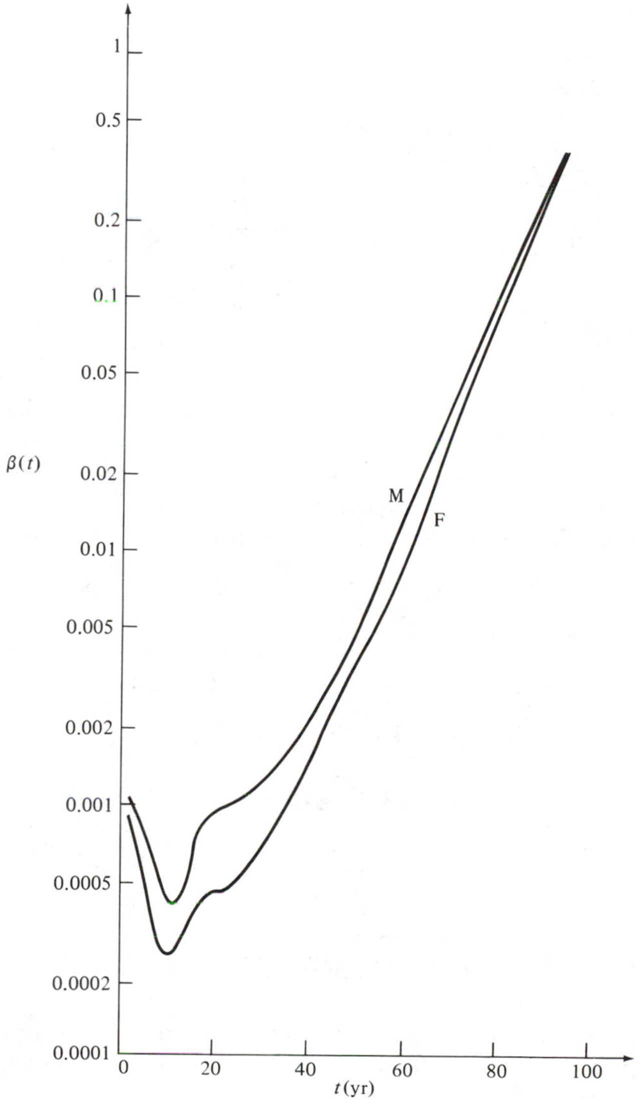

FIGURE 2.21 Conditional failure rate, human body.

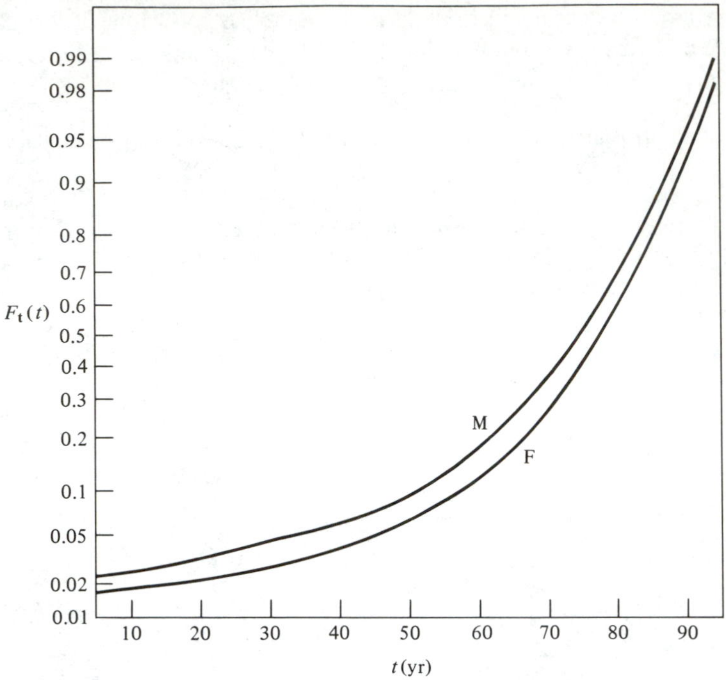

FIGURE 2.22 Cumulative failure-time distribution, human body.

functions $F_t(t)$ appear in Fig. 2.22. The values of the latter were taken from a Swedish mortality table for the period 1956–1960, the only data I had at hand.* Various portions of the function $-\ln[1 - F_t(t)]$ were fitted by orthogonal polynomials, and those were differentiated to obtain the failure rate $\beta(t)$. The rapidly rising portion for the older ages is fitted by the exponential dependence

$$\beta(t) \doteq b \exp[b(t - t_0)], \qquad t \geq 75 \text{ yr}.$$

with

$$b = 0.090825, \qquad t_0 = 79.15 \text{ yr (males)}$$

$$b = 0.10690, \qquad t_0 = 81.57 \text{ yr (females)}$$

Once that exponential term dominates the failure rate $\beta(t)$, our lives are snuffed out like so many candles.

2-3

FUNCTIONS OF A RANDOM VARIABLE

When a signal and accompanying random noise pass through a nonlinear device, such as a rectifier, the probability distribution of the combination of signal and

*_Statistisk Årsbok för Sverige_. Stockholm: Statistiska Centralbyrån, 1965, Table 46, p. 54.

noise is altered. In terms of its input x at any time t, the output of the device will be some function of x, such as

$$y = g(x) = A(e^{\mu x} - 1), \tag{2.61}$$

which is depicted in Fig. 2.23. A sample of the input at an arbitrary time t will be a random variable because of the noise; it might have a Gaussian probability density function $f_{\mathbf{x}}(x)$ like that in (2.19). The output $\mathbf{y}$ of the rectifier will then also be a random variable, and we may need to know its probability density function.

In this section we treat the problem of calculating the probability density function and the cumulative distribution function of a random variable $\mathbf{y}$ that is a given function $g(\mathbf{x})$ of a random variable $\mathbf{x}$ having a known probability density function $f_{\mathbf{x}}(x)$. We assume that the function $g(x)$ has a unique value for all values of x that might be taken on by the random variable $\mathbf{x}$.

A. Discrete Random Variable

The problem is simplest when the input random variable $\mathbf{x}$ is of discrete type, taking values in a countable set $x_1, x_2, \ldots$ with probabilities $\Pr(\mathbf{x} = x_k)$, $k = 1, 2, \ldots$, which sum to 1. Then the random variable $\mathbf{y}$ takes on only a countable set of values

$$y_1 = g(x_1), \quad y_2 = g(x_2), \quad \ldots, \quad y_k = g(x_k), \quad \ldots.$$

If $y = g(x)$ is a monotone function of x, as in Fig. 2.24(a), it has a unique inverse, which we write as

$$x = g^{-1}(y) \tag{2.62}$$

and which is depicted in Fig. 2.24(b). Then the probabilities of the various values of the random variable $\mathbf{y}$ are simply

$$\Pr(\mathbf{y} = y_k) = \Pr(\mathbf{x} = g^{-1}(y_k)).$$

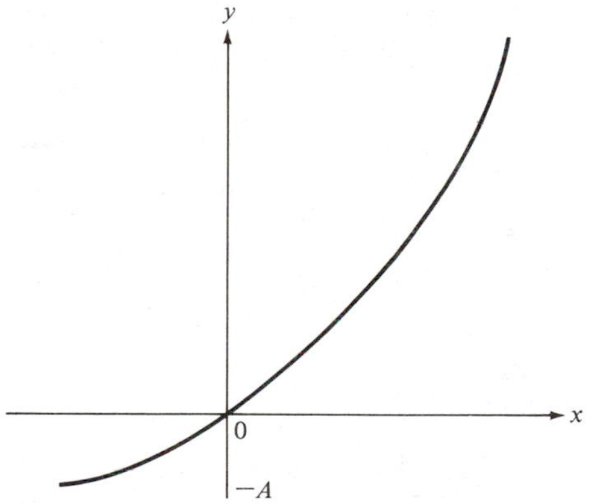

FIGURE 2.23 Rectifier characteristic.

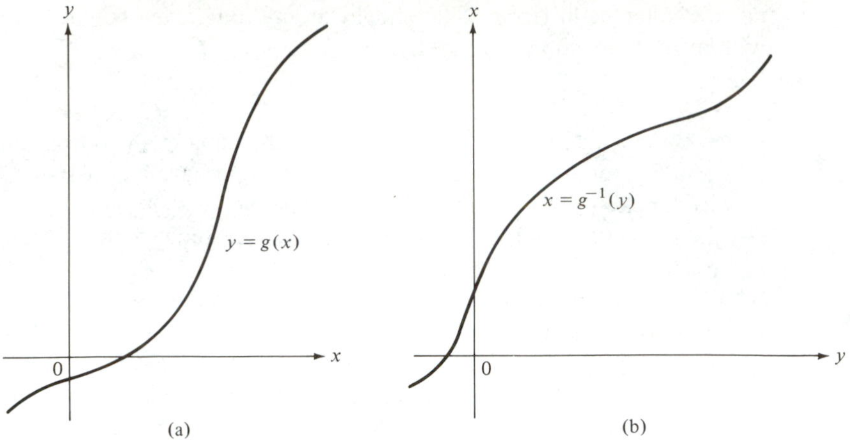

FIGURE 2.24 Monotone transformation and its inverse.

EXAMPLE 2.8 The random variable $\mathbf{x}$ has a Poisson distribution as in (1.27), and $\mathbf{y} = \mathbf{x}^3$. Then $\mathbf{y}$ takes on the integral values 0, 1, 8, 27, 81, . . . , and

$$\Pr(\mathbf{y} = k^3) = \frac{m^k e^{-m}}{k!}, \qquad k = 0, 1, 2, \ldots$$

most simply describes its distribution. ■

If $y = g(x)$ is not monotone, as when $y = x^2$, more than one value of x may go into the same value of y. One then adds the probabilities of those values of x. The procedure is best illustrated by an example.

EXAMPLE 2.9 Let $\mathbf{x}$ take on values -2, -1, 0, and 1 with probabilities

$$\Pr(\mathbf{x} = -2) = 0.1, \qquad \Pr(\mathbf{x} = -1) = 0.2,$$
$$\Pr(\mathbf{x} = 0) = 0.3, \qquad \Pr(\mathbf{x} = 1) = 0.4.$$

Then $\mathbf{y} = \mathbf{x}^2$ takes on values 0, 1, and 4 with probabilities

$$\Pr(\mathbf{y} = 0) = \Pr(\mathbf{x} = 0) = 0.3,$$
$$\Pr(\mathbf{y} = 1) = \Pr(\mathbf{x} = -1) + \Pr(\mathbf{x} = 1) = 0.2 + 0.4 = 0.6,$$
$$\Pr(\mathbf{y} = 4) = \Pr(\mathbf{x} = -2) = 0.1.$$ ■

Always check to be sure that the probabilities of the several values of the output random variable $\mathbf{y}$ add up to 1. The cumulative distribution functions of $\mathbf{x}$ and $\mathbf{y}$ in Example 2.9 are shown in Fig. 2.25.

B. Continuous Random Variable

1. MONOTONE TRANSFORMATIONS

Now let $\mathbf{x}$ be a continuous random variable with probability density function $f_{\mathbf{x}}(x)$ and let $\mathbf{y} = g(\mathbf{x})$ be a monotone function of $\mathbf{x}$, as in Fig. 2.26. For dx and dy infinitesimal, let

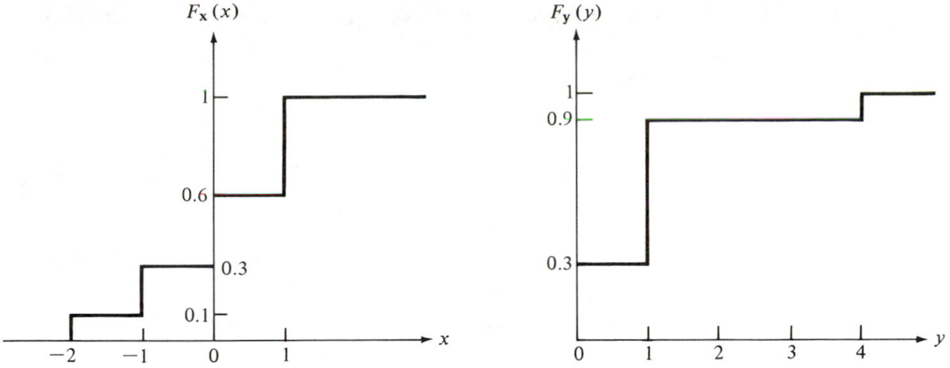

FIGURE 2.25 Transformation $y = x^2$ of discrete random variable.

$$
\begin{aligned}
y &= g(x), \\
y + dy &= g(x + dx) \qquad (2.63)\\
&\doteq g(x) + g'(x)\,dx,
\end{aligned}
$$

where the prime indicates the derivative,

$$g'(x) = \frac{dg}{dx}.$$

The event $(y < \mathbf{y} < y + dy)$ contains the same outcomes of our chance experiment E as the event $(x < \mathbf{x} < x + dx)$; we have merely relabeled them with values of **y** instead of with values of **x.** The probabilities of these two events must therefore be equal:

$$\Pr\,(y < \mathbf{y} < y + dy) = \Pr\,(x < \mathbf{x} < x + dx). \qquad (2.64)$$

In terms of the probability density functions of **x** and **y** these probabilities are

$$f_{\mathbf{y}}(y)|dy| = f_{\mathbf{x}}(x)|dx|. \qquad (2.65)$$

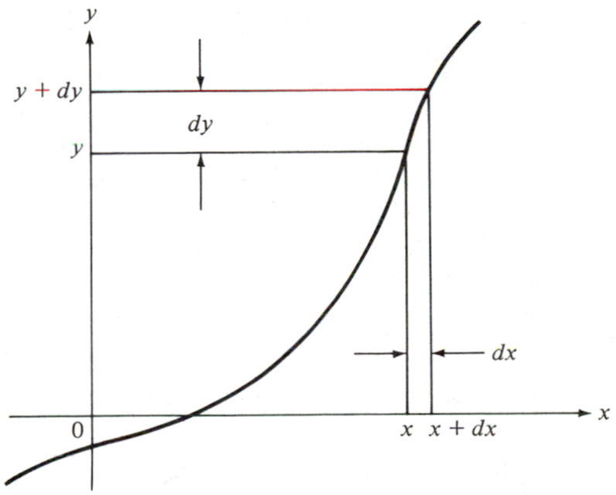

FIGURE 2.26. Monotone transformation.

Think of this important relation as expressing the "conservation of probability" in the transformation from $\mathbf{x}$ to $\mathbf{y} = g(\mathbf{x})$.

From (2.65) we obtain, by dividing by $|dy|$,

$$f_{\mathbf{y}}(y) = \frac{f_{\mathbf{x}}(x)}{|dy/dx|} = \left.\frac{f_{\mathbf{x}}(x)}{|g'(x)|}\right|_{x=g^{-1}(y)}. \tag{2.66}$$

Into (2.66) we must substitute x as a function of y in order to obtain the probability density function of the random variable $\mathbf{y}$ in terms of y; we indicate this by $x = g^{-1}(y)$. We have used the absolute value signs on the differentials in order to include monotone decreasing functions; it is only the magnitudes of dx and dy that matter in the probabilities in (2.65).

EXAMPLE 2.10 Let $\mathbf{x}$ be uniformly distributed on (0, 1),

$$f_{\mathbf{x}}(x) = \begin{cases} 1, & 0 < x < 1, \\ 0, & x < 0, \quad x > 1, \end{cases}$$

and let $\mathbf{y} = \sinh \mathbf{x}$ represent the transformation. Then

$$g'(x) = \cosh x,$$

and the output variable $\mathbf{y}$ takes on values only between 0 and sinh 1. Because

$$\cosh^2 x - \sinh^2 x = 1,$$

we find from (2.66) the probability density function of $\mathbf{y}$,

$$f_{\mathbf{y}}(y) = \begin{cases} 0, & y < 0, \quad y > \sinh 1, \\ \dfrac{1}{\cosh x} = (1 + y^2)^{-1/2}, & 0 < y < \sinh 1. \end{cases}$$ ■

The cumulative distribution function $F_{\mathbf{y}}(y)$ of the transformed variable can be obtained directly from the cumulative distribution function $F_{\mathbf{x}}(x)$. With $g(x)$ monotone increasing, the events

$$(\mathbf{y} \le y) \qquad \text{and} \qquad (\mathbf{x} \le g^{-1}(y))$$

are the same. Hence their probabilities are equal, and

$$F_{\mathbf{y}}(y) = F_{\mathbf{x}}(g^{-1}(y)). \tag{2.67}$$

EXAMPLE 2.10 (continued) Here the cumulative distribution function of the input variable is

$$F_{\mathbf{x}}(x) = \begin{cases} 0, & x < 0, \\ x, & 0 < x < 1, \\ 1, & x > 1, \end{cases}$$

and hence the cumulative distribution function of the output variable is

$$F_{\mathbf{y}}(y) = \begin{cases} 0, & y < 0, \\ \sinh^{-1} y, & 0 < y < \sinh 1, \\ 1, & y > \sinh 1. \end{cases}$$

The same result could be obtained, but more laboriously, by integrating the probability density function $f_{\mathbf{y}}(y)$. ■

By drawing a diagram, convince yourself that if the function $y = g(x)$ is monotone decreasing, the cumulative distribution function of **y** is

$$\begin{aligned} F_{\mathbf{y}}(y) &= \Pr(\mathbf{y} \leq y) \\ &= \Pr(\mathbf{x} \geq g^{-1}(y)) \\ &= 1 - \Pr(\mathbf{x} < g^{-1}(y)) \\ &= 1 - \Pr(\mathbf{x} \leq g^{-1}(y)) + \Pr(\mathbf{x} = g^{-1}(y)) \\ &= 1 - F_{\mathbf{x}}(g^{-1}(y)) + \Pr(\mathbf{x} = g^{-1}(y)), \end{aligned} \tag{2.68}$$

in which for the sake of completeness we have included the possibility that the random variable **x** is of discrete or mixed type. The last term would appear only for such values of y that the cumulative distribution function $F_{\mathbf{x}}(x)$ has a jump at $g^{-1}(y)$.

EXAMPLE 2.11 Linear Transformation (see Fig. 2.27)

When the transformation is linear,

$$\mathbf{y} = a\mathbf{x} + b,$$

with inverse

$$\mathbf{x} = g^{-1}(\mathbf{y}) = \frac{\mathbf{y} - b}{a},$$

we find immediately that

$$f_{\mathbf{y}}(y) = |a|^{-1} f_{\mathbf{x}}\left(\frac{y - b}{a}\right),$$

$$\begin{aligned} F_{\mathbf{y}}(y) &= F_{\mathbf{x}}\left(\frac{y - b}{a}\right), \qquad a > 0, \\ &= 1 - F_{\mathbf{x}}\left(\frac{y - b}{a}\right) + \Pr\left(\mathbf{x} = \frac{y - b}{a}\right), \qquad a < 0. \end{aligned}$$

■

2. MANY–ONE TRANSFORMATIONS

If several values of x can be transformed into one value of y, we say that $y = g(x)$ represents a *many–one transformation*. Thus with $\mathbf{y} = \mathbf{x}^2$, both x and $-x$ go into the same value of y. It is merely necessary to add the probabilities of all the "infinitesimal events" $(x < \mathbf{x} < x + dx)$ that go into the same event $(y < \mathbf{y} < y + dy)$.

In Fig. 2.28 we illustrate this point by a transformation that might, for instance, be given by a third-degree polynomial in x. Now the equations

$$y = g(x), \qquad y + dy = g(x + dx)$$

each have three roots, x_1, x_2, x_3 and $x_1 + dx_1, x_2 + dx_2, x_3 + dx_3$, respectively.

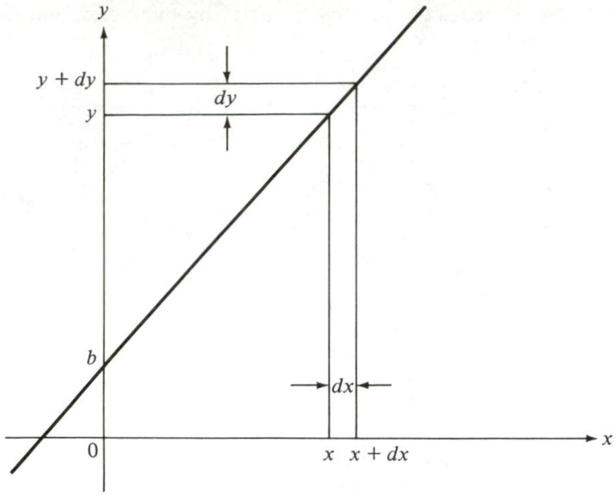

FIGURE 2.27 Linear transformation.

The event $(y < \mathbf{y} < y + dy)$ occurs when any one of three events involving $\mathbf{x}$ occurs,

$$(y < \mathbf{y} < y + dy) = (x_1 < \mathbf{x} < x_1 + dx_1) \cup (x_2 + dx_2 < \mathbf{x} < x_2) \cup (x_3 < \mathbf{x} < x_3 + dx_3).$$

These events will be mutually exclusive if dy is sufficiently small, and dy ultimately vanishes. Hence

$$\Pr(y < \mathbf{y} < y + dy) = \Pr(x_1 < \mathbf{x} < x_1 + dx_1) + \Pr(x_2 + dx_2 < \mathbf{x} < x_2) + \Pr(x_3 < \mathbf{x} < x_3 + dx_3),$$

and in terms of the probability density functions of $\mathbf{x}$ and $\mathbf{y}$,

$$f_{\mathbf{y}}(y)|dy| = f_{\mathbf{x}}(x_1)|dx_1| + f_{\mathbf{x}}(x_2)|dx_2| + f_{\mathbf{x}}(x_3)|dx_3|.$$

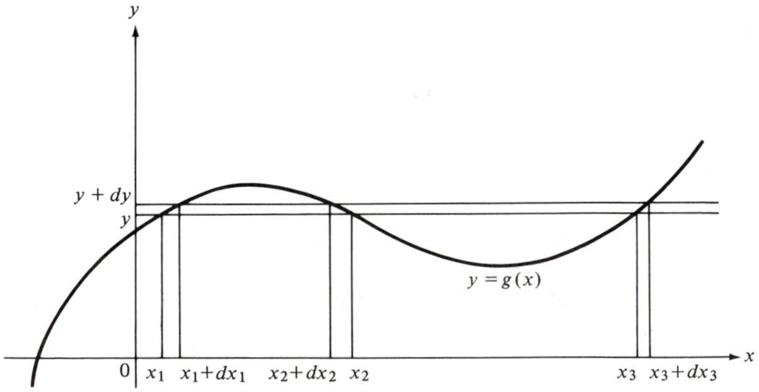

FIGURE 2.28 Many-one transformation.

(Observe that here the second component event is "backwards" since $dy/dx < 0$ at $x = x_2$. We must therefore use the absolute value $|dx_2|$ in the second term, and for generality we use the absolute values of the differentials in all three.) Dividing by $|dy|$, we find for the probability density function of the output $\mathbf{y}$

$$f_{\mathbf{y}}(y) = \left.\frac{f_{\mathbf{x}}(x_1)}{|g'(x_1)|}\right|_{x_1=g_1^{-1}(y)} + \left.\frac{f_{\mathbf{x}}(x_2)}{|g'(x_2)|}\right|_{x_2=g_2^{-1}(y)} + \left.\frac{f_{\mathbf{x}}(x_3)}{|g'(x_3)|}\right|_{x_3=g_3^{-1}(y)}, \tag{2.69}$$

where by $g_k^{-1}(y)$ we mean the kth root of the equation $y = g(x)$; $k = 1, 2, 3$ here. This solution is easily generalized to an arbitrary number of roots. Remember to add the probabilities $\Pr(x_k < \mathbf{x} < x_k + dx_k)$ of *all* the events that go into the same "output event" $(y < \mathbf{y} < y + dy)$.

EXAMPLE 2.12 Quadratic Transformation (see Fig. 2.29)

Let $y = a(x - c)^2 + b$, $a > 0$. Now values of y less than b are impossible, and $f_{\mathbf{y}}(y) \equiv 0$, $F_{\mathbf{y}}(y) \equiv 0$, $y < b$. For $y > b$, two values of x contribute to the same value of y, namely

$$x_1 = c - \left(\frac{y-b}{a}\right)^{1/2},$$

$$x_2 = c + \left(\frac{y-b}{a}\right)^{1/2}.$$

Since $g'(x) = 2a(x - c)$, we find for the probability density function of the output variable, as in (2.69),

$$f_{\mathbf{y}}(y) = \frac{1}{2}[a(y-b)]^{-1/2}\left[f_{\mathbf{x}}\left(c - \sqrt{\frac{y-b}{a}}\right) + f_{\mathbf{x}}\left(c + \sqrt{\frac{y-b}{a}}\right)\right]. \tag{2.70}$$

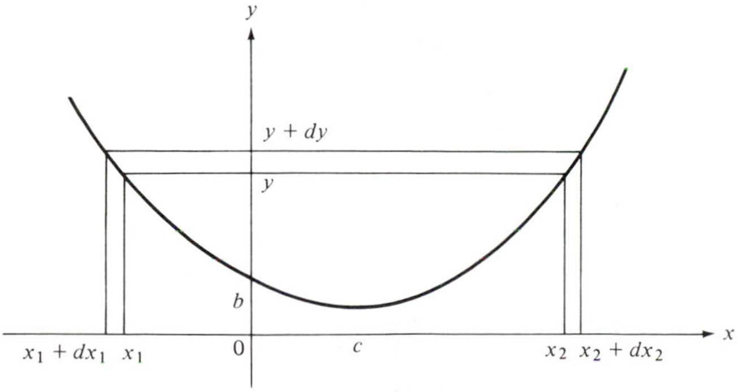

FIGURE 2.29 Quadratic transformation.

The cumulative distribution function of **y**, in terms of that of **x**, is

$$\begin{aligned} F_{\mathbf{y}}(y) &= \Pr(\mathbf{y} \le y) = \Pr(x_1 \le \mathbf{x} \le x_2) \\ &= F_{\mathbf{x}}(x_2) - F_{\mathbf{x}}(x_1) + \Pr(\mathbf{x} = x_1) \\ &= F_{\mathbf{x}}\left(c + \sqrt{\frac{y-b}{a}}\right) - F_{\mathbf{x}}\left(c - \sqrt{\frac{y-b}{a}}\right) \\ &\quad + \Pr\left(\mathbf{x} = c - \sqrt{\frac{y-b}{a}}\right). \end{aligned} \tag{2.71}$$

The last term occurs only for input random variables of discrete or mixed type; it is included for the sake of completeness.

Suppose, for instance, that the input random variable is Gaussian with mean m and variance σ^2, as in (2.19), and suppose for simplicity that $\mathbf{y} = \mathbf{x}^2$. Then

$$f_{\mathbf{y}}(y) = 0, \qquad y < 0,$$

and by (2.70)

$$\begin{aligned} f_{\mathbf{y}}(y) &= \tfrac{1}{2} y^{-1/2} (2\pi\sigma^2)^{-1/2} \left\{ \exp\left[-\frac{(y^{1/2}+m)^2}{2\sigma^2} \right] \right. \\ &\quad \left. + \exp\left[-\frac{(y^{1/2}-m)^2}{2\sigma^2} \right] \right\} U(y). \end{aligned}$$

■

C. Mixed Random Variable

If the input random variable **x** is of mixed type, the easiest way to find the distribution of the output random variable **y** is to treat the discrete portion, represented by jumps in the cumulative distribution function of **x** and by delta functions in the probability density function of **x**, by the method of part A and to treat the continuous portion by the method of part B. It suffices to illustrate the method by an example, and there is no point to writing down general formulas. Remember the bookkeeping principle: All the probability at the output must add up to 1.

EXAMPLE 2.13 Let the probability density function of the input be

$$f_{\mathbf{x}}(x) = \begin{cases} \frac{1}{4} + \frac{1}{2}\delta(x - \frac{1}{4}), & -1 < x < 1, \\ 0, & |x| > 1, \end{cases}$$

and let $\mathbf{y} = \mathbf{x}^2$. The cumulative distribution function of the input is

$$F_{\mathbf{x}}(x) = \begin{cases} 0, & x < -1, \\ \frac{1}{4}(x+1), & -1 < x < \frac{1}{4}, \\ \frac{1}{4}(x+1) + \frac{1}{2}, & \frac{1}{4} \le x < 1, \\ 1, & x \ge 1. \end{cases}$$

The probability density function and the cumulative distribution function of the input are sketched in Fig. 2.30. Be sure that you understand the relation between these two functions.

Because the random variable **x** takes on the value $\frac{1}{4}$ with probability $\frac{1}{2}$, the output variable **y** takes on the value $\frac{1}{16}$ with the same probability $\frac{1}{2}$, and the output probability

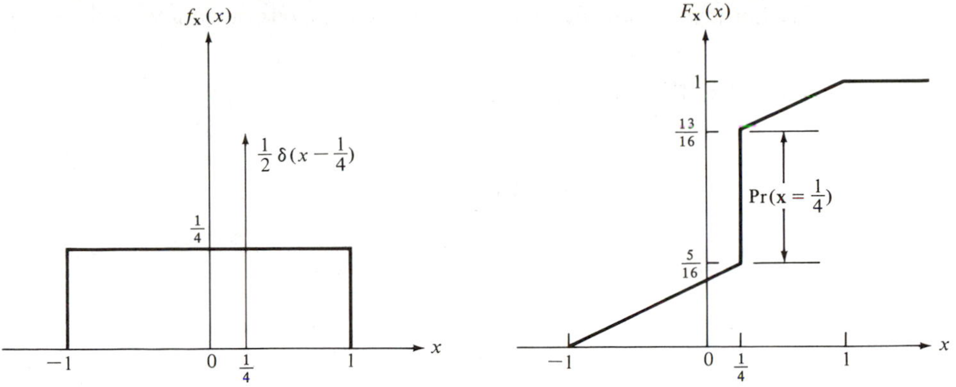

FIGURE 2.30 Distribution of mixed random variable at input.

density function will contain a term $\frac{1}{2}\delta(y - \frac{1}{16})$. That disposes of the discrete component of the input. For the continuous component, we use (2.70). Thus

$$f_y(y) = \begin{cases} 0, & y < 0, \quad y > 1, \\ \frac{1}{4}y^{-1/2} + \frac{1}{2}\delta(y - \frac{1}{16}), & 0 < y < 1, \end{cases}$$

the contributions from $+y^{1/2}$ and $-y^{1/2}$ being equal. This probability density function is sketched in Fig. 2.31(a).

To obtain the cumulative distribution function, we can either integrate the probability density function by (2.9) or use (2.71). The former is the simpler, and we obtain

$$F_y(y) = \begin{cases} 0, & y < 0, \\ \frac{1}{2}y^{1/2} + \frac{1}{2}U(y - \frac{1}{16}), & 0 \le y < 1, \\ 1, & y \ge 1, \end{cases}$$

which is illustrated in Fig. 2.31(b). ■

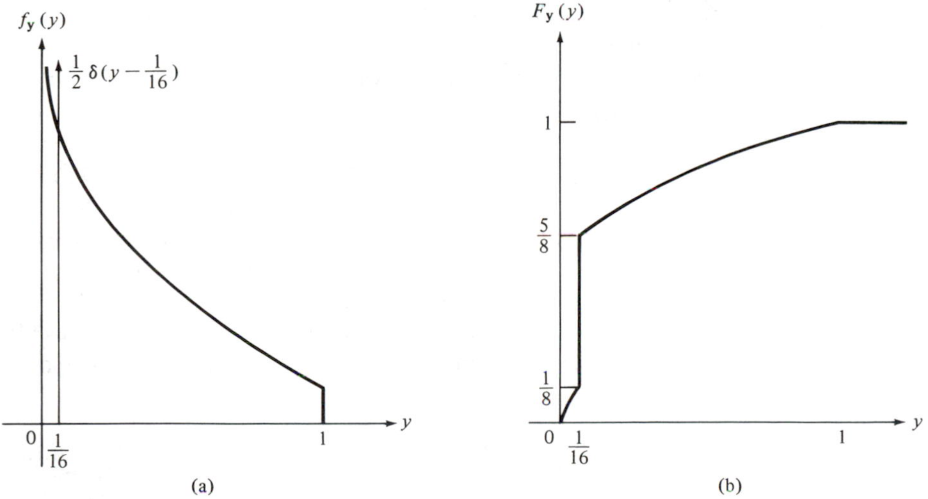

FIGURE 2.31 Distribution of mixed random variable at output of quadratic transformation.

A nonlinear transformation may convert a continuous random variable into a mixed random variable.

EXAMPLE 2.14 Halfwave Rectifier

The output $\mathbf{y}$ of a halfwave rectifier is defined by

$$\mathbf{y} = \mathbf{x}\, U(\mathbf{x}) = \begin{cases} \mathbf{x}, & \mathbf{x} \geq 0, \\ 0, & \mathbf{x} < 0, \end{cases}$$

in terms of the input $\mathbf{x}$. Let $\mathbf{x}$ be a continuous random variable with probability density function $f_{\mathbf{x}}(x)$. Because negative values of $\mathbf{y}$ are impossible,

$$f_{\mathbf{y}}(y) = 0, \qquad y < 0.$$

All negative values of $\mathbf{x}$ are transformed into $\mathbf{y} = 0$, and therefore

$$\Pr(\mathbf{y} = 0) = \Pr(\mathbf{x} \leq 0) = \int_{-\infty}^{0} f_{\mathbf{x}}(x)dx;$$

the probability density function of $\mathbf{y}$ must contain a delta-function at $y = 0$. For $\mathbf{x} > 0$, $\mathbf{y} = \mathbf{x}$ and $f_{\mathbf{y}}(y) = f_{\mathbf{x}}(y)$. The probability density function of the output $\mathbf{y}$ can be written as

$$f_{\mathbf{y}}(y) = f_{\mathbf{x}}(y)\, U(y) + \Pr(\mathbf{x} \leq 0)\delta(y),$$

and the output is a mixed random variable. ■

2-4 EXPECTED VALUES

A. The Mean

Given a random variable $\mathbf{x}$, one would like to know around what center its values cluster, and over how broad a range about that center they are, roughly speaking, dispersed. The former is specified by the mean, the latter by the standard deviation of the random variable. We turn our attention to the definitions of parameters such as these, which characterize the form and properties of probability distributions.

The *mean* or *expected value* of a random variable $\mathbf{x}$ is defined in terms of its probability density function by

$$E(\mathbf{x}) = \bar{x} = \int_{-\infty}^{\infty} x f_{\mathbf{x}}(x)\, dx. \tag{2.72}$$

If one thinks of the probability density function $f_{\mathbf{x}}(x)$ as the density of mass distributed along a line, the expected value $E(\mathbf{x})$ corresponds to the center of mass of the mass distribution, as defined in mechanics.

Equation (2.72) holds for both continuous and discrete random variables. If for a discrete random variable we substitute its probability density function in terms of delta functions from (2.15), we find for its expected value

$$E(\mathbf{x}) = \sum_{k=1}^{\infty} x_k \Pr(\mathbf{x} = x_k). \tag{2.73}$$

Each value of $\mathbf{x}$ is weighted with the probability of its occurrence, and the weighted values are summed to form the mean.

If the probability distribution function has a point of symmetry, such as μ in

$$f_{\mathbf{x}}(x) = \phi(x - \mu) = \phi(\mu - x),$$

then $E(\mathbf{x}) = \mu$, as follows easily from (2.72). Thus in the formula (2.19) for the Gaussian probability density function, the mean of the Gaussian random variable is equal to m.

Suppose that a large number N of independent trials of a chance experiment E are carried out, and that in the rth trial the discrete random variable $\mathbf{x}$ takes on the value ξ_r, which will be one of the numbers $x_1, x_2, \ldots, x_k, \ldots$ in (2.73). We are accustomed to defining the average of these numbers $\xi_1, \xi_2, \ldots, \xi_N$ by

$$\mathrm{Av}_N(\mathbf{x}) = \frac{1}{N} \sum_{r=1}^{N} \xi_r.$$

Suppose that in these N trials the random variable $\mathbf{x}$ takes on the value x_k $n_N(k)$ times, $k = 1, 2, \ldots$. Then our summation can be rearranged into the form

$$\mathrm{Av}_N(\mathbf{x}) = \sum_{k=1}^{\infty} \frac{x_k n_N(k)}{N}, \qquad \sum_{k=1}^{\infty} n_N(k) = N,$$

simply by grouping together the x_1's, the x_2's, and so on. From the standpoint of the relative-frequency concept of probability, discussed in Sec. 1-4B,

$$\frac{n_N(k)}{N} \simeq \Pr(\mathbf{x} = x_k)$$

when $N >> 1$. Our average then approaches the expected value in (2.73),

$$\mathrm{Av}_N(\mathbf{x}) \to E(\mathbf{x}) \qquad \text{as } N \to \infty.$$

The expected value as defined by (2.72) or (2.73) may turn out to be infinite. For the probability density function

$$f_{\mathbf{x}}(x) = \begin{cases} x^{-2}, & x > 1, \\ 0, & x \le 1, \end{cases}$$

for instance, we find that

$$E(\mathbf{x}) = \int_1^{\infty} x \frac{1}{x^2}\, dx = \infty.$$

This simply means that $E(\mathbf{x})$ is not a useful characterization of the distribution of this random variable. Mathematicians would say, "The mean of $\mathbf{x}$ does not exist."

B. Expected Value of a Function of a Random Variable

If $\mathbf{y} = g(\mathbf{x})$ is a function of a random variable of the kind treated in Sec. 2-3, its expected value

$$E(\mathbf{y}) = \int_{-\infty}^{\infty} y f_{\mathbf{y}}(y)\, dy$$

can be written

$$E[g(\mathbf{x})] = \overline{g(x)} = \int_{-\infty}^{\infty} g(x) f_{\mathbf{x}}(x)\, dx, \tag{2.74}$$

and it is usually simpler to calculate $E(\mathbf{y})$ by this formula than by first working out the probability density function of $\mathbf{y}$. Equation (2.74) generalizes the concept of expected value to an arbitrary function $g(\mathbf{x})$ of a random variable.

If the function $g(\mathbf{x})$ is monotone increasing, as in Fig. 2.26, (2.74) follows directly from the "conservation law,"

$$f_{\mathbf{y}}(y)\, dy = f_{\mathbf{x}}(x)\, dx.$$

If $g(x)$ is monotone decreasing, we must put

$$f_{\mathbf{y}}(y)\, dy = -f_{\mathbf{x}}(x)\, dx,$$

but now the integral over x will run from $+\infty$ to $-\infty$; the minus sign disappears when we change the integration variable from x to $x' = -x$, and we again obtain (2.74). If $g(x)$ represents a many–one transformation, as in part 2 of Sec. 2-3B, we need to account for the ranges of all the roots of $y = g(x)$, as in (2.69); we leave the details of verifying (2.74) to the reader.

C. Moments of a Random Variable

For the function $g(x) = x^n$, n a positive integer, the expected value

$$\begin{aligned} m_n = E(\mathbf{x}^n) &= \overline{x^n} \\ &= \int_{-\infty}^{\infty} x^n f_{\mathbf{x}}(x)\, dx \qquad \text{(continuous r.v.)} \qquad (2.75) \\ &= \sum_{k=1}^{\infty} x_k^n \Pr(\mathbf{x} = x_k) \qquad \text{(discrete r.v.)} \end{aligned}$$

is the nth *moment* of the random variable $\mathbf{x}$.

Of greater significance are the central moments

$$\begin{aligned} \mu_n = E[(\mathbf{x} - \bar{x})^n] &= \overline{(x - \bar{x})^n} \\ &= \int_{-\infty}^{\infty} (x - \bar{x})^n f_{\mathbf{x}}(x)\, dx \qquad \text{(continuous r.v.)} \qquad (2.76) \\ &= \sum_{k=1}^{\infty} (x_k - \bar{x})^n \Pr(\mathbf{x} = x_k) \qquad \text{(discrete r.v.)} \end{aligned}$$

with respect to the mean $\bar{x} = E(\mathbf{x})$ of the random variable $\mathbf{x}$. Most important of these is the second central moment

$$\begin{aligned} \mu_2 = \text{Var}\, \mathbf{x} &= E[(\mathbf{x} - \bar{x})^2] \\ &= \int_{-\infty}^{\infty} (x - \bar{x})^2 f_{\mathbf{x}}(x)\, dx \qquad \text{(continuous r.v.)} \qquad (2.77) \end{aligned}$$

$$= \sum_{k=1}^{\infty} (x_k - \bar{x})^2 \Pr(\mathbf{x} = x_k) \qquad \text{(discrete r.v.)}$$

which is called the *variance* of the random variable $\mathbf{x}$. The second moment m_2 corresponds to the moment of inertia of a mass distribution with respect to an axis through the origin $x = 0$. The variance $\mu_2 = \sigma^2$ corresponds to the moment of inertia with respect to an axis through the center of mass. The square root of the variance,

$$\sigma = \sqrt{\operatorname{Var} \mathbf{x}} \tag{2.78}$$

is called the *standard deviation* of the random variable $\mathbf{x}$ and corresponds to the radius of gyration in mechanics. It roughly measures over how broad a range the values of the random variable $\mathbf{x}$ are dispersed. If $\operatorname{Var} \mathbf{x} = \infty$, some other way of specifying the breadth of the distribution must be sought.

EXAMPLE 2.15 Variance of a Gaussian Random Variable

For the Gaussian probability density function in (2.19), the variance is given by

$$\operatorname{Var} \mathbf{x} = (2\pi\sigma^2)^{-1/2} \int_{-\infty}^{\infty} (x - m)^2 e^{-(x-m)^2/2\sigma^2}\, dx$$

$$= \sigma^2 (2\pi)^{-1/2} \int_{-\infty}^{\infty} y^2 e^{-y^2/2}\, dy$$

if we change the integration variable to $y = (x - m)/\sigma$. Integration by parts yields, with

$$u = y, \qquad dv = ye^{-y^2/2}\, dy,$$

$$du = dy, \qquad v = -e^{-y^2/2},$$

the result

$$\operatorname{Var} \mathbf{x} = \sigma^2 (2\pi)^{-1/2} \left[-ye^{-y^2/2} \Big|_{-\infty}^{\infty} + \int_{-\infty}^{\infty} e^{-y^2/2}\, dy \right] = \sigma^2$$

by virtue of the normalization integral for (2.19), or

$$\int_{-\infty}^{\infty} e^{-y^2/2}\, dy = (2\pi)^{1/2}.$$

Thus σ^2 in (2.19) is the variance of the Gaussian random variable, and σ is its standard deviation. ■

The variance is most conveniently calculated by a formula corresponding to the parallel-axes theorem for moments of inertia,

$$\operatorname{Var} \mathbf{x} = E(\mathbf{x}^2) - [E(\mathbf{x})]^2 = \overline{x^2} - \bar{x}^2. \tag{2.79}$$

To prove it, we expand the square in (2.77),

$$\operatorname{Var} \mathbf{x} = \int_{-\infty}^{\infty} (x^2 - 2\bar{x}x + \bar{x}^2) f_{\mathbf{x}}(x)\, dx$$

$$= E(\mathbf{x}^2) - 2\bar{x} \int_{-\infty}^{\infty} x f_{\mathbf{x}}(x)\, dx + \bar{x}^2 \int_{-\infty}^{\infty} f_{\mathbf{x}}(x)\, dx$$

$$= E(\mathbf{x}^2) - 2\bar{x}^2 + \bar{x}^2.$$

The variance Var $\mathbf{x}$ does not change with a shift of the mean of the random variable, as is indeed the case with all the central moments. Later we shall identify the mean of a random variable with a signal and fluctuations about the mean with the noise. The variance then measures the strength of the noise.

Scaling a random variable by a factor a multiplies its mean by a and its variance by a^2, as is easily verified from (2.72) and (2.77):

$$\begin{aligned} E(a\mathbf{x}) &= aE(\mathbf{x}), \\ \text{Var}\,(a\mathbf{x}) &= a^2\,\text{Var}\,\mathbf{x}. \end{aligned} \tag{2.80}$$

Indeed,

$$\text{Var}\,(a\mathbf{x} + b) = a^2\,\text{Var}\,\mathbf{x}$$

since the shift b does not affect the variance.

Statisticians call the quantity

$$\frac{\mu_3}{\sigma^3} = \frac{\overline{(x - \bar{x})^3}}{\sigma^3}$$

the *skewness* of the distribution; it vanishes if the probability density function is symmetrical about the mean $\bar{x}$. The quantity $(\mu_4 - 3\sigma^4)/\sigma^4$ is called the *kurtosis;* it vanishes if $\mathbf{x}$ is a normal (Gaussian) random variable. Quantities such as the mean, variance, skewness, and kurtosis can be estimated by taking averages over the outcomes of a long series of independent trials of the chance experiment E. They permit rough characterization of a probability density function whose true form is unknown.

EXAMPLE 2.16 For the discrete random variable $\mathbf{x}$ whose probability distribution is given in Example 2.9, the mean is

$$E(\mathbf{x}) = (-2)(0.1) + (-1)(0.2) + (0)(0.3) + (1)(0.4) = 0$$

and its variance is

$$\text{Var}\,\mathbf{x} = (-2)^2(0.1) + (-1)^2(0.2) + 0^2(0.3) + 1^2(0.4) = 1,$$

so that its standard deviation equals 1. ■

EXAMPLE 2.17 For random variables of mixed type, we use a combination of the formulas for discrete and continuous random variables. Thus for the random variable $\mathbf{x}$ of Example 2.13, we find that

$$E(\mathbf{x}) = \int_{-1}^{1} \frac{1}{4}x\,dx + \frac{1}{2}\cdot\frac{1}{4} = \frac{1}{8},$$

and

$$E(\mathbf{x}^2) = \int_{-1}^{1} \frac{1}{4}x^2\,dx + \frac{1}{2}\left(\frac{1}{4}\right)^2 = \frac{19}{96}.$$

Hence by (2.79),

$$\text{Var}\,\mathbf{x} = \frac{19}{96} - \frac{1}{64} = \frac{35}{192}$$

and the standard deviation is $\sqrt{\dfrac{35}{192}} = 0.427.$ ■

EXAMPLE 2.18 Halfwave Rectifier (continued)

The output of the halfwave rectifier introduced in Example 2.14 is $\mathbf{y} = \mathbf{x}\, U(\mathbf{x})$. Its expected value is

$$E(\mathbf{y}) = \int_{-\infty}^{\infty} x\, U(x)\, f_{\mathbf{x}}(x)\, dx = \int_{0}^{\infty} x\, f_{\mathbf{x}}(x)\, dx,$$

and because $[U(x)]^2 = U(x)$, its mean square value is

$$E(\mathbf{y}^2) = \int_{-\infty}^{\infty} x^2\, U(x)\, f_{\mathbf{x}}(x)\, dx = \int_{0}^{\infty} x^2\, f_{\mathbf{x}}(x)\, dx.$$

If, for instance, the input has a Gaussian distribution as in (2.19) with mean $m = 0$ and variance σ^2,

$$E(\mathbf{y}) = (2\pi\sigma^2)^{-1/2} \int_{0}^{\infty} x\, e^{-x^2/2\sigma^2}\, dx = (2\pi)^{-1/2}\, \sigma,$$

and from Example 2.15,

$$E(\mathbf{y}^2) = (2\pi\sigma^2)^{-1/2} \int_{0}^{\infty} x^2\, e^{-x^2/2\sigma^2}\, dx = \frac{1}{2}\sigma^2.$$

The variance of the output is then

$$\operatorname{Var} \mathbf{y} = \frac{\sigma^2}{2} - \frac{\sigma^2}{2\pi} = \frac{1}{2}(1 - \pi^{-1})\sigma^2 = 0.3408\, \sigma^2.$$

If the input is a sine wave of amplitude A and random phase, we can take

$$\mathbf{x} = A \sin \boldsymbol{\theta}$$

with the random variable $\boldsymbol{\theta}$ uniformly distributed over $(0, 2\pi)$,

$$f_{\boldsymbol{\theta}}(\theta) = (2\pi)^{-1}, \qquad 0 < \theta < 2\pi.$$

The output of the halfwave rectifier is

$$\mathbf{y} = \begin{cases} A \sin \boldsymbol{\theta}, & 0 < \boldsymbol{\theta} < \pi, \\ 0, & \pi < \boldsymbol{\theta} < 2\pi, \end{cases}$$

in terms of the random variable $\boldsymbol{\theta}$. The expected value of the output is

$$E(\mathbf{y}) = (2\pi)^{-1} \int_{0}^{\pi} A \sin \theta\, d\theta = \frac{A}{\pi},$$

and its mean square value is

$$E(\mathbf{y}^2) = (2\pi)^{-1} \int_{0}^{\pi} A^2 \sin^2 \theta\, d\theta = \frac{A^2}{4},$$

whereupon the variance of the output is

$$\operatorname{Var} \mathbf{y} = \frac{A^2}{4} - \left(\frac{A}{\pi}\right)^2 = 0.1487\, A^2.$$

Observe that these results can be obtained without first calculating the probability density function of the output $\mathbf{y}$. ■

D. Conditional Expected Values

Conditional probability density functions are probability density functions in their own right; the universal set S has merely been replaced by a restricted set of numerical or nonnumerical outcomes. Expected values with respect to conditional probability density functions are called *conditional expected values*. If M is a conditioning event,

$$E[g(\mathbf{x}) \mid M] = \int_{-\infty}^{\infty} g(x) f_{\mathbf{x}}(x \mid M)\, dx$$

is the conditional expected value of $g(\mathbf{x})$, given M.

In the problem of Sec. 2-2A involving life tests of transistors, for instance, the expected lifetime of those from supplier A is, by (2.42),

$$E(\mathbf{t} \mid A) = \int_{-\infty}^{\infty} t f_t(t \mid A)\, dt = k_a \int_0^{\infty} t e^{-k_a t}\, dt = \frac{1}{k_a},$$

and $E(\mathbf{t} \mid B) = 1/k_b$, $E(\mathbf{t} \mid C) = 1/k_c$.

The conditional variance of a random variable $\mathbf{x}$ with respect to a conditioning event M is, by (2.79),

$$\operatorname{Var}(\mathbf{x} \mid M) = E(\mathbf{x}^2 \mid M) - [E(\mathbf{x} \mid M)]^2. \tag{2.81}$$

Thus in this example,

$$E(\mathbf{t}^2 \mid A) = k_a \int_0^{\infty} t^2 e^{-k_a t}\, dt = \frac{2}{k_a^2}$$

and $\operatorname{Var}(\mathbf{t} \mid A) = 1/k_a^2$.

As before, the conditioning event could be the observation that the random variable $\mathbf{x}$ lies in a particular range of numerical outcomes of the experiment. Thus for a continuous random variable, by (2.53),

$$\begin{aligned} E(\mathbf{x} \mid a < \mathbf{x} < b) &= \int_{-\infty}^{\infty} x f_{\mathbf{x}}(x \mid a < \mathbf{x} < b)\, dx \\ &= [F_{\mathbf{x}}(b) - F_{\mathbf{x}}(a)]^{-1} \int_a^b x f_{\mathbf{x}}(x)\, dx. \end{aligned} \tag{2.82}$$

EXAMPLE 2.19 The random variable $\mathbf{x}$ is Gaussian with mean zero and variance σ^2. Find $E(\mathbf{x} \mid \mathbf{x} > 0)$ and $\operatorname{Var}(\mathbf{x} \mid \mathbf{x} > 0)$.

Now

$$f_{\mathbf{x}}(x) = (2\pi\sigma^2)^{-1/2} \exp\left(-\frac{x^2}{2\sigma^2}\right)$$

by (2.19), and by (2.53),

$$f_{\mathbf{x}}(x \mid \mathbf{x} > 0) = 2(2\pi\sigma^2)^{-1/2} e^{-x^2/2\sigma^2} U(x),$$

so that

$$E(\mathbf{x} \mid \mathbf{x} > 0) = 2(2\pi\sigma^2)^{-1/2} \int_0^{\infty} x e^{-x^2/2\sigma^2}\, dx = \sigma\left(\frac{2}{\pi}\right)^{1/2},$$

$$E(\mathbf{x}^2 \mid \mathbf{x} > 0) = 2(2\pi\sigma^2)^{-1/2} \int_0^{\infty} x^2 e^{-x^2/2\sigma^2}\, dx = \sigma^2$$

and by (2.81),

$$\text{Var}\,(\mathbf{x} \mid \mathbf{x} > 0) = \sigma^2\left(1 - \frac{2}{\pi}\right) = 0.3634\sigma^2. \qquad \blacksquare$$

EXAMPLE 2.20

$$E(\mathbf{x} \mid \mathbf{x} = a) = a,$$
$$\text{Var}\,(\mathbf{x} \mid \mathbf{x} = a) = 0. \qquad \blacksquare$$

E. The Chebyshev Inequality

Let $\bar{x}$ be the mean of the random variable **x**. We wish to bound the probability that **x** deviates from its mean $\bar{x}$ by more than a certain amount ε. That can be done in terms of the variance of **x** through the Chebyshev inequality,

$$\Pr\,(|\mathbf{x} - \bar{x}| \geq \varepsilon) \leq \frac{\text{Var}\,\mathbf{x}}{\varepsilon^2}. \tag{2.83}$$

Although the actual value of this probability may be much smaller than the bound, the bound is useful in theoretical studies, as we shall see.

To prove the Chebyshev inequality, we observe that the probability in question is

$$\Pr\,[(\mathbf{x} \leq \bar{x} - \varepsilon) \cup (\mathbf{x} \geq \bar{x} + \varepsilon)] = \int_{-\infty}^{\bar{x}-\varepsilon} f_{\mathbf{x}}(x)\,dx + \int_{\bar{x}+\varepsilon}^{\infty} f_{\mathbf{x}}(x)\,dx.$$

The variance, on the other hand, is

$$\begin{aligned}
\text{Var}\,\mathbf{x} &= \int_{-\infty}^{\infty} (x - \bar{x})^2 f_{\mathbf{x}}(x)\,dx \\
&\geq \int_{-\infty}^{\bar{x}-\varepsilon} (x - \bar{x})^2 f_{\mathbf{x}}(x)\,dx + \int_{\bar{x}+\varepsilon}^{\infty} (x - \bar{x})^2 f_{\mathbf{x}}(x)\,dx \\
&\geq \varepsilon^2 \int_{-\infty}^{\bar{x}-\varepsilon} f_{\mathbf{x}}(x)\,dx + \varepsilon^2 \int_{\bar{x}+\varepsilon}^{\infty} f_{\mathbf{x}}(x)\,dx = \varepsilon^2 \Pr\,(|\mathbf{x} - \bar{x}| \geq \varepsilon),
\end{aligned}$$

and dividing out ε^2 we obtain (2.83).

If Var $\mathbf{x} = 0$, the random variable **x** equals its mean value $\bar{x}$ "with probability 1."

F. Characteristic Functions

The characteristic function of the random variable **x** is, except for the sign of j, the Fourier transform of its probability density function,

$$\begin{aligned}
\Phi_{\mathbf{x}}(\omega) &= E(e^{j\omega\mathbf{x}}) = E(\cos \omega\mathbf{x}) + j\,E(\sin \omega\mathbf{x}) \\
&= \int_{-\infty}^{\infty} f_{\mathbf{x}}(x) e^{j\omega x}\,dx.
\end{aligned} \tag{2.84}$$

For a discrete random variable it is

$$\Phi_{\mathbf{x}}(\omega) = \sum_{k=1}^{\infty} \exp(j\omega x_k) \Pr(\mathbf{x} = x_k) \tag{2.85}$$

as follows by substituting (2.15). Putting $\omega = 0$, we obtain

$$\Phi_{\mathbf{x}}(0) = \int_{-\infty}^{\infty} f_{\mathbf{x}}(x)\, dx = 1.$$

Knowing the characteristic function, we can find the probability density function by the inverse Fourier transformation,

$$f_{\mathbf{x}}(x) = \int_{-\infty}^{\infty} \Phi_{\mathbf{x}}(\omega) e^{-j\omega x} \frac{d\omega}{2\pi}. \tag{2.86}$$

As we shall see later, the characteristic function of the sum of a number of independent random variables is simply related to the characteristic functions of each, and (2.86) is useful for calculating the probability density function of such a sum.

A principal use of the characteristic function is to calculate the moments of the random variable. Differentiating (2.84) or (2.85) with respect to $(j\omega)$ and setting $\omega = 0$, we obtain

$$\frac{d}{d(j\omega)} \Phi_{\mathbf{x}}(\omega) \bigg|_{\omega=0} = \int_{-\infty}^{\infty} x f_{\mathbf{x}}(x)\, dx = E(\mathbf{x}).$$

Continuing thus, we find for the nth moment

$$m_n = E(\mathbf{x}^n) = \frac{d^n}{d(j\omega)^n} \Phi_{\mathbf{x}}(\omega) \bigg|_{\omega=0}. \tag{2.87}$$

This process must stop as soon as an infinite value of the derivative is encountered at $\omega = 0$.

The Taylor series for the characteristic function $\Phi_{\mathbf{x}}(\omega)$ in the neighborhood of the origin is obtained from (2.87) or by substituting the power series for $e^{j\omega x}$ into (2.84) and integrating termwise,

$$\Phi_{\mathbf{x}}(\omega) = \sum_{k=0}^{\infty} \frac{(j\omega)^k}{k!} E(\mathbf{x}^k). \tag{2.88}$$

Thus an easy way to find the moments of a random variable is to expand its characteristic function about $\omega = 0$. This series must, however, be terminated by a remainder term just before the first infinite moment, if any such exist.

EXAMPLE 2.21 Cauchy Distribution

For a Cauchy random variable, the probability density function is

$$f_{\mathbf{x}}(x) = \frac{a/\pi}{x^2 + a^2}$$

and the characteristic function is

$$\Phi_{\mathbf{x}}(\omega) = e^{-a|\omega|}.$$

This Fourier relationship is most easily demonstrated by taking the inverse Fourier transform of the characteristic function and showing that the Cauchy probability density function ensues:

$$\int_{-\infty}^{\infty} \Phi_{\mathbf{x}}(\omega)e^{-j\omega x}\,\frac{d\omega}{2\pi} = \int_{-\infty}^{\infty} e^{-a|\omega|-j\omega x}\,\frac{d\omega}{2\pi}$$

$$= \int_{-\infty}^{0} e^{(a-jx)\omega}\,\frac{d\omega}{2\pi} + \int_{0}^{\infty} e^{-(a+jx)\omega}\,\frac{d\omega}{2\pi}$$

$$= \frac{1}{2\pi}\left(\frac{1}{a-jx} + \frac{1}{a+jx}\right) = \frac{a/\pi}{a^2+x^2},$$

as was to be shown.

The characteristic function $e^{-a|\omega|}$ of the Cauchy random variable has a discontinuous first derivative at $\omega = 0$ and cannot be expanded about $\omega = 0$ to give a power series of the form (2.88). The reason is that the moments of even order are infinite, as you can easily verify by attempting to calculate $E(\mathbf{x}^2)$. ■

EXAMPLE 2.22 Gaussian Random Variable

We first show that the characteristic function of a Gaussian random variable with mean m and variance σ^2 is

$$\Phi_{\mathbf{x}}(\omega) = \exp\,(jm\omega - \tfrac{1}{2}\sigma^2\omega^2). \tag{2.89}$$

From (2.19) and the definition in (2.84) we must evaluate

$$\Phi_{\mathbf{x}}(\omega) = (2\pi\sigma^2)^{-1/2}\int_{-\infty}^{\infty} \exp\left[-\frac{(x-m)^2}{2\sigma^2} + j\omega x\right] dx$$

$$= e^{jm\omega}\,(2\pi\sigma^2)^{-1/2}\int_{-\infty}^{\infty} \exp\left(-\frac{u^2}{2\sigma^2} + j\omega u\right) du$$

by putting $x = m + u$. Completing the square in the exponent, we find that

$$\Phi_{\mathbf{x}}(\omega) = \exp\,(jm\omega - \tfrac{1}{2}\sigma^2\omega^2)\cdot(2\pi\sigma^2)^{-1/2}\int_{-\infty}^{\infty} \exp\left[-\frac{(u-j\sigma^2\omega)^2}{2\sigma^2}\right] du.$$

Again changing the variable of integration, this time to $y = u - j\sigma^2\omega$, and utilizing the normalization integral, we obtain (2.89).

We can obtain the central moments of the Gaussian random variable by setting $m = 0$ in (2.89), in effect shifting the center of the probability density function to the origin. We then expand the characteristic function about $\omega = 0$,

$$\Phi_{\mathbf{x}}(\omega) = e^{-\sigma^2\omega^2/2} = \sum_{m=0}^{\infty} \frac{(-\frac{1}{2}\sigma^2)^m\omega^{2m}}{m!}.$$

The central moments of odd order are zero; those of even order follow from (2.88) by comparing terms with like powers of ω. Restoring the mean, we obtain for the central moments

$$E[(\mathbf{x} - m)^{2k}] = \frac{(2k)!}{2^k k!}\,\sigma^{2k} \tag{2.90}$$

for all positive integers k. Thus

$$E[(\mathbf{x} - m)^4] = \frac{4!\,\sigma^4}{2^2\cdot 2!} = 3\sigma^4, \tag{2.91}$$

and so on. ■

You should calculate the mean and the variance of the binomial and Poisson random variables by this method, although a simpler one appears in the next part.

G. Probability Generating Functions

When a random variable takes on values that are uniformly spaced, it is said to be a *lattice* variable. The commonest kind is one whose values are the non-negative integers, as in counting particles. A convenient tool for analyzing probability distributions of integer-valued random variables is the probability generating function defined by

$$h(z) = E(z^{\mathbf{x}}) = \sum_{k=0}^{\infty} z^k \Pr(\mathbf{x} = k) = \sum_{k=0}^{\infty} p_k z^k, \tag{2.92}$$

with $p_k = \Pr(\mathbf{x} = k)$. You may recognize this as the z-transform of the sequence of probabilities $\{p_k\}$, except that z^{-1} has been replaced by z. In particular,

$$h(1) = \sum_{k=0}^{\infty} p_k = 1.$$

Knowing the probability generating function we can find the probabilities p_k by differentiating and setting $z = 0$:

$$\Pr(\mathbf{x} = k) = p_k = (k!)^{-1} \frac{d^k}{dz^k} h(z) \bigg|_{z=0}. \tag{2.93}$$

The derivatives of the probability generating function at $z = 1$ yield the factorial moments:

$$\begin{aligned} c_n &= E[\mathbf{x}(\mathbf{x} - 1)(\mathbf{x} - 2) \cdots (\mathbf{x} - n + 1)] \\ &= \frac{d^n}{dz^n} h(z) \bigg|_{z=1}. \end{aligned} \tag{2.94}$$

In particular,

$$c_1 = E(\mathbf{x}) = m_1, \tag{2.95}$$

$$c_2 = E(\mathbf{x}^2 - \mathbf{x}) = m_2 - m_1, \tag{2.96}$$

so that

$$\operatorname{Var} \mathbf{x} = c_2 + c_1 - c_1^2. \tag{2.97}$$

The power-series expansion of $h(z)$ about the point $z = 1$ has coefficients $c_n/n!$,

$$h(z) = \sum_{n=0}^{\infty} \frac{c_n(z - 1)^n}{n!}. \tag{2.98}$$

It must stop with a remainder term before any term with c_n infinite.

EXAMPLE 2.23 Binomial Distribution

From (1.25) we find for the probability generating function of a binomial random variable

$$h(z) = \sum_{k=0}^{n} \binom{n}{k} p^k q^{n-k} z^k$$
$$= (pz + q)^n = [1 + p(z - 1)]^n$$

by the binomial theorem; $q = 1 - p$. Using the binomial theorem again, we find

$$h(z) = \sum_{r=0}^{n} \binom{n}{r} p^r (z - 1)^r, \tag{2.99}$$

so that the rth factorial moment is

$$c_r = \frac{n!}{(n - r)!} p^r, \qquad 0 \le r \le n. \tag{2.100}$$

In particular, the mean c_1 of a binomial random variable is

$$E(\mathbf{x}) = np,$$

and its variance is, by (2.97),

$$\text{Var } \mathbf{x} = n(n - 1)p^2 + np - n^2p^2 = np(1 - p). \tag{2.101}$$

These results will be applied in the next part. ■

EXAMPLE 2.24 Poisson Random Variable

From (1.27) we find for the probability generating function of a Poisson random variable

$$h(z) = \sum_{k=0}^{\infty} \frac{m^k}{k!} z^k e^{-m} = e^{m(z-1)} \tag{2.102}$$

by the power series for the exponential function. Expanding it in powers of $(z - 1)$ and using (2.98), we see that the factorial moments of the Poisson random variable are

$$c_r \equiv m^r.$$

Thus m is the mean of the Poisson random variable, and its variance is

$$\text{Var } \mathbf{x} = m. \tag{2.103}$$

The variance of a Poisson random variable equals its mean. ■

H. Law of Large Numbers

Looking for an event A in n trials of a chance experiment E, we define the relative frequency of its occurrence by

$$\mathbf{F} = \frac{\mathbf{k}}{n},$$

where $\mathbf{k}$ is the number of times A occurred in n trials and is a binomial random variable; $\mathbf{F}$ is a lattice random variable, taking values that are integral multiples of $1/n$, $0 \leq \mathbf{F} \leq 1$. Applying the scaling rules in Sec. 2-4C to the results of the previous part, with $p = \Pr(A)$, we find that

$$E(\mathbf{F}) = p,$$

$$\operatorname{Var} \mathbf{F} = \frac{p(1-p)}{n}.$$

Thus the larger n, the more closely the distribution of the random variable $\mathbf{F}$ clusters about its mean value p.

Let us apply the Chebyshev inequality (2.83) to the relative frequency $\mathbf{F}$,

$$\Pr(|\mathbf{F} - p| \geq \varepsilon) \leq \frac{p(1-p)}{n\varepsilon^2}. \tag{2.104}$$

This expresses the *law of large numbers*, which tells us that the larger the number n of trials, the less probable is a great deviation of the relative frequency $\mathbf{F}$ from the true probability p of the event A.

These n Bernoulli trials taken together constitute a chance experiment that we called E_n in Sec. 1-8. The probability $\Pr(|\mathbf{F} - p| \geq \varepsilon)$ in (2.104), viewed from the standpoint of the relative-frequency concept of probability described in Sec. 1-4B, must be taken to refer to a large number N of independent trials of experiment E_n, in each of which trials, n independent trials of experiment E are performed. As $N \to \infty$, the fraction of trials of E_n in which the relative frequency $\mathbf{F}$ deviates from $p = \Pr(A)$ by more than ε approaches something less than the right-hand side of (2.104); it will be the smaller, the larger the number n. This result cannot be taken as a justification of the relative-frequency concept, however, for to attempt to do so would involve us in an infinite regress, first of trials of E_n, then of trials of

$$\mathsf{E}_{nN} = \underbrace{\mathsf{E}_n \times \mathsf{E}_n \times \cdots \times \mathsf{E}_n}_{N \text{ times}},$$

then of trials of

$$\mathsf{E}_{nNN'} = \underbrace{\mathsf{E}_{nN} \times \mathsf{E}_{nN} \times \cdots \times \mathsf{E}_{nN}}_{N' \text{ times}},$$

and so on. The law of large numbers, (2.104), remains a deduction of the mathematical theory of probability and does not of itself furnish us with an interpretation of the concept of probability.

PROBLEMS

2-1 The random variable $\mathbf{x}$ has the probability density function

$$f_{\mathbf{x}}(x) = \begin{cases} c(a^2 - x^2), & |x| < a, \\ 0, & |x| > a. \end{cases}$$

In terms of a, find the constant c. Calculate the cumulative distribution function $F_{\mathbf{x}}(x)$ and the probability that

$$-\frac{a}{4} < \mathbf{x} < \frac{3a}{4}.$$

2-2 The random variable $\mathbf{x}$ has a probability density function of the form

$$f_{\mathbf{x}}(x) = ce^{-a|x|}$$

for all x. Determine the constant c and the cumulative distribution function $F_{\mathbf{x}}(x)$ of this random variable.

2-3 For c a positive odd multiple of $\frac{1}{2}$ (e.g., $c = \frac{1}{2}, \frac{3}{2}, \frac{5}{2}, \ldots$) derive an expression for the cumulative gamma distribution of (2.31) in terms of the error-function integral erfc $(\cdot)$ combined with e^{-x} and powers of x.

2-4 The random variable $\mathbf{x}$ has a Rayleigh distribution as in (2.24) with $b = 1$. Find the conditional cumulative distribution function of $\mathbf{x}$, given that $\mathbf{x} > a > 0$; that is, determine

$$F_{\mathbf{x}}(x \mid \mathbf{x} > a) = \Pr(\mathbf{x} < x \mid \mathbf{x} > a)$$

for all values of x.

2-5 The probability density function of the random variable $\mathbf{x}$ is

$$f_{\mathbf{x}}(x) = \begin{cases} 1 - |x|, & |x| < 1, \\ 0, & |x| \geq 1. \end{cases}$$

(a) Find and sketch the cumulative distribution function of $\mathbf{x}$.
(b) Calculate the probability that $|\mathbf{x}| > \frac{1}{2}$.
(c) Calculate the conditional probability density function $f_{\mathbf{x}}(x \mid |\mathbf{x}| > \frac{1}{2})$ of the random variable $\mathbf{x}$, given that $|\mathbf{x}| > \frac{1}{2}$, and sketch it.
(d) Calculate the conditional cumulative distribution function

$$F_{\mathbf{x}}\left(x \,\middle|\, |\mathbf{x}| > \frac{1}{2}\right) = \Pr\left(\mathbf{x} < x \,\middle|\, |\mathbf{x}| > \frac{1}{2}\right)$$

and sketch it.

2-6 For the probability density function in Problem 2-2, determine the conditional cumulative distribution functions $F_{\mathbf{x}}(x \mid \mathbf{x} > b)$ for both $b < 0$ and $b > 0$, and sketch them.

2-7 A communication receiver filters, amplifies, and rectifies the voltage across the terminals of an antenna. The final output of the receiver, sampled at a certain time t, is a random variable $\mathbf{x}$. When no signal, but only background noise is present at the input to the receiver, the probability density function of $\mathbf{x}$ is

$$f_0(x) = c_0 \exp(-c_0 x)\, U(x).$$

When a signal is also arriving, the probability density function of $\mathbf{x}$ is

$$f_1(x) = c_1 \exp(-c_1 x)\, U(x),$$

and $c_0 > c_1$. The prior probability that a signal is present equals $\frac{1}{2}$.

(a) Given that a particular value x of this random variable $\mathbf{x}$ has been observed at the output of the receiver, what is the posterior (conditional) probability that a signal is present? That is, calculate Pr (signal present $\mid \mathbf{x} = x$).
(b) If we wish to minimize the probability P_e of error in deciding whether a signal is present or not, over what range of values of x should we decide "signal present" and over what range should we decide "signal absent"?
(c) Calculate the minimum attainable probability P_e of error in terms of c_0 and c_1.

2-8 The conditional failure rate of a device is $\beta(t) = at^{1/2}$. Calculate the probability that it is still operating at t units of time after it is turned on.

2-9 The probability distribution of the discrete random variable $\mathbf{x}$ is given in the table below. Calculate the cumulative distribution functions of $\mathbf{x}$ and $\mathbf{y} = \mathbf{x}^4$.

$\mathbf{x}$	-2	-1	0	1	2
$\Pr(\mathbf{x} = x)$	0.05	0.15	0.20	0.35	0.25

2-10 The random variable $\mathbf{x}$ has the displaced Cauchy distribution, the form of whose probability density function is

$$f_{\mathbf{x}}(x) = \frac{a/\pi}{a^2 + (x - b)^2}.$$

A new random variable is defined by $\mathbf{y} = \cosh \mathbf{x}$. Find its probability density function and its cumulative distribution function.

2-11 The random variable $\mathbf{x}$ has a normal distribution

$$f_{\mathbf{x}}(x) = (2\pi\sigma^2)^{-1/2} \exp\left(-\frac{x^2}{2\sigma^2}\right).$$

Find the probability density function of the random variable

$$\mathbf{y} = \mathbf{x}^2 U(\mathbf{x}) = \begin{cases} \mathbf{x}^2, & \mathbf{x} \geq 0, \\ 0, & \mathbf{x} < 0. \end{cases}$$

2-12 Let the input, at any time t, to the rectifier described by (2.61) be a random variable $\mathbf{x}$ with the Gaussian probability density function in (2.19). Calculate the probability density function of the output of the rectifier.

2-13 The input to a limiter is a voltage $v(t)$ that at any time t is a Gaussian random variable $v(t) = \mathbf{v}$ with mean zero; its probability density function is the same as that in Problem 2-11. The output of the limiter is a random variable $\mathbf{w}$ given by

$$\mathbf{w} = \begin{cases} -b, & \mathbf{v} < -a, \\ \dfrac{b\mathbf{v}}{a}, & -a < \mathbf{v} < a, \\ b, & \mathbf{v} > a, \end{cases}$$

as shown in Fig. 2P-13. Determine the probability density function of the output $\mathbf{w}$.

2-14 The random variable $\mathbf{x}$ has a cumulative distribution function $F_{\mathbf{x}}(x)$. Define a new random variable $\mathbf{y}$ by

$$\mathbf{y} = g(\mathbf{x}) = F_{\mathbf{x}}(\mathbf{x}).$$

Show that $\mathbf{y}$ is uniformly distributed over (0, 1). Conversely, if we want a random variable that has a given probability density function $f_{\mathbf{x}}(x)$, and we know that $\mathbf{y}$ is uniformly distributed over (0, 1), we can take $\mathbf{x} = g^{-1}(\mathbf{y})$, where $g^{-1}(\mathbf{y})$ is the inverse function to that defined above, with

$$F_{\mathbf{x}}(x) = \int_{-\infty}^{x} f_{\mathbf{x}}(u)\, du$$

in terms of the desired probability density function. Which function $g^{-1}(\mathbf{y})$ yields a Cauchy random variable $\mathbf{x}$ as in (2.35)–(2.36) by transforming a uniformly distributed one? Which yields the arcsine law [(2.37)–(2.38)]? In numerical studies, samples of $\mathbf{y}$ are drawn from a table of random numbers or produced by a computer program called a "random-number generator."

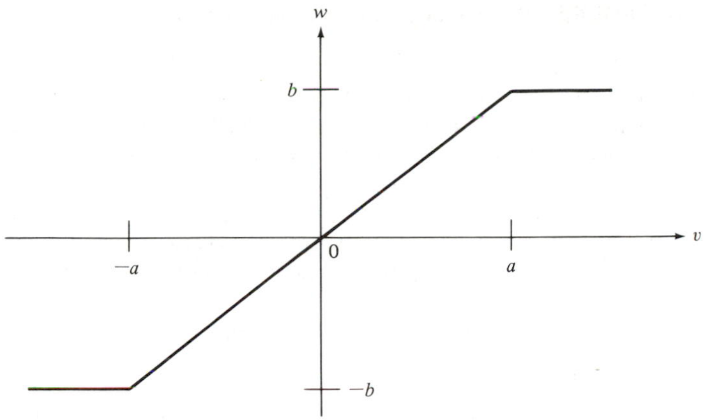

FIGURE 2P-13 Limiter characteristic.

2-15 The input to the limiter of Problem 2-13 is a sine wave

$$\mathbf{v} = A \cos(\omega t + \boldsymbol{\theta}), \qquad A > a,$$

whose phase $\boldsymbol{\theta}$ is random and uniformly distributed over $(0, 2\pi)$. The output w of the limiter is sampled at an arbitrary time t to obtain a random variable $\mathbf{w}$. Find the probability density function of the sampled output $\mathbf{w}$. [*Hint:* If $\boldsymbol{\theta}$ is uniformly distributed over $(0, 2\pi)$, so is $\boldsymbol{\psi} = (\omega t + \boldsymbol{\theta}) \pmod{2\pi}$, and we can put $\mathbf{v} = A \cos \boldsymbol{\psi}$.]

2-16 The probability density function of the random variable $\mathbf{x}$ is

$$f_{\mathbf{x}}(x) = \begin{cases} \frac{1}{2}(1 - |x|) + \frac{1}{4}\delta(x + \frac{1}{2}) + \frac{1}{4}\delta(x - \frac{1}{2}), & -1 < x < 1, \\ 0, & x < -1, \quad x > 1. \end{cases}$$

Find the probability density function of the random variable

$$\mathbf{y} = \cos\left(\frac{\pi \mathbf{x}}{2}\right).$$

2-17 Calculate the mean and the variance of the random variable of Problem 2-1.

2-18 Calculate the mean and the variance of the random variable of Problem 2-2.

2-19 Calculate the means and the variances of the random variables $\mathbf{x}$ and $\mathbf{y}$ of Problem 2-9.

2-20 The random variable $\mathbf{x}$ has expected value 2 and variance 5. Find the expected value of $\mathbf{y} = (3\mathbf{x} - 2)^2$.

2-21 Calculate the variance of the random variable $\mathbf{y}$ of Problem 2-11.

2-22 Calculate the variance Var $\mathbf{w}$ of the output $\mathbf{w}$ of the limiter in Problem 2-13 in terms of a, b, σ^2, and the function erfc $(\cdot)$.

2-23 The random variable $\mathbf{x}$ has the probability density function

$$f_{\mathbf{x}}(x) = \begin{cases} 6a^{-3}(ax - x^2), & 0 < x < a, \\ 0, & x < 0, \quad x > a. \end{cases}$$

Calculate $E(\mathbf{x} \mid \mathbf{x} > a/3)$.

2-24 Calculate the variance Var $\mathbf{w}$ of the output of the limiter in Problem 2-15, the input to which is a sine wave of random phase.

2-25 The probability density function of the random variable $\mathbf{x}$ is

$$f_{\mathbf{x}}(x) = e^{-x}U(x),$$

where $U(x)$ is the unit step function. Find the characteristic function of $\mathbf{x}$ and from it determine the first three moments of $\mathbf{x}$, $E(\mathbf{x}^k)$, $k = 1, 2, 3$. Calculate Var $\mathbf{x}$.

2-26 Calculate the mean and the variance of the Rayleigh distribution in eq. (2.24). Also calculate $E(\mathbf{x}^{2k})$ for all integers $k > 0$.

2-27 Calculate the mean and the variance of the gamma distribution in eq. (2.26). Calculate $E(\mathbf{x}^k)$ for all integers $k > 0$. (*Hint:* First determine the characteristic function of $\mathbf{x}$.)

2-28 Calculate the mean time to failure, $E(\mathbf{t})$, for the device in Problem 2-8, expressing the result in terms of the gamma function.

2-29 The integer-valued random variable $\mathbf{n}$ has the probability distribution

$$\Pr(\mathbf{n} = k) = (1 - v)v^k, \qquad 0 < v < 1, \quad 0 \le k < \infty.$$

Calculate the probability generating function, the mean, and the variance of this random variable.

2-30 The input to a quantizer is a random positive voltage $\mathbf{v}$ whose probability density function is

$$f_{\mathbf{v}}(v) = ae^{-av}U(v).$$

The output of the quantizer is equal to the integer k when

$$k < \mathbf{v} \le k + 1, \qquad 0 \le k < \infty.$$

Find the expected value of the output of the quantizer.

2-31 A nonnegative integer-valued random variable $\mathbf{n}$ has the probability generating function

$$h_{\mathbf{n}}(z) = \left(\frac{1 - v}{1 - vz}\right)^M,$$

where M is a positive integer. Calculate the probability distribution $\Pr(\mathbf{n} = k)$ of $\mathbf{n}$, for all $k \ge 0$. Also calculate the mean $E(\mathbf{n})$ and the variance Var $\mathbf{n}$ of this random variable.

CHAPTER 3

Bivariate Probability Distributions

3-1

JOINT DISTRIBUTIONS OF PAIRS OF RANDOM VARIABLES

The outcome of a chance experiment is often labeled not by only one, but by two or more numbers. We measure, for instance, the voltage between a point in a circuit and ground at two times τ_1 and τ_2; the outcome is the pair of voltages (v_1, v_2), $v_1 = v(\tau_1)$, $v_2 = v(\tau_2)$. Or the voltages at two points in the circuit are measured at the same or different times. With light from a common source falling on two adjacent photodetectors, we count the numbers n_1 and n_2 of photoelectrons emitted by each during an interval $(0, T)$. The components of the velocity of a turbulent fluid are measured at two or more points. A satellite contains sensors determining the magnetic field intensity and the charged-particle flux at regular intervals. Events in such experiments are sets of numerically identified outcomes, and when a probability measure has been assigned to the events, an outcome specified by, say, n numbers becomes n-tuple of random variables. We begin by treating the probabilistic description of pairs of random variables. The concepts required are simple extensions of those introduced in Chapter 2 for describing single random variables. Once they are understood, their generalization to cover outcomes specified by more than two random variables will be easy. We defer it to the next chapter.

It is convenient to visualize an outcome (**x**, **y**) consisting of a pair of numbers **x** and **y** as a point with coordinates (**x**, **y**) in the *XY*-plane. An event then corresponds to a region or a group of disjoint regions in that plane. The probability measure assigned to the events of the chance experiment can be thought of as a distribution of mass over the plane. The total amount of mass is equal to 1. As with a single random variable, there are two principal ways of specifying the probability measure; they involve the joint cumulative distribution function and the joint probability density function of **x** and **y**.

A. Joint Cumulative Distribution Function

For any pair of numbers x and y, the event $(\mathbf{x} \le x, \mathbf{y} \le y)$ is represented by a quadrant of the *XY*-plane having its vertex at the point (x, y) and opening to the lower left, as in Fig. 3.1. The probability of this event is denoted by

$$F_{\mathbf{xy}}(x, y) = \Pr(\mathbf{x} \le x, \mathbf{y} \le y) \tag{3.1}$$

and is called the *joint cumulative distribution function* of **x** and **y**.

In order to specify a probability measure consistent with the axioms of Sec. 1-3, the function $F_{\mathbf{xy}}(x, y)$ must possess certain properties. It must be nonnegative,

$$F_{\mathbf{xy}}(x, y) \ge 0.$$

Because the event $\mathbf{x} \le -\infty, \mathbf{y} \le -\infty$ is empty,

$$F_{\mathbf{xy}}(-\infty, -\infty) = 0,$$

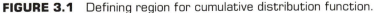

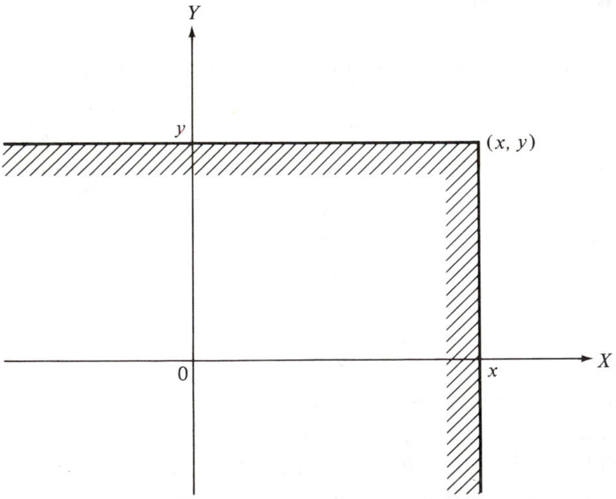

FIGURE 3.1 Defining region for cumulative distribution function.

and because all outcomes $(\mathbf{x}, \mathbf{y})$ lie somewhere in the plane, which constitutes the universal set S,

$$F_{\mathbf{xy}}(\infty, \infty) = 1.$$

The probability assigned to the rectangle in Fig. 3.2 is given, according to the axioms, by

$$\begin{aligned}
&\Pr(x_1 < \mathbf{x} \le x_2, y_1 < \mathbf{y} \le y_2) \\
&\quad = \Pr(x_1 < \mathbf{x} \le x_2, \mathbf{y} \le y_2) - \Pr(x_1 < \mathbf{x} \le x_2, \mathbf{y} \le y_1) \\
&\quad = \Pr(\mathbf{x} \le x_2, \mathbf{y} \le y_2) - \Pr(\mathbf{x} \le x_1, \mathbf{y} \le y_2) \qquad (3.2) \\
&\qquad - [\Pr(\mathbf{x} \le x_2, \mathbf{y} \le y_1) - \Pr(\mathbf{x} \le x_1, \mathbf{y} \le y_1)] \\
&\quad = F_{\mathbf{xy}}(x_2, y_2) - F_{\mathbf{xy}}(x_1, y_2) - F_{\mathbf{xy}}(x_2, y_1) + F_{\mathbf{xy}}(x_1, y_1) \ge 0.
\end{aligned}$$

As this quantity is a probability, it must always be nonnegative, and that means that the cumulative distribution function $F_{\mathbf{xy}}(x, y)$ must generally rise as the point (x, y) moves upward and to the right, although it may remain constant over certain portions of the plane and rise abruptly along certain lines or curves. When $F_{\mathbf{xy}}(x, y)$ is a continuous function of both x and y, the random variables $\mathbf{x}$ and $\mathbf{y}$ are of continuous type and are called *continuous* random variables.

The event $(\mathbf{x} \le x)$ is represented by all that part of the XY-plane lying to the left of the vertical line cutting the x-axis at point x, as in Fig. 3.3. The probability of this event is

$$F_{\mathbf{x}}(x) = \Pr(\mathbf{x} \le x) = \Pr(\mathbf{x} \le x, \mathbf{y} \le \infty) = F_{\mathbf{xy}}(x, \infty) \qquad (3.3)$$

and constitutes the cumulative distribution function of the random variable $\mathbf{x}$ in the sense of Sec. 2-1. In this context, $F_{\mathbf{x}}(x)$ is called the *marginal cumulative distribution function* of $\mathbf{x}$. In the same way,

$$F_{\mathbf{y}}(y) = F_{\mathbf{xy}}(\infty, y) \qquad (3.4)$$

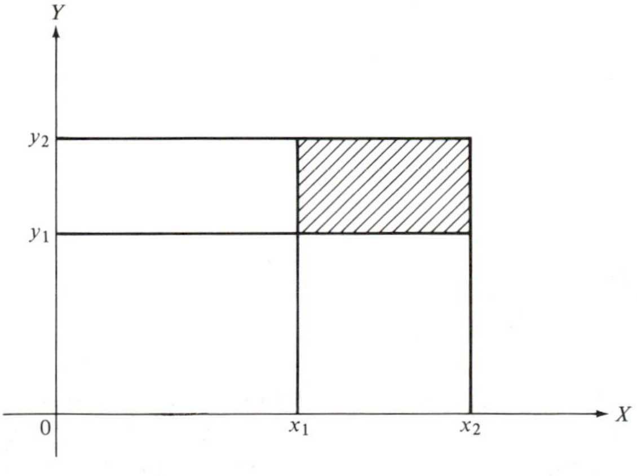

FIGURE 3.2 Region for eq. (3.2).

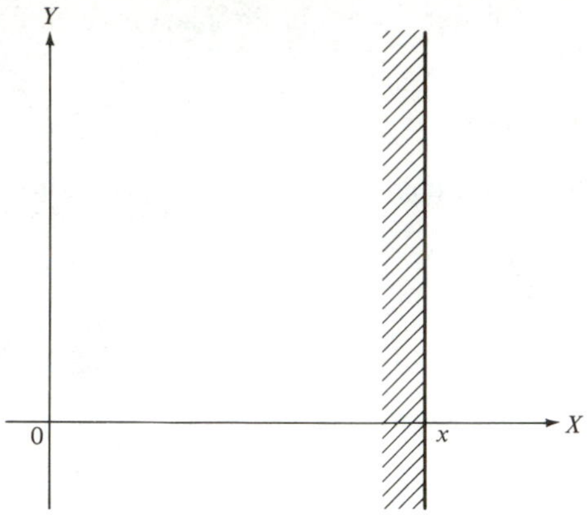

FIGURE 3.3 Region defining marginal distribution.

is the marginal cumulative distribution function of the random variable $\mathbf{y}$. Both of these marginal cumulative distribution functions can thus be derived from the joint cumulative distribution function $F_{\mathbf{xy}}(x, y)$, but in general $F_{\mathbf{x}}(x)$ and $F_{\mathbf{y}}(y)$ do *not* suffice to determine $F_{\mathbf{xy}}(x, y)$.

EXAMPLE 3.1. Let

$$F_{\mathbf{xy}}(x, y) = \begin{cases} 0, & x < 0 \text{ or } y < 0 \text{ or both}, \\ xy, & 0 \le x < 1,\ 0 \le y < 1, \\ x, & 0 \le x < 1,\ y \ge 1, \\ y, & x \ge 1,\ 0 \le y \le 1, \\ 1, & x \ge 1,\ y \ge 1. \end{cases}$$

Then the marginal cumulative distribution functions of $\mathbf{x}$ and $\mathbf{y}$ are

$$F_{\mathbf{x}}(x) = \begin{cases} 0, & x < 0 \\ x, & 0 \le x < 1, \\ 1, & x \ge 1, \end{cases}$$

$$F_{\mathbf{y}}(y) = \begin{cases} 0, & y < 0, \\ y, & 0 \le y < 1, \\ 1, & y \ge 1, \end{cases}$$

and $\mathbf{x}$ and $\mathbf{y}$ are each uniformly distributed over (0, 1). ■

B. Joint Probability Density Function: Continuous Random Variables

As Example 3.1 shows, the expressions for particular joint cumulative distribution functions can be cumbersome, and we need a simpler way of describing a distribution of probability over the *XY*-plane. We find it in the joint probability density function, which is defined in terms of the cumulative distribution function by

$$f_{\mathbf{xy}}(x, y) = \frac{\partial^2 F_{\mathbf{xy}}(x, y)}{\partial x \, \partial y} \tag{3.5}$$

provided that the cumulative distribution function $F_{\mathbf{xy}}(x, y)$ is continuous and differentiable. The function $f_{\mathbf{xy}}(x, y)$ is sometimes called a *bivariate probability density function*.

In Example 3.1, for instance,

$$f_{\mathbf{xy}}(x, y) = \begin{cases} 1, & 0 \le x < 1, \quad 0 \le y < 1, \\ 0, & \text{elsewhere}. \end{cases}$$

This is the uniform probability density function for a pair of random variables.

From (3.2) the probability assigned to the infinitesimal rectangle $(x < \mathbf{x} < x + dx, y < \mathbf{y} < y + dy)$ is

$$\begin{aligned} \Pr\,(x < \mathbf{x} < x + dx, \ y < \mathbf{y} < y + dy) \\ &= [F_{\mathbf{xy}}(x + dx, y + dy) - F_{\mathbf{xy}}(x, y + dy)] \\ &\quad - [F_{\mathbf{xy}}(x + dx, y) - F_{\mathbf{xy}}(x, y)] \\ &\doteq \frac{\partial F_{\mathbf{xy}}(x, y + dy)}{\partial x} dx - \frac{\partial F_{\mathbf{xy}}(x, y)}{\partial x} dx \\ &\doteq \frac{\partial^2 F_{\mathbf{xy}}(x, y)}{\partial x \, \partial y} dx \, dy = f_{\mathbf{xy}}(x, y) \, dx \, dy. \end{aligned}$$

Thus the probability that $\mathbf{x}$ lies between x and $x + dx$ and $\mathbf{y}$ lies between y and $y + dy$ equals the probability density function $f_{\mathbf{xy}}(x, y)$ times the area $dx\,dy$ of the infinitesimal rectangle. Because this probability must be nonnegative,

$$f_{\mathbf{xy}}(x, y) \ge 0 \tag{3.6}$$

for all (x, y).

The cumulative distribution function can be recovered from the probability density function by integrating (3.5) over the quadrant in Fig. 3.1,

$$F_{\mathbf{xy}}(x, y) = \int_{-\infty}^{x} dx' \int_{-\infty}^{y} dy' \, f_{\mathbf{xy}}(x', y'). \tag{3.7}$$

Because in particular $F_{\mathbf{xy}}(\infty, \infty) = 1$,

$$\int_{-\infty}^{\infty} \int_{-\infty}^{\infty} f_{\mathbf{xy}}(x, y) \, dx \, dy = 1. \tag{3.8}$$

This is the normalization integral for the joint probability density function $f_{\mathbf{xy}}(x, y)$.

It is helpful to visualize the joint probability density function $f_{\mathbf{xy}}(x, y)$ as the height of a pile of sand lying on the XY-plane. Then (3.8) states that there is a unit amount of sand in all. The quantity of sand in a rectilinear column standing over the rectangle $(x < \mathbf{x} < x + dx,\ y < \mathbf{y} < y + dy)$ equals

$$f_{\mathbf{xy}}(x, y)\, dx\, dy = \Pr(x < \mathbf{x} < x + dx,\ y < \mathbf{y} < y + dy).$$

Let the event $(\mathbf{x}, \mathbf{y}) \in \mathcal{A}$ be represented by a region $\mathcal{A}$ of arbitrary form in the XY-plane, as in Fig. 3.4. The amount of probability assigned to this event is

$$\Pr((\mathbf{x}, \mathbf{y}) \in \mathcal{A}) = \iint_{\mathcal{A}} f_{\mathbf{xy}}(x, y)\, dx\, dy \tag{3.9}$$

corresponding to the mass of sand in a column over the region $\mathcal{A}$.

A probability measure defined as in (3.9) clearly satisfies the first three axioms of probability as stated in Sec. 1-3 when the function $f_{\mathbf{xy}}(x, y)$ is nonnegative and normalized as in (3.8). All probabilities $\Pr((\mathbf{x}, \mathbf{y}) \in \mathcal{A})$ are then nonnegative by (3.9) (Axiom I). The universal set S is represented by the entire XY-plane, and (3.8) corresponds to Axiom II. Two mutually exclusive events $(\mathbf{x}, \mathbf{y}) \in \mathcal{A}$ and $(\mathbf{x}, \mathbf{y}) \in \mathcal{B}$ are represented by disjoint regions $\mathcal{A}$ and $\mathcal{B}$ in the plane, whence

$$\begin{aligned}
&\Pr((\mathbf{x}, \mathbf{y}) \in \mathcal{A} \cup \mathcal{B}) \\
&\quad = \iint_{\mathcal{A}\cup\mathcal{B}} f_{\mathbf{xy}}(x, y)\, dx\, dy = \iint_{\mathcal{A}} f_{\mathbf{xy}}(x, y)\, dx\, dy + \iint_{\mathcal{B}} f_{\mathbf{xy}}(x, y)\, dx\, dy \\
&\quad = \Pr((\mathbf{x}, \mathbf{y}) \in \mathcal{A}) + \Pr((\mathbf{x}, \mathbf{y}) \in \mathcal{B}),
\end{aligned}$$

which is Axiom III. The reader should now translate Corollaries 1 to 4 of Sec. 1-3 into this context. That Axiom IV is also satisfied derives from the properties of integrals like (3.9), as developed in the theory of integration.

The marginal probability density functions of $\mathbf{x}$ and $\mathbf{y}$ are obtained by differentiating (3.3) and (3.4) after substituting (3.7),

$$\begin{aligned}
f_{\mathbf{x}}(x) &= \frac{dF_{\mathbf{x}}(x)}{dx} = \frac{d}{dx} F_{\mathbf{xy}}(x, \infty) \\
&= \frac{d}{dx} \int_{-\infty}^{x} dx' \int_{-\infty}^{\infty} dy\, f_{\mathbf{xy}}(x', y) \\
&= \int_{-\infty}^{\infty} f_{\mathbf{xy}}(x, y)\, dy,
\end{aligned} \tag{3.10}$$

and similarly,

$$f_{\mathbf{y}}(y) = \int_{-\infty}^{\infty} f_{\mathbf{xy}}(x, y)\, dx. \tag{3.11}$$

The marginal probability density function of one random variable is obtained by integrating the joint probability density function over the range of the other variable. We can think of $f_{\mathbf{x}}(x)\, dx$ as the probability assigned to the vertical strip lying between x and $x + dx$, and $f_{\mathbf{y}}(y)\, dy$ as the probability assigned to the horizontal strip lying between y and $y + dy$ (see Fig. 3.5):

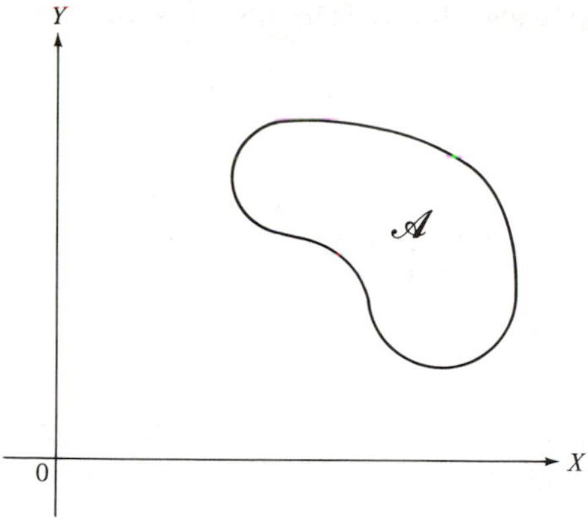

FIGURE 3.4 Region defining event.

$$
\begin{aligned}
f_{\mathbf{x}}(x)\,dx &= \Pr\,(x < \mathbf{x} < x + dx,\ -\infty < \mathbf{y} < \infty),\\
f_{\mathbf{y}}(y)\,dy &= \Pr\,(-\infty < \mathbf{x} < \infty,\ y < \mathbf{y} < y + dy).
\end{aligned} \tag{3.12}
$$

EXAMPLE 3.2. The joint probability density function of $(\mathbf{x}, \mathbf{y})$ is given by

$$f_{\mathbf{xy}}(x, y) = C(1 - x - y)$$

for values of x and y for which (x, y) lies within the triangle $\triangle$ as shown in Fig. 3.6; it is bounded by the lines $x = 0$, $y = 0$, $x + y = 1$. Outside the triangle $\triangle$, $f_{\mathbf{xy}}(x, y) \equiv 0$.

The constant C is determined from the condition that the total probability is equal to 1:

$$
\begin{aligned}
1 &= \iint_{\triangle} f_{\mathbf{xy}}(x, y)\,dx\,dy = \int_0^1 dx \int_0^{1-x} C(1 - x - y)\,dy\\
&= \tfrac{1}{2}C \int_0^1 (1 - x)^2\,dx = \tfrac{1}{6}C,
\end{aligned}
$$

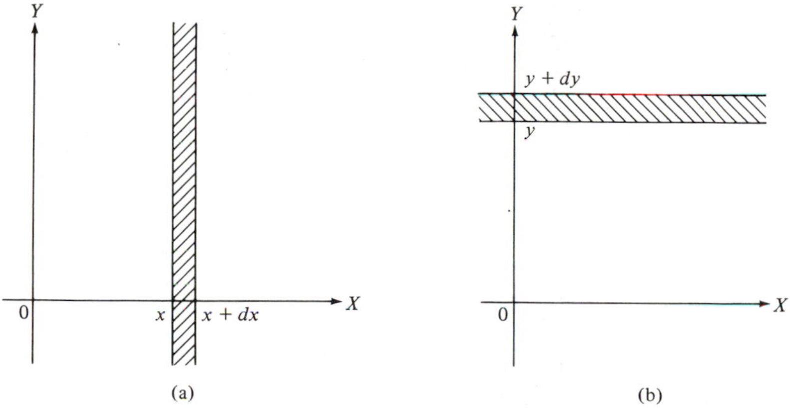

FIGURE 3.5 Regions defining marginal probability density functions.

whence $C = 6$. The marginal probability density function of $\mathbf{x}$ is, by (3.10),

$$f_{\mathbf{x}}(x) = \begin{cases} 6\int_0^{1-x} (1 - x - y)\,dy = 3(1-x)^2, & 0 \le x < 1, \\ 0, & x < 0, \quad x \ge 1, \end{cases}$$

when we integrate over the strip lying between x and $x + dx$, the portions of the strip for $y > 1 - x$ and $y < 0$ contributing zero probability. Similarly,

$$f_{\mathbf{y}}(y) = \begin{cases} 3(1-y)^2, & 0 \le y < 1, \\ 0, & y < 0, \quad y \ge 1. \end{cases}$$

The joint cumulative distribution function of $(\mathbf{x}, \mathbf{y})$ equals 0 for $x < 0$ or $y < 0$ or both. When the point (x, y) lies inside the triangle Δ in Fig. 3.6, the cumulative distribution function is

$$F_{\mathbf{xy}}(x, y) = 6\int_0^x dx' \int_0^y dy'\,(1 - x' - y'), \qquad \begin{cases} 0 < x < 1 \\ 0 < y < 1 \\ 0 < x + y < 1 \end{cases}$$

If $0 < y < 1$, $x > 1 - y$, the point (x, y) lies outside the triangle to its right. For $1 - y \le x < 1$,

$$F_{\mathbf{xy}}(x, y) = 6\int_0^{1-x} dy' \int_0^x dx'\,(1 - x' - y') + 6\int_{1-x}^{y} dy' \int_0^{1-y'} dx'\,(1 - x' - y').$$

For $x > 1$, $0 < y < 1$, on the other hand,

$$F_{\mathbf{xy}}(x, y) = 6\int_0^y dy' \int_0^{1-y'} dx'\,(1 - x' - y').$$

For $y > 1$, $0 < x < 1$,

$$F_{\mathbf{xy}}(x, y) = 6\int_0^x dx' \int_0^{1-x'} dy'\,(1 - x' - y'),$$

and for $x > 1$, $y > 1$, $F_{\mathbf{xy}}(x, y) = 1$. We leave the evaluation of these double integrals to the reader and urge him or her to verify the limits of integration by drawing a diagram for each case. ■

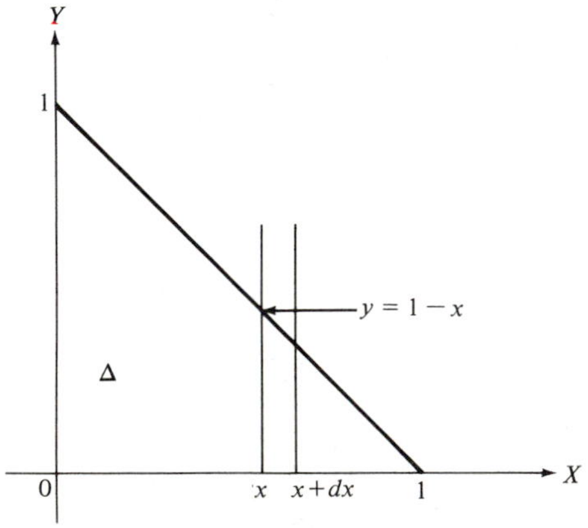

FIGURE 3.6 Region over which density function in Example 3.2 is defined.

EXAMPLE 3.3. The following joint probability density function vanishes outside a circle of radius a,

$$f_{\mathbf{xy}}(x, y) = \begin{cases} C(a^2 - x^2 - y^2), & 0 < x^2 + y^2 < a^2, \\ 0, & x^2 + y^2 > a^2. \end{cases}$$

The constant C is most easily evaluated by carrying out the normalization integral in polar coordinates,

$$\int_{-\infty}^{\infty}\int_{-\infty}^{\infty} f_{\mathbf{xy}}(x, y)\, dx\, dy = C\int_0^a r\, dr \int_0^{2\pi} d\theta\, (a^2 - r^2)$$

$$= 2\pi C\int_0^a (a^2 r - r^3)\, dr = \tfrac{1}{2}\pi C a^4,$$

so that $C = (2/\pi)a^{-4}$. The marginal probability density function of $\mathbf{x}$ is now

$$f_{\mathbf{x}}(x) = C\int_{-\sqrt{a^2 - x^2}}^{\sqrt{a^2 - x^2}} (a^2 - x^2 - y^2)\, dy = \tfrac{4}{3}\, C(a^2 - x^2)^{3/2}$$

$$= \frac{8}{3\pi}\, a^{-4}(a^2 - x^2)^{3/2}, \qquad -a < x < a,$$

$$f_{\mathbf{x}}(x) = 0, \qquad |x| > a.$$

The marginal probability density function of $\mathbf{y}$ has the same form. The joint cumulative distribution function is neither simple nor instructive, and we forbear evaluating it.

The probability that $\mathbf{x}^2 + \mathbf{y}^2 < b^2$ equals the integral of the joint probability density function $f_{\mathbf{xy}}(x, y)$ over a circle of radius b. For $0 < b < a$ we find, again using polar coordinates,

$$\Pr(\mathbf{x}^2 + \mathbf{y}^2 < b^2) = \frac{b^2(2a^2 - b^2)}{a^4}, \qquad 0 < b < a. \quad ■$$

C. Pairs of Discrete Random Variables

Suppose that a fair cubical die is cast. The winnings of two persons watching the outcome are given by $\mathbf{x}$ and $\mathbf{y}$ as in the following table:

ζ	1	2	3	4	5	6
$\mathbf{x}$	−2	1	5	−1	−3	4
$\mathbf{y}$	1	−4	−3	2	−6	7

Each of the six pairs of numbers $(-2, 1), (1, -4), \ldots, (4, 7)$ has probability $\frac{1}{6}$. We can imagine point masses equal to $\frac{1}{6}$ located at the corresponding six points of the XY-plane, as shown in Fig. 3.7; each point has been labeled with the face ζ of the die that evokes that pair of winnings $(\mathbf{x}, \mathbf{y})$. These random variables $\mathbf{x}$ and $\mathbf{y}$ are discrete. Their cumulative distribution function $F_{\mathbf{xy}}(x, y)$ can be calculated by adding the probabilities attached to points $(\mathbf{x}, \mathbf{y})$ lying in the quadrant $(\mathbf{x} \le x, \mathbf{y} \le y)$. It will consist of flat steps, rising along horizontal and vertical lines through the six point masses. The probability $\Pr((\mathbf{x}, \mathbf{y}) \in \mathscr{A})$

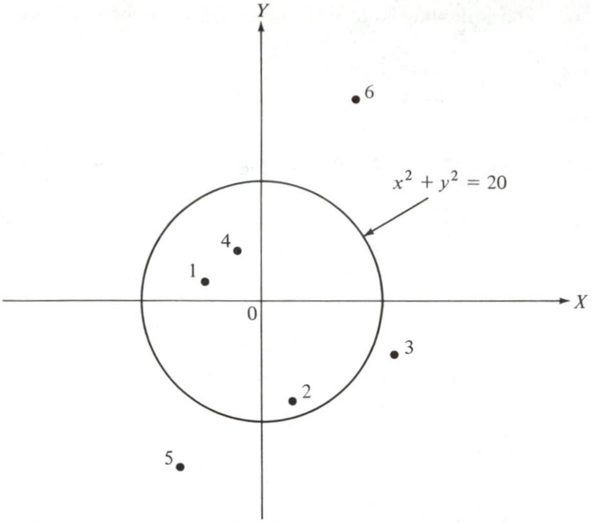

FIGURE 3.7 Points specifying discrete random variables.

that the point $(\mathbf{x}, \mathbf{y})$ lies in an arbitrary region $\mathscr{A}$ of the XY-plane is simply the sum of the masses contained within $\mathscr{A}$. Thus

$$\Pr(\mathbf{x}^2 + \mathbf{y}^2 < 20) = \Pr(\mathbf{x} = -2, \mathbf{y} = 1) + \Pr(\mathbf{x} = 1, \mathbf{y} = -4) + \Pr(\mathbf{x} = -1, \mathbf{y} = 2) = \tfrac{1}{2}.$$

In general, the probability distribution of a pair of discrete random variables can be presented most simply by specifying the probability attached to each of a countable number of such point masses,

$$\Pr(\mathbf{x} = x_k, \mathbf{y} = y_k) = P(x_k, y_k), \qquad k = 1, 2, \ldots, \tag{3.13}$$

$$\sum_{k=1}^{\infty} P(x_k, y_k) = 1.$$

The marginal distribution of $\mathbf{x}$ is obtained by summing the probabilities of all points having the same x-coordinate. The marginal distribution of $\mathbf{y}$ is obtained by summing the probabilities of all points having the same y-coordinate.

EXAMPLE 3.4. For all nonnegative integers k, m, let

$$\Pr(\mathbf{x} = k, \mathbf{y} = m) = (1 - v_1)(1 - v_2)v_1^k v_2^m;$$

$\mathbf{x}$ and $\mathbf{y}$ might be the numbers of photoelectrons counted at two photodetectors. Then the marginal distribution of $\mathbf{x}$ is

$$\Pr(\mathbf{x} = k) = \sum_{m=0}^{\infty} \Pr(\mathbf{x} = k, \mathbf{y} = m) = (1 - v_1)v_1^k,$$

and that of $\mathbf{y}$ is

$$\Pr(\mathbf{y} = m) = (1 - v_2)v_2^m.$$

The probability that for any positive integer p,

$$\mathbf{x} + \mathbf{y} \le p,$$

is obtained by summing the probabilities attached to mass points in that region of the XY-plane (see Fig. 3.8).

$$\begin{aligned}
\Pr(\mathbf{x} + \mathbf{y} \le p) &= \sum_{k=0}^{p} \sum_{m=0}^{p-k} (1 - v_1)(1 - v_2) v_1^k v_2^m \\
&= \sum_{k=0}^{p} (1 - v_1) v_1^k (1 - v_2^{p-k+1}) \\
&= 1 - v_1^{p+1} - (1 - v_1) v_2^{p+1} \sum_{k=0}^{p} \left(\frac{v_1}{v_2}\right)^k \\
&= 1 - v_1^{p+1} - (1 - v_1) v_2^{p+1} \left[\frac{1 - (v_1/v_2)^{p+1}}{1 - (v_1/v_2)}\right] \\
&= 1 - \left(\frac{1 - v_1}{v_2 - v_1}\right) v_2^{p+2} + \left(\frac{1 - v_2}{v_2 - v_1}\right) v_1^{p+2}
\end{aligned}$$

provided that $v_1 \neq v_2$. Here we used the common formula for summing a geometrical progression,

$$\sum_{k=0}^{n} a^k = \frac{1 - a^{n+1}}{1 - a}.$$

■

Distributions of discrete random variables can be expressed in terms of joint probability density functions by means of two-dimensional delta functions, which are simply products of ordinary Dirac delta functions,

$$\delta(x - a, y - b) = \delta(x - a)\, \delta(y - b),$$

If $g(x, y)$ is any function of x and y continuous at (a, b),

$$\begin{aligned}
&\iint_{R(a,b)} g(x, y)\, \delta(x - a, y - b)\, dx\, dy \\
&\quad = \iint_{R(a,b)} g(x, y)\, \delta(x - a)\, \delta(y - b)\, dx\, dy = g(a, b),
\end{aligned}$$

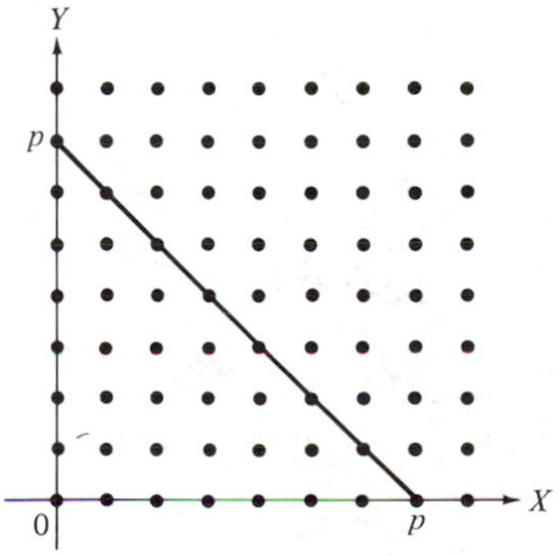

FIGURE 3.8 Locus of discrete random variables in Example 3.4.

where $R(a, b)$ is any region of the XY-plane containing the point (a, b) in its interior.

If the discrete random variables $\mathbf{x}$ and $\mathbf{y}$ take on a countable number of pairs of values (x_k, y_k), with probabilities as in (3.13), their joint probability density function is

$$f_{\mathbf{xy}}(x, y) = \sum_{k=1}^{\infty} P(x_k, y_k)\delta(x - x_k)\delta(y - y_k). \tag{3.14}$$

Thus for the example at the beginning of this part,

$$\begin{aligned} f_{\mathbf{xy}}(x, y) &= \tfrac{1}{6}\,\delta(x + 2)\delta(y - 1) + \tfrac{1}{6}\,\delta(x - 1)\delta(y + 4) \\ &\quad + \tfrac{1}{6}\,\delta(x - 5)\delta(y + 3) + \tfrac{1}{6}\,\delta(x + 1)\delta(y - 2) \\ &\quad + \tfrac{1}{6}\,\delta(x + 3)\delta(y + 6) + \tfrac{1}{6}\,\delta(x - 4)\delta(y - 7). \end{aligned}$$

The joint probability density function of the random variables $\mathbf{x}$ and $\mathbf{y}$ in Example 3.4 is

$$f_{\mathbf{xy}}(x, y) = (1 - v_1)(1 - v_2) \sum_{k=0}^{\infty} \sum_{m=0}^{\infty} v_1^k v_2^m \delta(x - k)\delta(y - m).$$

D. Mixed Random Variables and Line Masses

The joint probability density function of a pair of random variables might consist of both a continuous portion and a countable set of two-dimensional delta functions; the latter correspond to point masses located here and there in the XY-plane, the former to a continuous distribution of mass. Such random variables are of the mixed type. One must be sure that the sum of all the point masses and the integral over the continuous portion of the distribution comes out equal to 1.

Probability might be distributed along a curve in the XY-plane, much like mass distributed along a curved wire, as illustrated in Fig. 3.9. Let the equation of the curve be $y = g(x)$, and assume it to be monotone increasing, as in the figure. Let the joint probability density function of the random variables $\mathbf{x}$ and $\mathbf{y}$ be

$$f_{\mathbf{xy}}(x, y) = f_{\mathbf{x}}(x)\delta(y - g(x)), \tag{3.15}$$

where

$$\int_{-\infty}^{\infty} f_{\mathbf{x}}(x)\, dx = 1, \qquad f_{\mathbf{x}}(x) \geq 0.$$

Integration over $-\infty < y < \infty$ shows that $f_{\mathbf{x}}(x)$ is the marginal probability density function of the random variable $\mathbf{x}$. The marginal probability density function of $\mathbf{y}$ is

$$f_{\mathbf{y}}(y) = \int_{-\infty}^{\infty} f_{\mathbf{x}}(x)\delta(y - g(x))\, dx.$$

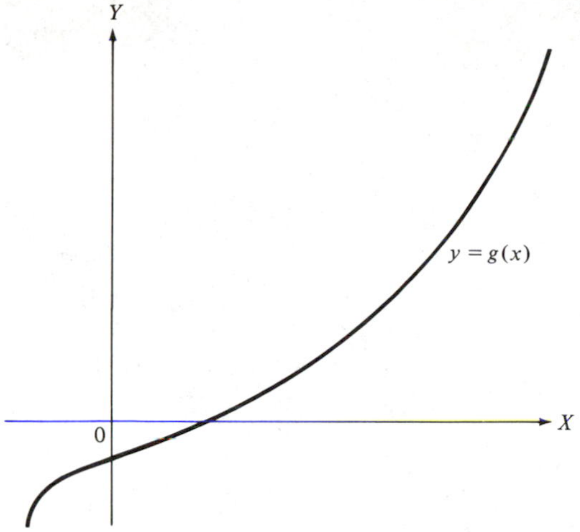

FIGURE 3.9 Distribution of probability along a line.

To evaluate this integral, we change the integration variable to

$$z = g(x), \qquad dz = g'(x)\,dx,$$

obtaining

$$f_{\mathbf{y}}(y) = \int_{-\infty}^{\infty} f_{\mathbf{x}}(g^{-1}(z))\delta(y - z)\,\frac{dz}{g'(g^{-1}(z))},$$

where $x = g^{-1}(z)$ is the inverse function to $z = g(x)$. Integrating over z, we find that

$$f_{\mathbf{y}}(y) = \frac{f_{\mathbf{x}}(g^{-1}(y))}{g'(g^{-1}(y))} = \left.\frac{f_{\mathbf{x}}(x)}{g'(x)}\right|_{x = g^{-1}(y)}, \tag{3.16}$$

which is the same as (2.66) for a monotone increasing function $y = g(x)$.

EXAMPLE 3.5. Let the probability be uniformly distributed around the unit circle, with density $1/(2\pi)$. What is the marginal probability density function of $\mathbf{x}$? (See Fig. 3.10.)

The probability that $\mathbf{x}$ lies between x and $x + dx$ equals the masses of the two arc segments lying in the strip

$$x < \mathbf{x} < x + dx.$$

The length of each segment is $|d\theta|$, where $x = \cos\theta$, $y = \sin\theta$. Because $dx = -\sin\theta\,d\theta$,

$$|d\theta| = \frac{dx}{|\sin\theta|} = \frac{dx}{\sqrt{1 - x^2}},$$

and the probability is

$$\Pr(x < \mathbf{x} < x + dx) = 2\left(\frac{1}{2\pi}\right)\frac{dx}{\sqrt{1 - x^2}},$$

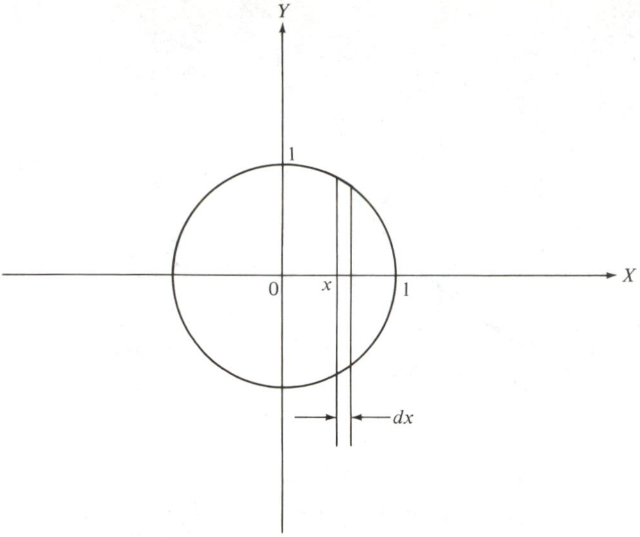

FIGURE 3.10 Probability distribution on a circle.

whence the marginal probability density function of **x** is

$$f_{\mathbf{x}}(x) = \pi^{-1}(1 - x^2)^{-1/2}.$$

The random variable **x** is distributed according to the arcsine law. Its cumulative distribution $F_{\mathbf{x}}(x) = \Pr(\mathbf{x} < x)$ equals $(2\pi)^{-1}$ times the length of the part of the circle to the left of the vertical line at x and is easily seen to be $1 - \theta/\pi$, which reduces to (2.38) for $a = 1$. ■

3-2

CONDITIONAL DISTRIBUTIONS OF PAIRS OF RANDOM VARIABLES

A. Nonnumerical Attributes

The outcome of an experiment may be specified by both a pair of numbers **x** and **y** and some nonnumerical attribute. The probability measure for the experiment could be given in terms of joint probability density functions of the random variables **x** and **y**, conditioned on the status of the nonnumerical attribute, much as a single random variable with a nonnumerical attribute was treated in Sec. 2-2A. An example should suffice, and we take it from hypothesis testing.

EXAMPLE 3.6. Detection of a Random Signal

As in Sec. 1-11, let H_0 denote the hypothesis, "Only noise is present at the input to a receiver," and let H_1 denote the hypothesis, "A random signal is present in addition to the noise." The random variables **x** and **y** are samples of the input voltage at two times t_1 and t_2. Under hypothesis H_0 these have the joint probability density function

$$f_{\mathbf{xy}}(x, y \mid H_0) = (2\pi\sigma_0^2)^{-1} \exp\left(-\frac{x^2 + y^2}{2\sigma_0^2}\right).$$

Under hypothesis H_1 their joint probability density function is

$$f_{\mathbf{xy}}(x, y \mid H_1) = (2\pi\sigma_1^2)^{-1} \exp\left(-\frac{x^2 + y^2}{2\sigma_1^2}\right),$$

with $\sigma_1 > \sigma_0$. Let the prior probabilities of the hypotheses be

$$\Pr(H_0) = p_0, \qquad \Pr(H_1) = p_1, \qquad p_0 + p_1 = 1.$$

Here the presence or the absence of a signal is a nonnumerical attribute, designated by H_0 or H_1, and the outcome of the experiment is specified by $(\mathbf{x}, \mathbf{y}, H_0)$ or $(\mathbf{x}, \mathbf{y}, H_1)$, as the case may be. The probability that $x < \mathbf{x} < x + dx$ and $y < \mathbf{y} < y + dy$ and no signal is present is

$$\begin{aligned}\Pr(x < \mathbf{x} < x + dx, y < \mathbf{y} < y + dy, H_0) \\ &= \Pr(H_0) f_{\mathbf{xy}}(x, y \mid H_0)\, dx\, dy \\ &= p_0(2\pi\sigma_0^2)^{-1} \exp\left(-\frac{x^2 + y^2}{2\sigma_0^2}\right) dx\, dy.\end{aligned}$$

The total probability that $\mathbf{x}$ lies between x and $x + dx$ and $\mathbf{y}$ lies between y and $y + dy$ is

$$\Pr(x < \mathbf{x} < x + dx, y < \mathbf{y} < y + dy) = f_{\mathbf{xy}}(x, y)\, dx\, dy,$$

where

$$f_{\mathbf{xy}}(x, y) = \Pr(H_0) f_{\mathbf{xy}}(x, y \mid H_0) + \Pr(H_1) f_{\mathbf{xy}}(x, y \mid H_1)$$

by the principle of total probability.

By Bayes's theorem, the probability that a signal is present when it has been observed that $x < \mathbf{x} < x + dx$ and $y < \mathbf{y} < y + dy$ is

$$\begin{aligned}&\Pr(H_1 \mid x < \mathbf{x} < x + dx, y < \mathbf{y} < y + dy) \\ &= \frac{\Pr(H_1) f_{\mathbf{xy}}(x, y \mid H_1)}{f_{\mathbf{xy}}(x, y)} \\ &= \frac{p_1(2\pi\sigma_1^2)^{-1} \exp[-(x^2 + y^2)/2\sigma_1^2]}{p_0(2\pi\sigma_0^2)^{-1} \exp[-(x^2 + y^2)/2\sigma_0^2] + p_1(2\pi\sigma_1^2)^{-1} \exp[-(x^2 + y^2)/2\sigma_1^2]},\end{aligned}$$

the infinitesimal area $dx\, dy$ canceling from numerator and denominator. It is customary to write this probability as $\Pr(H_1 \mid x, y)$ in the limit $dx \to 0$, $dy \to 0$. The probability $\Pr(H_0 \mid x, y)$ is similarly defined.

A receiver might be designed to indicate the presence of a signal whenever $\Pr(H_1 \mid x, y) > \Pr(H_0 \mid x, y)$, that is, whenever

$$p_1(2\pi\sigma_1^2)^{-1} \exp\left(-\frac{x^2 + y^2}{2\sigma_1^2}\right) > p_0(2\pi\sigma_0^2)^{-1} \exp\left(-\frac{x^2 + y^2}{2\sigma_0^2}\right)$$

or whenever

$$x^2 + y^2 > R^2 = \frac{2\sigma_0^2\sigma_1^2}{\sigma_1^2 - \sigma_0^2} \ln\left(\frac{\sigma_1^2 p_0}{\sigma_0^2 p_1}\right).$$

Otherwise the receiver decides no signal is present. By virtue of what we said about Bayes's rule in Sec. 1-11, this receiver attains the minimum probability P_e of error with respect to the universal set of all possible outcomes $(\mathbf{x}, \mathbf{y}, H_k)$, $k = 0, 1$, of the experiment.

The *false-alarm probability*, that is, the probability that the receiver declares a signal is present when it is not, will be

$$\Pr(\mathbf{x}^2 + \mathbf{y}^2 > R^2 \mid H_0) = \iint_{x^2+y^2>R^2} f_{\mathbf{xy}}(x, y \mid H_0)\, dx\, dy$$

$$= (2\pi\sigma_0^2)^{-1} \int_R^\infty r\, dr \int_0^{2\pi} d\theta \exp\left(-\frac{r^2}{2\sigma_0^2}\right)$$

$$= \exp\left(-\frac{R^2}{2\sigma_0^2}\right),$$

when we convert our integrand to polar coordinates. Similarly, the *probability of detection,* that is, the probability of declaring a signal present when one has really come in, is

$$\Pr(\mathbf{x}^2 + \mathbf{y}^2 > R^2 \mid H_1) = \exp\left(-\frac{R^2}{2\sigma_1^2}\right).$$

The probability of error is now

$$P_e = p_0 \Pr(\mathbf{x}^2 + \mathbf{y}^2 > R^2 \mid H_0) + p_1 \Pr(\mathbf{x}^2 + \mathbf{y}^2 < R^2 \mid H_1)$$

$$= p_0 \exp\left(-\frac{R^2}{2\sigma_0^2}\right) + p_1\left[1 - \exp\left(-\frac{R^2}{2\sigma_1^2}\right)\right],$$

with R^2 defined as above.

The marginal probability density function of the random variable $\mathbf{x}$, given that a signal is present, is by (3.10)

$$f_{\mathbf{x}}(x \mid H_1) = \int_{-\infty}^{\infty} f_{\mathbf{xy}}(x,y \mid H_1)\, dy = (2\pi\sigma_1^2)^{-1/2} \exp\left(-\frac{x^2}{2\sigma_1^2}\right).$$

■

Conditional joint probability density functions are treated in the same way as ordinary joint probability density functions and are subject to all the same rules.

B. Numerical Conditioning Events

Suppose we are told that the pair of random variables $\mathbf{x}$ and $\mathbf{y}$ has been observed to lie in a certain region $\mathcal{A}$ of the XY-plane. What is their joint probability distribution when conditioned on that event $(\mathbf{x}, \mathbf{y}) \in \mathcal{A}$?

Let $R(x, y)$ be that portion of the quadrant $\{\mathbf{x}, \mathbf{y} : \mathbf{x} < x, \mathbf{y} < y\}$ that intersects the region $\mathcal{A}$, as shown in Fig. 3.11,

$$R(x, y) = \{\mathbf{x}, \mathbf{y} : \mathbf{x} < x, \mathbf{y} < y, (\mathbf{x}, \mathbf{y}) \in \mathcal{A}\}$$

$$= \{\mathbf{x}, \mathbf{y} : \mathbf{x} < x, \mathbf{y} < y\} \cap \mathcal{A}.$$

Then the conditional joint cumulative distribution function of $\mathbf{x}$ and $\mathbf{y}$, given the event $(\mathbf{x}, \mathbf{y}) \in \mathcal{A}$, is

$$F_{\mathbf{xy}}(x, y \mid (\mathbf{x}, \mathbf{y}) \in \mathcal{A}) = \frac{\Pr((\mathbf{x}, \mathbf{y}) \in R(x, y))}{\Pr((\mathbf{x}, \mathbf{y}) \in \mathcal{A})} \tag{3.17}$$

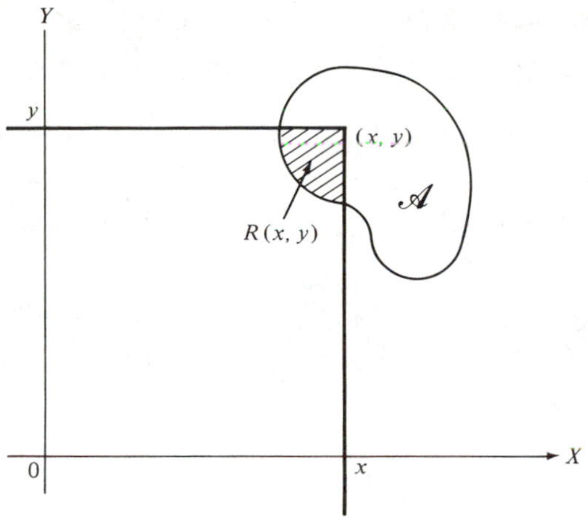

FIGURE 3.11 Region for numerical conditioning event.

by the definition of conditional probability in (1.9). In terms of the joint probability density function of $\mathbf{x}$ and $\mathbf{y}$, the numerator is

$$\Pr((\mathbf{x}, \mathbf{y}) \in R(x, y)) = \iint_{R(x,y)} f_{\mathbf{xy}}(x, y)\, dx\, dy \tag{3.18}$$

and the denominator is

$$\Pr((\mathbf{x}, \mathbf{y}) \in \mathcal{A}) = \iint_{\mathcal{A}} f_{\mathbf{xy}}(x, y)\, dx\, dy. \tag{3.19}$$

If the region $R(x, y)$ does not intersect $\mathcal{A}$ at all, the conditional cumulative distribution function in (3.17) vanishes. If $R(x, y)$ includes all of the region $\mathcal{A}$, $F_{\mathbf{xy}}(x, y \mid (\mathbf{x}, \mathbf{y}) \in \mathcal{A}) = 1$. When $\mathbf{x}$ and $\mathbf{y}$ are discrete random variables, the integrals in (3.18) and (3.19) are replaced by summations over the point masses in the defining regions.

If we now apply the definition of a joint probability density function in (3.5), we obtain the joint conditional probability density function of $\mathbf{x}$ and $\mathbf{y}$, given $(\mathbf{x}, \mathbf{y}) \in \mathcal{A}$,

$$f_{\mathbf{xy}}(x, y \mid (\mathbf{x}, \mathbf{y}) \in \mathcal{A}) = \begin{cases} 0, & (x, y) \notin \mathcal{A}, \\ \dfrac{f_{\mathbf{xy}}(x, y)}{\Pr((\mathbf{x}, \mathbf{y}) \in \mathcal{A})}, & (x, y) \in \mathcal{A}. \end{cases} \tag{3.20}$$

This also follows from applying the definition in (1.9) to the probability

$$\Pr(x < \mathbf{x} < x + dx, y < \mathbf{y} < y + dy \mid (\mathbf{x}, \mathbf{y}) \in \mathcal{A})$$
$$= \frac{\Pr((\mathbf{x}, \mathbf{y}) \in \Delta_{xy} \cap \mathcal{A})}{\Pr((\mathbf{x}, \mathbf{y}) \in \mathcal{A})},$$

where $\triangle_{xy}$ is the infinitesimal rectangle

$$\triangle_{xy} = \{\mathbf{x}, \mathbf{y} : (x < \mathbf{x} < x + dx, \ y < \mathbf{y} < y + dy)\}$$

at (x, y) in the XY-plane. If $(x, y) \in \mathcal{A}$, the rectangle $\triangle_{xy}$ lies inside the region $\mathcal{A}$, and

$$\Pr\left((\mathbf{x}, \mathbf{y}) \in \triangle_{xy} \cap \mathcal{A}\right) = \Pr\left((\mathbf{x}, \mathbf{y}) \in \triangle_{xy}\right) = f_{\mathbf{xy}}(x, y)\, dx\, dy;$$

otherwise $\Pr\left((\mathbf{x}, \mathbf{y}) \in \triangle_{xy} \cap \mathcal{A}\right) = 0$, whence (3.20). For $(x, y) \in \mathcal{A}$ the conditional probability density function in (3.20) is proportional to the unconditional joint probability density function $f_{\mathbf{xy}}(x, y)$, but increased by just such a factor $1/\Pr((\mathbf{x}, \mathbf{y}) \in \mathcal{A})$ that

$$\int_{-\infty}^{\infty}\int_{-\infty}^{\infty} f_{\mathbf{xy}}(x, y \mid (\mathbf{x}, \mathbf{y}) \in \mathcal{A})\, dx\, dy = 1. \tag{3.21}$$

The portions of the plane outside the region $\mathcal{A}$ contribute zero to this integral because of (3.20). The universal set S—the entire XY-plane—has been replaced by the set of points lying in the region $\mathcal{A}$.

EXAMPLE 3.7. Let the joint probability density function of $\mathbf{x}$ and $\mathbf{y}$ be

$$f_{\mathbf{xy}}(x, y) = (2\pi\sigma^2)^{-1} \exp\left(-\frac{x^2 + y^2}{2\sigma^2}\right).$$

We seek the joint conditional probability density function

$$f_{\mathbf{xy}}(x, y \mid \mathbf{x}^2 + \mathbf{y}^2 < b^2).$$

Denote the circle $\mathbf{x}^2 + \mathbf{y}^2 < b^2$ by C. Then

$$\begin{aligned}\Pr(\mathbf{x}^2 + \mathbf{y}^2 < b^2) &= \Pr((\mathbf{x}, \mathbf{y}) \in C)\\ &= (2\pi\sigma^2)^{-1} \iint_C \exp\left(-\frac{x^2 + y^2}{2\sigma^2}\right) dx\, dy.\end{aligned}$$

This can most easily be evaluated by changing to polar coordinates:

$$\begin{aligned}\Pr((\mathbf{x}, \mathbf{y}) \in C) &= (2\pi\sigma^2)^{-1} \int_0^b r\, dr \int_0^{2\pi} d\theta \exp\left(-\frac{r^2}{2\sigma^2}\right)\\ &= 1 - \exp\left(-\frac{b^2}{2\sigma^2}\right).\end{aligned}$$

Hence by (3.20),

$$f_{\mathbf{xy}}(x, y \mid \mathbf{x}^2 + \mathbf{y}^2 < b^2) =$$
$$\begin{cases} 0, & x^2 + y^2 > b^2, \\ (2\pi\sigma^2)^{-1}\left[1 - \exp\left(-\dfrac{b^2}{2\sigma^2}\right)\right]^{-1} \exp\left(-\dfrac{x^2 + y^2}{2\sigma^2}\right), & x^2 + y^2 < b^2. \end{cases}$$ ■

EXAMPLE 3.8. Refer to Example 3.4. The conditional joint probability distribution of the numbers $\mathbf{x}$ and $\mathbf{y}$ of electrons counted, given that the total number of electrons is no greater than p, is specified by

$$\Pr(\mathbf{x} = k, \mathbf{y} = m \mid \mathbf{x} + \mathbf{y} \le p) = \left\{1 - \left(\frac{1 - v_1}{v_2 - v_1}\right)v_2^{p+2} + \left(\frac{1 - v_2}{v_2 - v_1}\right)v_1^{p+2}\right\}^{-1}$$
$$\times (1 - v_1)(1 - v_2)v_1^k v_2^m,$$
$$k + m \le p, \quad k \ge 0, \quad m \ge 0,$$
$$= 0, \qquad k + m > p \text{ or } k < 0 \text{ or } m < 0. \qquad ■$$

Marginal conditional probability density functions can be defined as in (3.10) and (3.11); one integrates out the superfluous variable:

$$f_{\mathbf{x}}(x \mid (\mathbf{x}, \mathbf{y}) \in \mathcal{A}) = \int_{-\infty}^{\infty} f_{\mathbf{xy}}(x, y \mid (\mathbf{x}, \mathbf{y}) \in \mathcal{A})\, dy, \tag{3.22}$$

$$f_{\mathbf{y}}(y \mid (\mathbf{x}, \mathbf{y}) \in \mathcal{A}) = \int_{-\infty}^{\infty} f_{\mathbf{xy}}(x, y \mid (\mathbf{x}, \mathbf{y}) \in \mathcal{A})\, dx. \tag{3.23}$$

The integrand will vanish over those ranges of the variable of integration where the point (x, y) lies outside the region $\mathcal{A}$.

EXAMPLE 3.7 (continued). The marginal conditional probability density function of $\mathbf{x}$ in this example is

$$f_{\mathbf{x}}(x \mid \mathbf{x}^2 + \mathbf{y}^2 < b^2) = (2\pi\sigma^2)^{-1}\left[1 - \exp\left(-\frac{b^2}{2\sigma^2}\right)\right]^{-1}$$
$$\times \exp\left(-\frac{x^2}{2\sigma^2}\right)\int_{-(b^2-x^2)^{1/2}}^{(b^2-x^2)^{1/2}} \exp\left(-\frac{y^2}{2\sigma^2}\right) dy$$
$$= (2\pi\sigma^2)^{-1/2}\left[1 - \exp\left(-\frac{b^2}{2\sigma^2}\right)\right]^{-1}$$
$$\times \left[1 - 2\,\text{erfc}\left(\frac{(b^2 - x^2)^{1/2}}{\sigma}\right)\right]\exp\left(-\frac{x^2}{2\sigma^2}\right),$$
$$-b < x < b.$$

The limits of integration are $\pm(b^2 - x^2)^{1/2}$ because the joint conditional probability density function $f_{\mathbf{xy}}(x, y \mid \mathbf{x}^2 + \mathbf{y}^2 < b^2)$ vanishes for $|y| > (b^2 - x^2)^{1/2}$. The function erfc x was defined in (2.20). Furthermore,

$$f_{\mathbf{x}}(x \mid \mathbf{x}^2 + \mathbf{y}^2 < b^2) = 0, \qquad |x| > b.$$

The marginal conditional probability density function of $\mathbf{y}$ has a similar form. ■

Let the defining event be $a < \mathbf{x} < b$. Then by (3.20),

$$f_{\mathbf{xy}}(x, y \mid a < \mathbf{x} < b) = \begin{cases} 0, & x < a,\ x > b, \\ \dfrac{f_{\mathbf{xy}}(x, y)}{\Pr(a < \mathbf{x} < b)}, & a < x < b, \end{cases} \tag{3.24}$$

where

$$\Pr(a < \mathbf{x} < b) = \int_a^b f_{\mathbf{x}}(x)\, dx = \int_a^b dx \int_{-\infty}^{\infty} dy\, f_{\mathbf{xy}}(x, y). \tag{3.25}$$

The marginal conditional probability density function of $\mathbf{y}$ is

$$f_{\mathbf{y}}(y \mid a < \mathbf{x} < b) = \frac{\int_a^b f_{\mathbf{xy}}(x', y)\,dx'}{\Pr(a < \mathbf{x} < b)}, \tag{3.26}$$

for the integrand in (3.23) vanishes for $x < a$, $x > b$.

Now we suppose the interval (a, b) very narrow, putting

$$a = x - \tfrac{1}{2}\varepsilon, \qquad b = x + \tfrac{1}{2}\varepsilon,$$

where ε is a positive number that we are going to allow to vanish. From (3.25),

$$\Pr(a < \mathbf{x} < b) \doteq \varepsilon f_{\mathbf{x}}(x),$$

and the numerator in (3.26) is

$$\int_{x-(1/2)\varepsilon}^{x+(1/2)\varepsilon} f_{\mathbf{xy}}(x', y)\,dx' \doteq \varepsilon f_{\mathbf{xy}}(x, y).$$

Hence the marginal probability density function of $\mathbf{y}$ becomes

$$f_{\mathbf{y}}(y \mid x - \tfrac{1}{2}\varepsilon < \mathbf{x} < x + \tfrac{1}{2}\varepsilon) \doteq \frac{f_{\mathbf{xy}}(x, y)}{f_{\mathbf{x}}(x)},$$

the factor ε canceling from numerator and denominator. Letting ε go to zero, we obtain the conditional probability density function of the random variable $\mathbf{y}$, given that $\mathbf{x}$ was observed to take the value x,

$$f_{\mathbf{y}}(y \mid \mathbf{x} = x) = \frac{f_{\mathbf{xy}}(x, y)}{f_{\mathbf{x}}(x)} = \frac{f_{\mathbf{xy}}(x, y)}{\int_{-\infty}^{\infty} f_{\mathbf{xy}}(x, y')\,dy'}. \tag{3.27}$$

This important conditional probability density function is often designated by $f_{\mathbf{y}}(y \mid x)$. Similarly,

$$\begin{aligned} f_{\mathbf{x}}(x \mid \mathbf{y} = y) = f_{\mathbf{x}}(x \mid y) &= \frac{f_{\mathbf{xy}}(x, y)}{f_{\mathbf{y}}(y)} \\ &= \frac{f_{\mathbf{xy}}(x, y)}{\int_{-\infty}^{\infty} f_{\mathbf{xy}}(x', y)\,dx'}. \end{aligned} \tag{3.28}$$

These conditional probability density functions have the usual properties of probability density functions,

$$f_{\mathbf{y}}(y \mid x) \geq 0, \qquad \int_{-\infty}^{\infty} f_{\mathbf{y}}(y \mid x)\,dy = 1. \tag{3.29}$$

Needless to say, the conditional probability density function $f_{\mathbf{y}}(y \mid x)$ has no meaning for values of x such that $f_{\mathbf{x}}(x) = 0$. Conditional cumulative distributions can be found by integration,

$$F_{\mathbf{y}}(y \mid x) = \int_{-\infty}^{y} f_{\mathbf{y}}(y' \mid x)\, dy'.$$

From (3.27) and (3.28) we find that

$$f_{\mathbf{y}}(y \mid x) = \frac{f_{\mathbf{x}}(x \mid y)\, f_{\mathbf{y}}(y)}{f_{\mathbf{x}}(x)} \tag{3.30}$$

which is a form of Bayes's theorem.

EXAMPLE 3.9. Refer to Example 3.2. Now the conditional probability density function of $\mathbf{y}$, given $\mathbf{x} = x$, is for $0 < x < 1$

$$f_{\mathbf{y}}(y \mid x) = \begin{cases} \dfrac{2(1 - x - y)}{(1 - x)^2}, & 0 < y < 1 - x, \\ 0, & y < 0, \quad y > 1 - x; \end{cases}$$

it is meaningless for $x < 0$, $x > 1$. Similarly, for $0 < y < 1$,

$$f_{\mathbf{x}}(x \mid y) = \begin{cases} \dfrac{2(1 - y - x)}{(1 - y)^2}, & 0 < x < 1 - y, \\ 0, & x < 0, \quad x < 1 - y. \end{cases}$$

Verifying (3.29), we have

$$\int_{-\infty}^{\infty} f_{\mathbf{y}}(y \mid x)\, dy = 2(1 - x)^{-2} \int_{0}^{1-x} (1 - x - y)\, dy = 1.$$

The conditional cumulative distribution function of $\mathbf{y}$, given $\mathbf{x} = x$, is

$$\begin{aligned} F_{\mathbf{y}}(y \mid x) &= 2(1 - x)^{-2} \int_{0}^{y} (1 - x - y')\, dy' \\ &= \frac{(1 - x)^2 - (1 - x - y)^2}{(1 - x)^2}, \qquad 0 < y < 1 - x, \\ &= 0, \qquad y < 0, \\ &= 1, \qquad y > 1 - x. \end{aligned}$$

■

EXAMPLE 3.10. Parameter Estimation

Signals arrive with random amplitudes $\boldsymbol{\mu}$ that have a Gaussian distribution with mean 0 and variance b^2,

$$f_{\boldsymbol{\mu}}(\mu) = (2\pi b^2)^{-1/2} \exp\left(-\frac{\mu^2}{2b^2}\right). \tag{3.31}$$

We observe not the signal amplitude μ, but its sum with noise of unknown instantaneous value; that is, we observe a random variable $\mathbf{x}$ having a conditional probability density function

$$f_{\mathbf{x}}(x \mid \mu) = (2\pi s^2)^{-1/2} \exp\left[-\frac{(x - \mu)^2}{2s^2}\right], \tag{3.32}$$

where the variance s^2 represents the strength of the noise. What is the conditional probability density function of the signal amplitude $\boldsymbol{\mu}$, given that we observed a particular value of $\mathbf{x}$, say $\mathbf{x} = x$? To apply Bayes's theorem in (3.30), we must calculate the marginal probability density function of **x**:

$$f_{\mathbf{x}}(x) = \int_{-\infty}^{\infty} f_{\mathbf{x}\boldsymbol{\mu}}(x, \mu)\, d\mu = \int_{-\infty}^{\infty} f_{\mathbf{x}}(x \mid \mu) f_{\boldsymbol{\mu}}(\mu)\, d\mu$$

$$= (2\pi s^2)^{-1/2}(2\pi b^2)^{-1/2} \int_{-\infty}^{\infty} \exp\left[-\frac{(x-\mu)^2}{2s^2} - \frac{\mu^2}{2b^2}\right] d\mu \qquad (3.33)$$

$$= [2\pi(s^2 + b^2]^{-1/2} \exp\left[-\frac{x^2}{2(s^2+b^2)}\right].$$

The integration can be carried out by expanding the exponent, completing the square, and using the normalization integral for the Gaussian probability density function. Alternatively, one recognizes (3.33) as the convolution of two Gaussian density functions with variances s^2 and b^2, respectively; the convolution theorem with (2.89) for $m = 0$ shows that the result must be a Gaussian density function with variance $s^2 + b^2$.

Now we can apply Bayes's theorem in (3.30) to write, after some algebra,

$$f_{\boldsymbol{\mu}}(\mu \mid x) = (2\pi s'^2)^{-1/2} \exp\left[-\left(\mu - \frac{b^2 x}{s^2+b^2}\right)^2 \Big/ 2s'^2\right],$$
$$s'^2 = \frac{s^2 b^2}{s^2 + b^2}. \qquad (3.34)$$

This is sometimes called the *posterior probability density function* of the signal amplitude $\boldsymbol{\mu}$, given the observation $\mathbf{x} = x$, and $f_{\boldsymbol{\mu}}(\mu)$ in (3.31) is the *prior probability density function* of $\boldsymbol{\mu}$.

The principle of maximum likelihood, which is equivalent to Bayes's rule, tells us that the best estimate of the value of the signal amplitude $\boldsymbol{\mu}$, given our observation that $\mathbf{x} = x$, is the value for which the conditional probability density function $f_{\boldsymbol{\mu}}(\mu \mid x)$ is largest. Denoting this *maximum-likelihood estimate* by $\hat{\mu}$, we find from (3.34) that

$$\hat{\boldsymbol{\mu}} = \frac{b^2 \mathbf{x}}{s^2 + b^2}. \qquad (3.35)$$

Substituting into (3.34) we find that the conditional probability density function of the true amplitude $\boldsymbol{\mu}$, given that our estimate $\hat{\boldsymbol{\mu}}$ has taken a particular value $\hat{\mu}$, is

$$f_{\boldsymbol{\mu}}(\mu \mid \hat{\mu}) = (2\pi s'^2)^{-1/2} \exp\left[-\frac{(\mu - \hat{\mu})^2}{2s'^2}\right]. \qquad (3.36)$$

Now $\mu - \hat{\mu}$ is the error in our estimate, and (3.36) shows that the variance of the error, sometimes called the *mean-square error,* equals

$$s'^2 = \frac{s^2 b^2}{s^2 + b^2}.$$

We see that if the noise is very strong, $s^2 >> b^2$, $\hat{\mu} \simeq 0$ and $s'^2 \simeq b^2$, and our observation of **x** has not told us much about the strength $\boldsymbol{\mu}$ of the signal. On the other hand, if the noise is weak, $s^2 << b^2$, $\hat{\mu} \simeq x$, and $s'^2 \simeq s^2$, and our estimate gives little weight to our prior knowledge about the signal strength, embodied in the prior probability density function $f_{\boldsymbol{\mu}}(\mu)$ in (3.31). ■

C. Independent Random Variables

If the conditional probability density function $f_{\mathbf{y}}(y \mid x)$ does not depend on x,

$$f_{\mathbf{y}}(y \mid x) = f_{\mathbf{y}}(y), \qquad (3.37)$$

observation of the random variable $\mathbf{x}$ gives us no information about the value of $\mathbf{y}$, and the random variables $\mathbf{x}$ and $\mathbf{y}$ are said to be *statistically independent*. Equivalently, by (3.27), $\mathbf{x}$ and $\mathbf{y}$ are independent if their joint probability density function factors into the product of their marginal probability density functions,

$$f_{\mathbf{xy}}(x, y) = f_{\mathbf{x}}(x) f_{\mathbf{y}}(y), \tag{3.38}$$

It is easy to show that when $\mathbf{x}$ and $\mathbf{y}$ are independent, their joint cumulative distribution function also factors,

$$F_{\mathbf{xy}}(x, y) = F_{\mathbf{x}}(x) F_{\mathbf{y}}(y), \tag{3.39}$$

into the product of their marginal cumulative distribution functions.

In Sec. 1-7C we introduced the concept of independent chance experiments. Let the random variable $\mathbf{x}$ be defined in experiment E_1, let the random variable $\mathbf{y}$ be defined in experiment E_2, and let experiments E_1 and E_2 be independent. The pair $(\mathbf{x}, \mathbf{y})$ of random variables is now defined in the compound experiment $\mathsf{E} = \mathsf{E}_1 \times \mathsf{E}_2$.

As in Fig. 1.9 and as in (1.19) let A be the event $(x < \mathbf{x} < x + dx)$ in E_1, and let B be the event $(y < \mathbf{y} < y + dy)$ in E_2. Then in $\mathsf{E}_1 \times \mathsf{E}_2$ the event

$$(x < \mathbf{x} < x + dx) = (x < \mathbf{x} < x + dx,\ -\infty < \mathbf{y} < \infty)$$

is designated by $A \times S_2$, and

$$(y < \mathbf{y} < y + dy) = (-\infty < \mathbf{x} < \infty,\ y < \mathbf{y} < y + dy)$$

is designated by $S_1 \times B$. The event forming their intersection is now

$$(x < \mathbf{x} < x + dx,\ y < \mathbf{y} < y + dy) = A \times B = (A \times S_2) \cap (S_1 \times B),$$

and the independence of the experiments E_1 and E_2 implies as in (1.22) that

$$f_{\mathbf{xy}}(x, y)\, dx\, dy = f_{\mathbf{x}}(x)\, dx \cdot f_{\mathbf{y}}(y)\, dy,$$

from which (3.38) follows. Random variables defined in independent chance experiments are statistically independent in the sense of (3.37) and (3.38).

As an example, the joint Gaussian density function used in Example 3.7 factors as in (3.38), with

$$f_{\mathbf{x}}(x) = (2\pi\sigma^2)^{-1/2} \exp\left(-\frac{x^2}{2\sigma^2}\right),$$

$$f_{\mathbf{y}}(y) = (2\pi\sigma^2)^{-1/2} \exp\left(-\frac{y^2}{2\sigma^2}\right),$$

whence $\mathbf{x}$ and $\mathbf{y}$ are independent. By Example 3.9, on the other hand, the joint probability density function $f_{\mathbf{xy}}(x, y)$ given in Example 3.2 does not factor as in (3.38), and there $\mathbf{x}$ and $\mathbf{y}$ are not independent random variables.

JOINTLY GAUSSIAN RANDOM VARIABLES

The bivariate normal (Gaussian) probability density function is

$$f_{\mathbf{xy}}(x, y) = (2\pi\sigma_x\sigma_y)^{-1}(1 - r^2)^{-1/2}$$

$$\times \exp\left[- \frac{\dfrac{(x - \bar{x})^2}{\sigma_x^2} - \dfrac{2r(x - \bar{x})(y - \bar{y})}{\sigma_x\sigma_y} + \dfrac{(y - \bar{y})^2}{\sigma_y^2}}{2(1 - r^2)} \right], \quad (3.40)$$

where $\bar{x} = E(\mathbf{x})$, $\bar{y} = E(\mathbf{y})$, $\sigma_x^2 = \text{Var } \mathbf{x}$, and $\sigma_y^2 = \text{Var } \mathbf{y}$. The parameter r is called the *correlation coefficient* of the random variables $\mathbf{x}$ and $\mathbf{y}$ for reasons that will be brought out later; $-1 < r < 1$. This function of x and y is maximum at the point $(\bar{x}, \bar{y})$. Taking $\bar{x} = 0$ and $\bar{y} = 0$ simply translates the whole function so that its summit lies above the origin, without changing its shape, and we now do so in order to avoid needlessly complicating our discussion of this important probability density function. We can further simplify matters by adopting scaled random variables: $\mathbf{x}' = \mathbf{x}/\sigma_x$, $\mathbf{y}' = \mathbf{y}/\sigma_y$, which according to (2.80) have unit variance. Equivalently, we put $\sigma_x = \sigma_y = 1$. Dropping the primes, we write our joint probability density function as

$$f_{\mathbf{xy}}(x, y) = (2\pi)^{-1}(1 - r^2)^{-1/2} \exp\left[- \frac{x^2 - 2rxy + y^2}{2(1 - r^2)} \right]. \quad (3.41)$$

That this joint probability density function—like (3.40)—is properly normalized can be proved by integrating it over the XY-plane. We first integrate on y, writing in the exponent

$$x^2 - 2rxy + y^2 = (y - rx)^2 + (1 - r^2)x^2.$$

The marginal probability density function of $\mathbf{x}$ is then, as in (3.10),

$$\begin{aligned} f_{\mathbf{x}}(x) &= \int_{-\infty}^{\infty} f_{\mathbf{xy}}(x, y)\, dy \\ &= (2\pi)^{-1}(1 - r^2)^{-1/2} \\ &\quad \times \int_{-\infty}^{\infty} \exp\left[- \frac{(y - rx)^2 + (1 - r^2)x^2}{2(1 - r^2)} \right] dy \\ &= (2\pi)^{-1/2} \exp\left(- \frac{x^2}{2} \right) \end{aligned} \quad (3.42)$$

when we change the variable of integration to $y' = y - rx$ and use the normalization integral for the Gaussian probability density function of (2.19). Now integrating over x, we find that

$$\int_{-\infty}^{\infty} f_{\mathbf{x}}(x)\, dx = \int_{-\infty}^{\infty}\int_{-\infty}^{\infty} f_{\mathbf{xy}}(x, y)\, dx\, dy = 1.$$

It is helpful to visualize the joint probability density function $f_{\mathbf{xy}}(x, y)$ as a campaniform heap of sand lying on the XY-plane. The lines of constant probability density $f_{\mathbf{xy}}(x, y)$, that is, the contours of equal height on our sandpile, are ellipses,

$$x^2 - 2rxy + y^2 = C^2, \tag{3.43}$$

as sketched in Fig. 3.12. Differentiating (3.43) with respect to x we find that

$$(y - rx)\frac{dy}{dx} = ry - x.$$

Hence the ellipses have zero slope where intersected by the line $x = ry$ and infinite (vertical) slope where intersected by the line $y = rx$, as shown in the figure.

The conditional probability density function of $\mathbf{y}$, given that the random variable $\mathbf{x}$ has taken on the value x, is by (3.27), after dividing (3.41) by (3.42),

$$f_{\mathbf{y}}(y \mid x) = [2\pi(1 - r^2)]^{-1/2} \exp\left[-\frac{(y - rx)^2}{2(1 - r^2)}\right]. \tag{3.44}$$

Its peak lies on the line $y = rx$. It corresponds to a vertical slice through our sandpile made by a plane perpendicular to the X-axis at the point x, but so magnified that its area equals 1. All such slices have the familiar Gaussian shape.

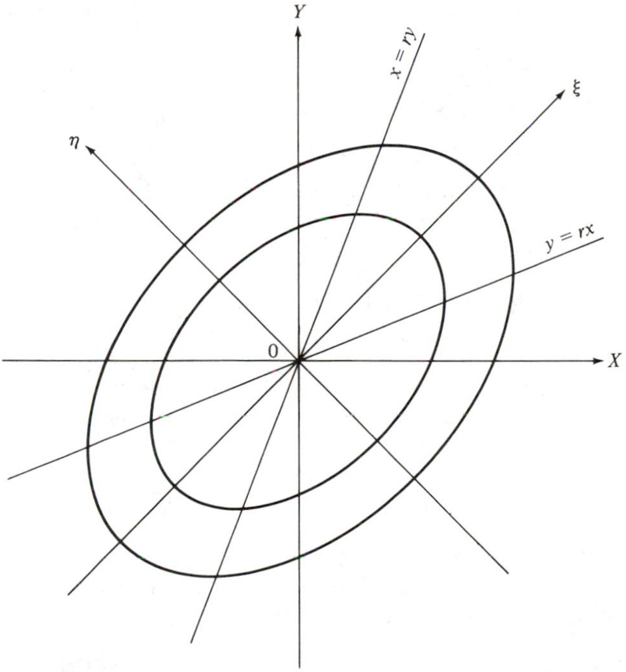

FIGURE 3.12 Contour lines of Gaussian probability density function.

Similarly,

$$f_{\mathbf{x}}(x \mid y) = [2\pi(1 - r^2)]^{-1/2} \exp\left[-\frac{(x - ry)^2}{2(1 - r^2)}\right]. \tag{3.45}$$

and its peak lies on the line $x = ry$.

When $r = 0$, $f_{\mathbf{y}}(y \mid x) = f_{\mathbf{y}}(y)$, and the Gaussian random variables **x** and **y** are statistically independent. The contour lines in Fig. 3.12 are then circles, the variances of **x** and **y** having been taken equal. For positive values of the correlation coefficient r, the ellipses are elongated with their major axes along the line $y = x$ and their minor axes along the line $y = -x$. To show this, we rotate the coordinate axes through 45° by introducing new variables

$$\xi = \frac{y + x}{\sqrt{2}}, \qquad \eta = \frac{y - x}{\sqrt{2}}.$$

Then the equation of the ellipse in (3.43) becomes

$$\frac{\xi^2}{1 + r} + \frac{\eta^2}{1 - r} = \frac{C}{1 - r^2} = b^2,$$

which is the equation of an ellipse with semimajor axis $b(1 + r)^{1/2}$ and semiminor axis $b(1 - r)^{1/2}$. As the correlation coefficient r approaches 1, the ellipses become more and more elongated along the line $x = y$. The lines $x = ry$ and $y = rx$ in Fig. 3.12 approach nearer and nearer to that line. The total mass under our joint Gaussian probability density function must remain equal to 1, and we see that as $r \to 1$,

$$\begin{aligned} f_{\mathbf{xy}}(x, y) &\to (2\pi\sigma^2)^{-1/2} e^{-x^2/2} \delta(y - x) \\ &= (2\pi\sigma^2)^{-1/2} e^{-y^2/2} \delta(y - x) \qquad (r = 1). \end{aligned}$$

At $r = 1$ all the probability lies along the line $x = y$, and the random variables **x** and **y** are equal. (The mathematician appends the qualifying phrase "with probability 1.") Now the conditional probability density functions in (3.44) and (3.45) are delta functions,

$$f_{\mathbf{y}}(y \mid x) = f_{\mathbf{x}}(x \mid y) = \delta(x - y),$$

for their variances $1 - r^2$ have shrunk to zero.

For negative values of the correlation coefficient r, on the other hand, the elliptical contours of constant probability density stretch into the second and fourth quadrants, and as r approaches -1,

$$f_{\mathbf{xy}}(x, y) \to (2\pi\sigma^2)^{-1/2} e^{-x^2/2} \delta(x + y) \qquad (r = -1).$$

When $r = -1$, $\mathbf{x} = -\mathbf{y}$ "with probability 1."

For $0 < r < 1$ the values of **x** and **y** more often have the same sign than not. The probability of the event "**x** and **y** have the same sign" is

$$\Pr\left[\{\mathbf{x}, \mathbf{y} : \mathbf{x} > 0, \mathbf{y} > 0\} + \{\mathbf{x}, \mathbf{y} : \mathbf{x} < 0, \mathbf{y} < 0\}\right]$$

and is calculated by integrating the joint probability density function in (3.41) over the first and third quadrants. Both quadrants contribute equally and

$$P = \Pr(\mathbf{x} \text{ and } \mathbf{y} \text{ have the same sign})$$

$$= \pi^{-1}(1 - r^2)^{-1/2} \int_0^\infty \int_0^\infty \exp\left[-\frac{x^2 - 2rxy + y^2}{2(1 - r^2)}\right] dx\, dy,$$

which is perhaps most easily evaluated by changing to polar coordinates. We put

$$x = p \cos \theta, \qquad y = p \sin \theta, \qquad dx\, dy = p\, dp\, d\theta,$$

and obtain

$$P = \pi^{-1}(1 - r^2)^{-1/2} \int_0^{\pi/2} d\theta \int_0^\infty p\, dp$$

$$\times \exp\left[-\frac{p^2(\cos^2\theta + \sin^2\theta - 2r\cos\theta\sin\theta)}{2(1 - r^2)}\right]$$

$$= \pi^{-1}(1 - r^2)^{1/2} \int_0^{\pi/2} \frac{d\theta}{1 - r\sin 2\theta}$$

$$= \tfrac{1}{2} + \pi^{-1}\sin^{-1} r.$$

This probability is greater than $\frac{1}{2}$ for $0 < r < 1$ and less than $\frac{1}{2}$ for $-1 < r < 0$.

3-4

FUNCTIONS OF TWO RANDOM VARIABLES

If two random variables $\mathbf{x}$ and $\mathbf{y}$ are defined in the same chance experiment, any function of them

$$\mathbf{z} = g(\mathbf{x}, \mathbf{y})$$

defines a new random variable $\mathbf{z}$ in the same experiment. In Example 3.6, for instance, we found that the decision about the presence or absence of a signal could be based on the value of $\mathbf{x}^2 + \mathbf{y}^2 = \mathbf{z}$. The sum or the product of two random variables may be significant in certain situations. The problem arises of finding the probability density functions of such functions $g(\mathbf{x}, \mathbf{y})$ of a pair of random variables. There are several methods, and now one, now another may be simplest to apply.

METHOD 1

Calculate the cumulative distribution function of $\mathbf{z} = g(\mathbf{x}, \mathbf{y})$ by determining the probability

$$F_{\mathbf{z}}(z) = \Pr(\mathbf{z} \le z) = \Pr[g(\mathbf{x}, \mathbf{y}) \le z]. \tag{3.46}$$

The event $g(\mathbf{x}, \mathbf{y}) \le z$ is represented by a certain region in the XY-plane, which is bounded by the curve $g(x, y) = z$. Call that region $R(z)$. In some problems it may consist of a number of disconnected parts. The probability in (3.46) is found by integrating the joint probability density function $f_{\mathbf{xy}}(x, y)$ over the region $R(z)$,

$$F_{\mathbf{z}}(z) = \iint\limits_{R(z)} f_{\mathbf{xy}}(x, y)\, dx\, dy. \tag{3.47}$$

The probability density function $f_{\mathbf{z}}(z)$ of $\mathbf{z} = g(\mathbf{x}, \mathbf{y})$ is obtained by differentiating this cumulative distribution function with respect to z.

EXAMPLE 3.11. $\mathbf{z} = \mathbf{x}^2 + \mathbf{y}^2$

This random variable $\mathbf{z}$ is never negative, and

$$F_{\mathbf{z}}(z) = 0, \qquad z < 0.$$

For $z > 0$, the region $R(z)$ is a circle of radius $z^{1/2}$ about the origin. If the joint probability density function of $\mathbf{x}$ and $\mathbf{y}$ is

$$f_{\mathbf{xy}}(x, y) = (2\pi\sigma^2)^{-1} \exp\left(-\frac{x^2 + y^2}{2\sigma^2}\right),$$

we find by changing to polar coordinates

$$\begin{aligned} F_{\mathbf{z}}(z) &= (2\pi\sigma^2)^{-1} \iint\limits_{R(z)} \exp\left(-\frac{x^2 + y^2}{2\sigma^2}\right) dx\, dy \\ &= (2\pi\sigma^2)^{-1} \int_0^{z^{1/2}} r\, dr \int_0^{2\pi} d\theta \exp\left(-\frac{r^2}{2\sigma^2}\right) \\ &= (1 - e^{-z/2\sigma^2})U(z), \end{aligned}$$

and the probability density function of $\mathbf{z} = \mathbf{x}^2 + \mathbf{y}^2$ is

$$f_{\mathbf{z}}(z) = \frac{dF_{\mathbf{z}}(z)}{dz} = (2\sigma^2)^{-1} e^{-z/2\sigma^2} U(z).$$

■

EXAMPLE 3.12. $\mathbf{z} = \max(\mathbf{x}, \mathbf{y})$

This random variable $\mathbf{z}$ is the larger of $\mathbf{x}$ and $\mathbf{y}$. Now $\mathbf{z} \le z$ if and only if both $\mathbf{x} \le z$ and $\mathbf{y} \le z$. The region $R(z)$ is sketched in Fig. 3.13. Hence in terms of the joint cumulative distribution function of $\mathbf{x}$ and $\mathbf{y}$,

$$\begin{aligned} F_{\mathbf{z}}(z) &= \Pr(\mathbf{z} \le z) \\ &= \Pr(\mathbf{x} \le z, \mathbf{y} \le z) = F_{\mathbf{xy}}(z, z) \\ &= \int_{-\infty}^{z} \int_{-\infty}^{z} f_{\mathbf{xy}}(x, y)\, dx\, dy, \end{aligned}$$

and the probability density function of $\mathbf{z}$ is, by differentiation,

$$f_{\mathbf{z}}(z) = \int_{-\infty}^{z} f_{\mathbf{xy}}(z, y)\, dy + \int_{-\infty}^{z} f_{\mathbf{xy}}(x, z)\, dx. \tag{3.48}$$

■

EXAMPLE 3.13. $\mathbf{z} = a\mathbf{x} + b\mathbf{y}$

When $a > 0$, $b > 0$, the region $R(z)$ lies to the lower left of the line

$$z = ax + by,$$

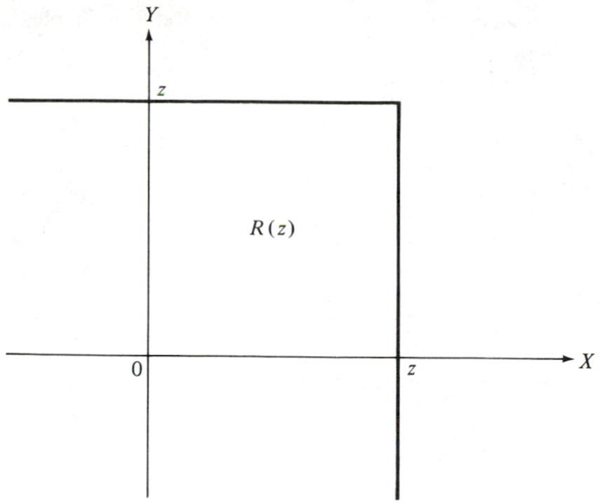

FIGURE 3.13 Region max(**x**, **y**) $\leq z$.

as shown in Fig. 3.14. The cumulative distribution function of $\mathbf{z} = a\mathbf{x} + b\mathbf{y}$ is

$$F_{\mathbf{z}}(z) = \int_{-\infty}^{\infty} dy \int_{-\infty}^{(z-by)/a} dx\, f_{\mathbf{xy}}(x, y). \tag{3.49}$$

Differentiating with respect to z, we find the probability density function

$$f_{\mathbf{z}}(z) = a^{-1} \int_{-\infty}^{\infty} dy\, f_{\mathbf{xy}}\left(\frac{z - by}{a}, y\right). \tag{3.50}$$

If **x** and **y** are independent random variables, and $a = b = 1$, the probability density function of $\mathbf{z} = \mathbf{x} + \mathbf{y}$ is

$$f_{\mathbf{z}}(z) = \int_{-\infty}^{\infty} f_{\mathbf{x}}(z - y) f_{\mathbf{y}}(y)\, dy; \tag{3.51}$$

that is, it is the convolution of the probability density functions of **x** and **y**. This important result and its generalization for the sum of an arbitrary number of independent random variables are useful in many problems. ■

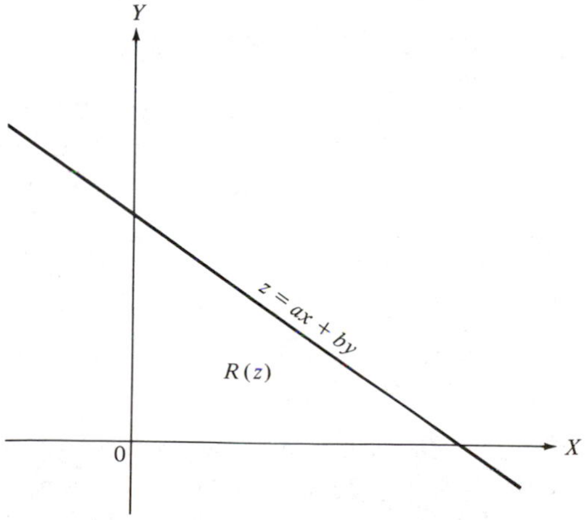

FIGURE 3.14 Region $a\mathbf{x} + b\mathbf{y} < z$.

EXAMPLE 3.14. Let **x** and **y** be independent discrete random variables with probability distributions

$$\Pr(\mathbf{x} = k) = (1 - v_1)v_1^k, \qquad k = 0, 1, 2, \ldots,$$

$$\Pr(\mathbf{y} = m) = (1 - v_2)v_2^m, \qquad m = 0, 1, 2, \ldots,$$

as in Example 3.4. The cumulative distribution of $\mathbf{z} = \mathbf{x} + \mathbf{y}$ is found as in that example,

$$\begin{aligned} F_{\mathbf{z}}(z) &= \Pr(\mathbf{x} + \mathbf{y} \le [z]) \\ &= 1 - \left(\frac{1 - v_1}{v_2 - v_1}\right)v_2^{[z]+2} + \left(\frac{1 - v_2}{v_2 - v_1}\right)v_1^{[z]+2}, \qquad z \ge 0, \end{aligned}$$

where $[z]$ equals the greatest integer in z. Then for all nonnegative integers p, $v_1 \ne v_2$,

$$\begin{aligned} \Pr(\mathbf{z} = p) &= F_{\mathbf{z}}(p) - F_{\mathbf{z}}(p - 1) \\ &= \frac{(1 - v_1)(1 - v_2)}{v_2 - v_1}(v_2^{p+1} - v_1^{p+1}) \\ &= \sum_{m=0}^{p} \Pr(\mathbf{x} = p - m) \Pr(\mathbf{y} = m), \end{aligned}$$

in analogy with the convolution in (3.51). This convolution could be evaluated by means of z-transforms, or equivalently by means of the probability generating functions introduced in Sec. 2-4G. ■

METHOD 2

Fix one of the two random variables, say **y**, at a particular one of its possible values y. Then $\mathbf{z} = g(\mathbf{x}, y)$ is random only through its dependence on the single random variable **x**. We have a transformation from the random variable **x** to the random variable **z**, in which y is now only a parameter that is held fixed. We can therefore apply the technique of Sec. 2-3 to determining the conditional probability density function $f_{\mathbf{z}}(z \mid y)$ of **z**, given $\mathbf{y} = y$. The probability density function of **z** is then found from

$$f_{\mathbf{z}}(z) = \int_{-\infty}^{\infty} f_{\mathbf{z}}(z \mid y) f_{\mathbf{y}}(y)\, dy, \tag{3.52}$$

where $f_{\mathbf{y}}(y)$ is the marginal probability density function of **y**.

If $z = g(x, y)$ is a monotone function of x, for fixed y, let its inverse be $x = g^{-1}(z, y)$. Then by (2.66), the conditional probability density function of **z** is

$$f_{\mathbf{z}}(z \mid y) = \frac{f_{\mathbf{x}}(x \mid y)}{\left|\dfrac{\partial g(x, y)}{\partial x}\right|}\Bigg|_{x = g^{-1}(z,y)} \tag{3.53}$$

Observe that unless **x** and **y** are independent random variables, it is necessary to use the conditional probability density function $f_{\mathbf{x}}(x \mid y) = f_{\mathbf{x}}(x \mid \mathbf{y} = y)$, defined as in (3.28). If for fixed y the equation $z = g(x, y)$ has several roots when solved for x as a function of z, we must sum (3.53) over all such roots, as in (2.69).

EXAMPLE 3.15. $\mathbf{z} = \mathbf{x}/\mathbf{y}$

Here it is simplest to fix $\mathbf{y}$ at some value y. Then $\mathbf{z} = \mathbf{x}/y$ is a simple scaling of the random variable $\mathbf{x}$ by the factor $1/y$, $g^{-1}(z, y) = yz$, and

$$\frac{\partial g(x, y)}{\partial x} = \frac{1}{y}.$$

Hence by (3.53),

$$f_{\mathbf{z}}(z \mid y) = |y|\, f_{\mathbf{x}}(yz \mid y).$$

Using (3.52), we find that

$$\begin{aligned} f_{\mathbf{z}}(z) &= \int_{-\infty}^{\infty} |y|\, f_{\mathbf{x}}(yz \mid y) f_{\mathbf{y}}(y)\, dy \\ &= \int_{-\infty}^{\infty} |y|\, f_{\mathbf{xy}}(yz, y)\, dy. \end{aligned} \tag{3.54}$$

If, for instance, $\mathbf{x}$ and $\mathbf{y}$ are independent with exponential distributions,

$$f_{\mathbf{x}}(x) = ae^{-ax}U(x), \qquad a > 0,$$

$$f_{\mathbf{y}}(y) = be^{-by}U(y), \qquad b > 0,$$

we find that

$$f_{\mathbf{z}}(z) = ab \int_0^{\infty} ye^{-ayz-by}U(yz)U(y)\, dy = \frac{ab}{(az + b)^2}\, U(z).$$

■

EXAMPLE 3.16. $\mathbf{z} = \mathbf{xy}$

Again fixing $\mathbf{y}$ at y, we find that $\mathbf{z} = y\mathbf{x}$ is a scaled version of the random variable $\mathbf{x}$, and by (3.53),

$$f_{\mathbf{z}}(z \mid y) = |y|^{-1} f_{\mathbf{x}}\left(\frac{z}{y} \mid y\right),$$

whence the probability density function of $\mathbf{z} = \mathbf{xy}$ is

$$\begin{aligned} f_{\mathbf{z}}(z) &= \int_{-\infty}^{\infty} \frac{dy}{|y|}\, f_{\mathbf{x}}\left(\frac{z}{y} \mid y\right) f_{\mathbf{y}}(y) \\ &= \int_{-\infty}^{\infty} \frac{dy}{|y|}\, f_{\mathbf{xy}}\left(\frac{z}{y}, y\right). \end{aligned} \tag{3.55}$$

Suppose, for instance, that the random variables $\mathbf{x}$ and $\mathbf{y}$ are jointly Gaussian, with means zero, variances 1, and correlation coefficient r, their joint probability density function given by (3.41). Then the probability density function of their product $\mathbf{z}$ is

$$\begin{aligned} f_{\mathbf{z}}(z) &= (2\pi)^{-1}(1 - r^2)^{-1/2} \int_{-\infty}^{\infty} \frac{dy}{|y|} \exp\left[-\frac{(z/y)^2 - 2r(z/y)y + y^2}{2(1 - r^2)}\right] \\ &= \pi^{-1}(1 - r^2)^{-1/2} \exp\left(\frac{rz}{1 - r^2}\right) \int_0^{\infty} \frac{dy}{y} \exp\left[-\frac{y^2 + z^2/y^2}{2(1 - r^2)}\right] \\ &= (2\pi)^{-1}(1 - r^2)^{-1/2} \exp\left(\frac{rz}{1 - r^2}\right) \int_0^{\infty} \frac{du}{u} \exp\left[-\frac{u + z^2/u}{2(1 - r^2)}\right], \end{aligned}$$

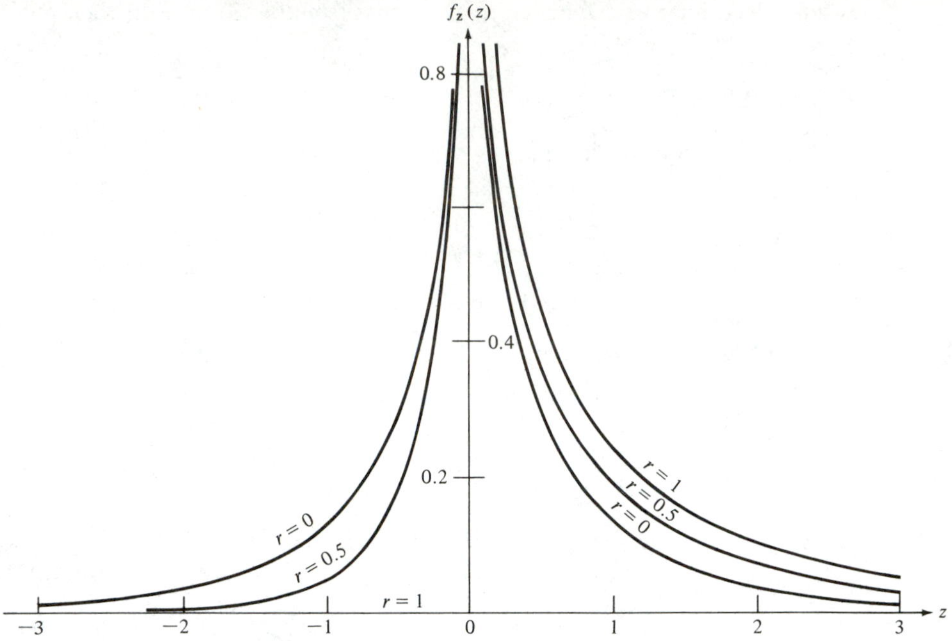

FIGURE 3.15 Probability density function: product of two correlated Gaussian random variables. Curves are indexed with the correlation coefficient r.

where we put $u = y^2$. This last integral is recognized as a Laplace transform, and we can look it up in a table,* obtaining

$$f_{\mathbf{z}}(z) = \pi^{-1}(1 - r^2)^{-1/2} \exp\left(\frac{rz}{1 - r^2}\right) K_0\left(\frac{|z|}{1 - r^2}\right), \qquad -1 < r < 1, \tag{3.56}$$

in terms of the modified Bessel function

$$K_0(z) = \sum_{k=0}^{\infty} \frac{s_k(z^2/4)^k}{(k!)^2},$$

$$s_0 = -\gamma - \ln\left(\frac{z}{2}\right), \qquad s_k = s_{k-1} + \frac{1}{k}, \qquad k > 0,$$

$$\gamma = \text{Euler's constant} = 0.5772156649.$$

In Fig. 3.15 we plot this probability density function for $r = 0, \frac{1}{2}$, and 1. For $r = 1$, $\mathbf{x} = \mathbf{y}$ and $\mathbf{z} = \mathbf{x}^2$, for which an easy calculation by the method of Sec. 2-3 shows that

$$f_{\mathbf{z}}(z) = (2\pi z)^{-1/2} e^{-z/2} U(z). \tag{3.57}$$

If $0 < r < 1$, the product $\mathbf{z} = \mathbf{xy}$ is more likely to be positive than negative, for as we saw at the end of Sec. 3-3, $\mathbf{x}$ and $\mathbf{y}$ are more likely to have the same sign than opposite signs. The probability density function of $\mathbf{z} = \mathbf{xy}$ therefore grows for $z > 0$ and diminishes for $z < 0$ as the correlation coefficient r increases from 0 to 1. ■

*A. Erdélyi, Ed., *Tables of Integral Transforms,* Vol. 1. New York: McGraw-Hill, 1954, p. 146, eq. (2.9).

METHOD 3

The third method for finding the probability density function of $\mathbf{z} = g(\mathbf{x}, \mathbf{y})$ uses the definition

$$\begin{aligned} \Pr(z < \mathbf{z} < z + dz) &= f_{\mathbf{z}}(z)\, dz \\ &= \Pr(z < g(\mathbf{x}, \mathbf{y}) < z + dz). \end{aligned} \tag{3.58}$$

The region $z < g(\mathbf{x}, \mathbf{y}) < z + dz$ is a curved strip in the XY-plane, possibly in several pieces. One calculates the probability measure of this strip by integrating the joint probability density function $f_{\mathbf{xy}}(x, y)$ over it.

EXAMPLE 3.17. The random variables $\mathbf{x}$ and $\mathbf{y}$ are independent and Gaussian with means m_x and m_y and equal variances σ^2. We seek the probability density function of

$$\mathbf{z} = (\mathbf{x}^2 + \mathbf{y}^2)^{1/2}$$

The probability in (3.58) is the measure of the annulus A of radius z and width dz, shown in Fig. 3.16. The probability sought is then

$$f_{\mathbf{z}}(z)\, dz = \iint_A (2\pi\sigma^2)^{-1} \exp\left[-\frac{(x - m_x)^2 + (y - m_y)^2}{2\sigma^2}\right] dx\, dy. \tag{3.59}$$

We evaluate the integral by converting the integrand to polar coordinates,

$$x = r\cos\theta, \qquad y = r\sin\theta, \qquad dx\, dy = r\, dr\, d\theta,$$

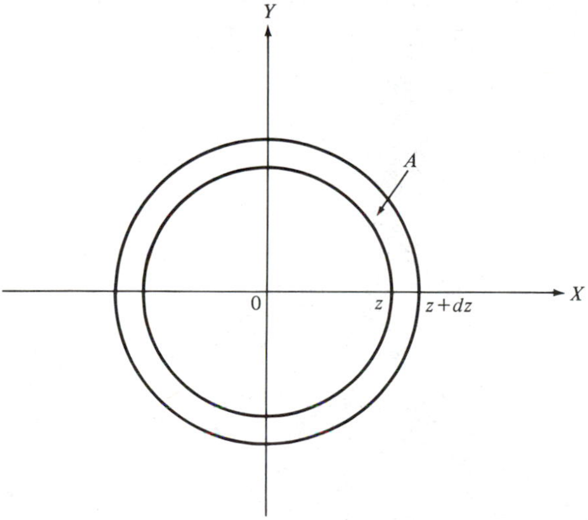

FIGURE 3.16 Annulus of constant z, Example 3.17.

whence

$$
\begin{aligned}
f_{\mathbf{z}}(z)\,dz &= (2\pi\sigma^2)^{-1}\int_z^{z+dz} r\,dr \int_0^{2\pi} d\theta \\
&\quad \times \exp\left[-\frac{(r\cos\theta - m_x)^2 + (r\sin\theta - m_y)^2}{2\sigma^2}\right] \\
&= (2\pi\sigma^2)^{-1}\int_z^{z+dz} r\,dr \int_0^{2\pi} d\theta \\
&\quad \times \exp\left[-\frac{r^2 - 2r(m_x\cos\theta + m_y\sin\theta) + m_x^2 + m_y^2}{2\sigma^2}\right].
\end{aligned}
\tag{3.60}
$$

Let us put

$$m_x = m\cos\alpha, \qquad m_y = m\sin\alpha, \qquad m_x\cos\theta + m_y\sin\theta = m\cos(\theta - \alpha).$$

Since dz is infinitesimal, the integral over r yields dz times the integrand with r replaced by z. Hence the probability density function of z is

$$f_{\mathbf{z}}(z) = (2\pi\sigma^2)^{-1} z \exp\left(-\frac{z^2 + m^2}{2\sigma^2}\right)\int_0^{2\pi} \exp\left[\frac{mz\cos(\theta - \alpha)}{\sigma^2}\right] d\theta.$$

As the integrand is periodic in θ, the result will be independent of α, and we find that

$$f_{\mathbf{z}}(z) = \frac{z}{\sigma^2}\exp\left(-\frac{z^2 + m^2}{2\sigma^2}\right) I_0\left(\frac{mz}{\sigma^2}\right), \tag{3.61}$$

in terms of the modified Bessel function

$$
\begin{aligned}
I_0(x) &= \frac{1}{2\pi}\int_0^{2\pi} e^{x\cos\theta}\,d\theta \\
&= \sum_{k=0}^{\infty}\frac{(\frac{1}{4}x^2)^k}{(k!)^2}.
\end{aligned}
\tag{3.62}
$$

The probability density function in (3.61) is called the *Rayleigh–Rice distribution*. It is the probability density function of the output of a linear rectifier whose input is both a sinusoidal signal and Gaussian noise that has been filtered by a narrowband pass filter

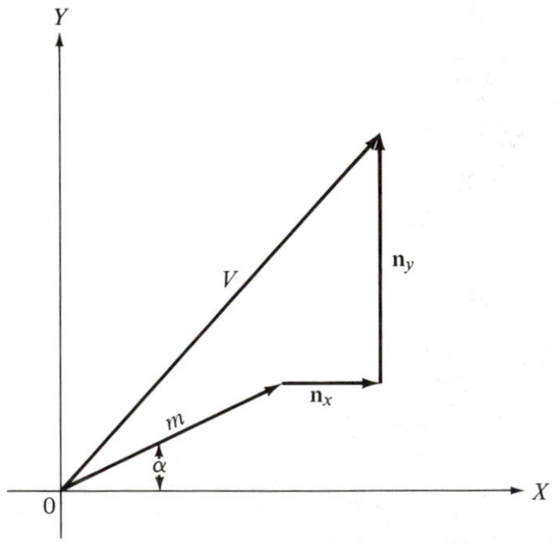

FIGURE 3.17 Phasor of signal plus noise.

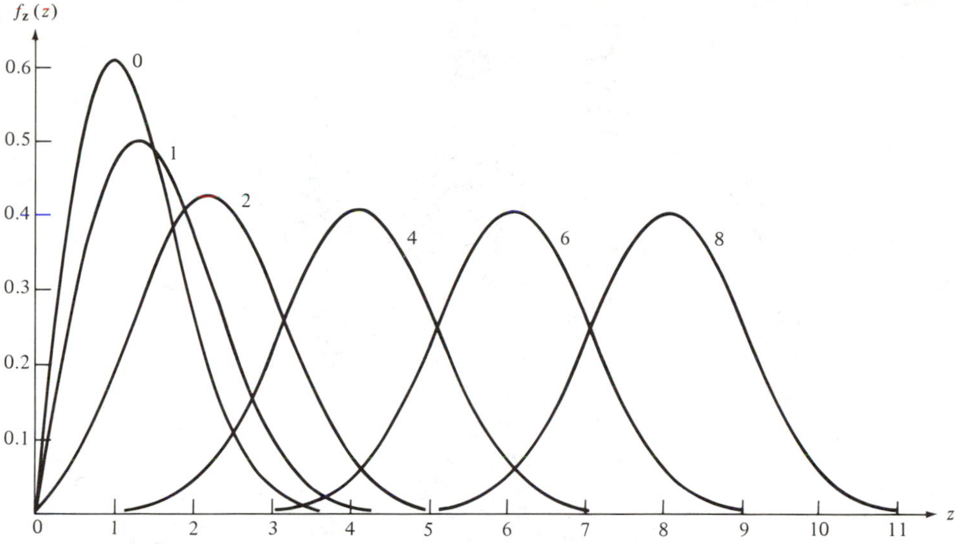

FIGURE 3.18 Rayleigh–Rice probability density function (3.61). Curves are indexed with the value of m; σ = 1.

passing frequencies in the neighborhood of the frequency $\Omega/2\pi$ of the signal. The input to the rectifier is then

$$v_{\text{in}}(t) = \text{Re}\,(Ve^{j\Omega t})$$
$$= \text{Re}\,[me^{j\Omega t + j\alpha} + (\mathbf{n}_x + j\mathbf{n}_y)e^{j\Omega t}]$$

where $\mathbf{n}_x$ and $\mathbf{n}_y$ are independent Gaussian random variables of mean zero and equal variances σ^2. The output of the linear rectifier equals the length $|V|$ of the phasor V,

$$\mathbf{z} = |me^{j\alpha} + \mathbf{n}_x + j\mathbf{n}_y|,$$

as shown in Fig. 3.17. When $m = 0$, $f_{\mathbf{z}}(z)$ reduces to the Rayleigh distribution as in (2.24). The density function $f_{\mathbf{z}}(z)$ is plotted in Fig. 3.18 for various values of m; $\sigma = 1$. ■

3-5

TRANSFORMATION OF A PAIR OF RANDOM VARIABLES

A. Continuous Random Variables: Joint Density Function

In Example 3.17 we might have wanted not the probability density function of $\mathbf{z} = (\mathbf{x}^2 + \mathbf{y}^2)^{1/2}$ alone, but the joint probability density function of both the amplitude $\mathbf{z}$ and the phase $\boldsymbol{\theta} = \arg(\mathbf{x} + j\mathbf{y})$. The joint probability that $\mathbf{z}$ lies between z and $z + dz$ and $\boldsymbol{\theta}$ lies between θ and $\theta + d\theta$,

$$f_{\mathbf{z}\boldsymbol{\theta}}(z, \theta)\, dz\, d\theta = \Pr(z < \mathbf{z} < z + dz, \theta < \boldsymbol{\theta} < \theta + d\theta),$$

equals the joint probability density function $f_{\mathbf{xy}}(x, y)$ times the area $dA = z\, dz\, d\theta$ of the infinitesimal element of the XY-plane where the annulus in Fig. 3.19

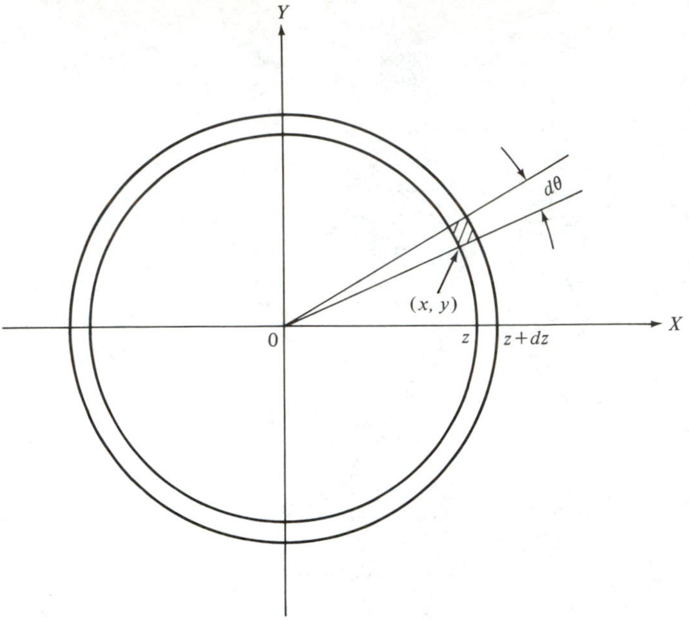

FIGURE 3.19 Region for definition of density function of amplitude and phase.

is cut by the two rays making angles θ and θ + $d\theta$ with the x-axis. From (3.59) this probability is

$$f_{\mathbf{z\theta}}(z, \theta)\,dz\,d\theta = f_{\mathbf{xy}}(x,y)\,dA$$

$$= z\,dz\,d\theta\,(2\pi\sigma^2)^{-1}\exp\left[-\frac{(x-m_x)^2+(y-m_y)^2}{2\sigma^2}\right],$$

where (x, y) are the coordinates of the lower corner of the infinitesimal element shown in the figure. Substituting

$$x = z\cos\theta, \qquad y = z\sin\theta$$

as before, we find that the joint probability density function of **z** and **θ** is

$$f_{\mathbf{z\theta}}(z, \theta) = (2\pi\sigma^2)^{-1}z\exp\left[-\frac{z^2-2mz\cos(\theta-\alpha)+m^2}{2\sigma^2}\right]U(z), \qquad (3.63)$$

with $m\cos\alpha = m_x$, $m\sin\alpha = m_y$.

This example instantiates the general problem of calculating the joint probability density function of random variables $\mathbf{z}_1$ and $\mathbf{z}_2$ defined by the pair of functions

$$\mathbf{z}_1 = g_1(\mathbf{x}_1, \mathbf{x}_2), \qquad \mathbf{z}_2 = g_2(\mathbf{x}_1, \mathbf{x}_2) \qquad (3.64)$$

where g_1 and g_2 are single-valued functions of the random variables $\mathbf{x}_1$ and $\mathbf{x}_2$. These equations (3.64) define a transformation from the X_1X_2-plane to the

Z_1Z_2-plane, whereby each point $(\mathbf{x}_1, \mathbf{x}_2)$ in the former goes into a unique point $(\mathbf{z}_1, \mathbf{z}_2)$ of the latter.

When the transformation in eq. (3.64) is one-to-one, those equations can be solved for $\mathbf{x}_1$ and $\mathbf{x}_2$ to yield

$$\mathbf{x}_1 = h_1(\mathbf{z}_1, \mathbf{z}_2), \qquad \mathbf{x}_2 = h_2(\mathbf{z}_1, \mathbf{z}_2), \tag{3.65}$$

and each point $(\mathbf{z}_1, \mathbf{z}_2)$ in the Z_1Z_2-plane arises from a unique point $(\mathbf{x}_1, \mathbf{x}_2)$ in the X_1X_2-plane. All points in the infinitesimal rectangle A_z shown in Fig. 3.20(b) arise from points in the infinitesimal parallelogram A_x shown in Fig. 3.20(a), which lies between the two curves defined by

$$z_1 = g_1(x_1, x_2) \qquad \text{and} \qquad z_1 + dz_1 = g_1(x_1, x_2)$$

and between the two curves defined by

$$z_2 = g_2(x_1, x_2) \qquad \text{and} \qquad z_2 + dz_2 = g_2(x_1, x_2).$$

Each outcome of the chance experiment specified by a point in the infinitesimal parallelogram A_x in Fig. 3.20(a) is, by the transformation in (3.64), relabeled by the coordinates of a point in the infinitesimal rectangle A_z in Fig. 3.20(b). The probability Pr $[(\mathbf{z}_1, \mathbf{z}_2) \in A_z]$ of the event "the relabeled outcome $(\mathbf{z}_1, \mathbf{z}_2)$ lies in A_z," therefore equals the probability Pr $[(\mathbf{x}_1, \mathbf{x}_2) \in A_x]$ of the event "the outcome $(\mathbf{x}_1, \mathbf{x}_2)$ lies in A_x." In terms of the joint probability density functions of the random variables, these probabilities are

$$f_{\mathbf{x}_1\mathbf{x}_2}(x_1, x_2) \cdot \text{Area } (A_x) = f_{\mathbf{z}_1\mathbf{z}_2}(z_1, z_2) \cdot \text{Area } (A_z).$$

The problem of calculating the joint probability density function $f_{\mathbf{z}_1\mathbf{z}_2}(z_1, z_2)$ thus reduces to finding the ratio of the areas of A_x and A_z. It is shown in texts on the

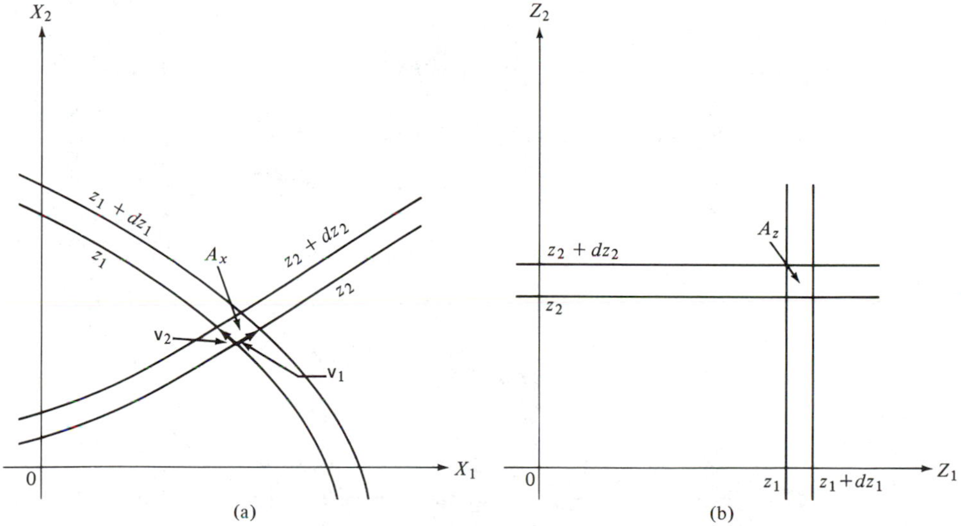

FIGURE 3.20 Transformation of a pair of random variables.

calculus that this ratio is

$$\frac{\text{Area }(A_x)}{\text{Area }(A_z)} = \left| J\begin{pmatrix} x_1 & x_2 \\ z_1 & z_2 \end{pmatrix} \right| = \left| J\begin{pmatrix} z_1 & z_2 \\ x_1 & x_2 \end{pmatrix} \right|^{-1}, \tag{3.66}$$

where the "Jacobian" J is defined by the determinant

$$J\begin{pmatrix} x_1 & x_2 \\ z_1 & z_2 \end{pmatrix} = \begin{vmatrix} \dfrac{\partial x_1}{\partial z_1} & \dfrac{\partial x_1}{\partial z_2} \\ \\ \dfrac{\partial x_2}{\partial z_1} & \dfrac{\partial x_2}{\partial z_2} \end{vmatrix} \tag{3.67}$$

and

$$J\begin{pmatrix} z_1 & z_2 \\ x_1 & x_2 \end{pmatrix} = \begin{vmatrix} \dfrac{\partial z_1}{\partial x_1} & \dfrac{\partial z_1}{\partial x_2} \\ \\ \dfrac{\partial z_2}{\partial x_1} & \dfrac{\partial z_2}{\partial x_2} \end{vmatrix} \tag{3.68}$$

By $\partial x_1/\partial z_1$ we mean $\partial h_1(z_1, z_2)/\partial z_1$, and by $\partial z_1/\partial x_1$ we mean $\partial g_1(x_1, x_2)/\partial x_1$, and so on. Hence we find for the joint probability density function of the two new random variables

$$\begin{aligned} f_{\mathbf{z}_1\mathbf{z}_2}(z_1,z_2) &= \frac{f_{\mathbf{x}_1\mathbf{x}_2}(x_1, x_2)}{\left| J\begin{pmatrix} z_1 & z_2 \\ x_1 & x_2 \end{pmatrix} \right|} \\ &= \left| J\begin{pmatrix} x_1 & x_2 \\ z_1 & z_2 \end{pmatrix} \right| f_{\mathbf{x}_1\mathbf{x}_2}(x_1,x_2). \end{aligned} \tag{3.69}$$

Into the right-hand side we must substitute x_1 and x_2 in terms of z_1 and z_2 as determined by (3.65).

In the example at the beginning of this section, with

$$x = z \cos \theta, \qquad y = z \sin \theta,$$

the Jacobian is

$$J\begin{pmatrix} x & y \\ z & \theta \end{pmatrix} = \begin{vmatrix} \dfrac{\partial x}{\partial z} & \dfrac{\partial y}{\partial z} \\ \\ \dfrac{\partial x}{\partial \theta} & \dfrac{\partial y}{\partial \theta} \end{vmatrix} = \begin{vmatrix} \cos\theta & \sin\theta \\ -z\sin\theta & z\cos\theta \end{vmatrix} = z$$

and (3.63) then easily follows from the second form in (3.69).

For the sake of completeness we shall derive the ratio of infinitesimal areas in (3.66). The area of a parallelogram defined as in Fig. 3.21 by two vectors

$$\mathsf{v}_1 = (v_{11}, v_{12}), \qquad \mathsf{v}_2 = (v_{21}, v_{22})$$

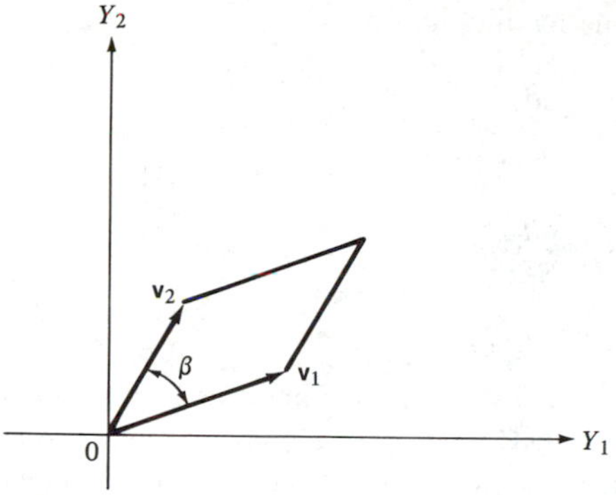

FIGURE 3.21 Parallelogram based on vectors v_1, v_2.

from the origin of a plane with rectangular coordinates y_1, y_2 equals

$$
\begin{aligned}
|v_1| \cdot |v_2| \sin \beta &= [v_1^2 v_2^2 - v_1^2 v_2^2 \cos^2 \beta]^{1/2} \\
&= [v_1^2 v_2^2 - (v_1 \cdot v_2)^2]^{1/2} \\
&= [(v_{11}^2 + v_{12}^2)(v_{21}^2 + v_{22}^2) - (v_{11}v_{21} + v_{12}v_{22})^2]^{1/2} \qquad (3.70) \\
&= |v_{11}v_{22} - v_{12}v_{21}| = |\det v| = \left| \begin{vmatrix} v_{11} & v_{12} \\ v_{21} & v_{22} \end{vmatrix} \right|,
\end{aligned}
$$

where by $|v_k| = (v_k^2)^{1/2}$ we mean the length of the vector v_k, and det v stands for the determinant of the components of v_1 and v_2.

In Fig. 3.20(a) the parallelogram A_x is spanned by the vectors v_1 and v_2. The vector v_1 goes from the point

$$(x_1, x_2) = (h_1(z_1, z_2), h_2(z_1, z_2))$$

to the point

$$(h_1(z_1 + dz_1, z_2), h_2(z_1 + dz_1, z_2)) = \left(x_1 + \frac{\partial h_1}{\partial z_1} dz_1, x_2 + \frac{\partial h_2}{\partial z_1} dz_1\right)$$

and hence has the components

$$v_1 = \left(\frac{\partial h_1}{\partial z_1} dz_1, \frac{\partial h_2}{\partial z_1} dz_1\right).$$

Similarly, the vector v_2 goes from (x_1, x_2) to the point

$$(h_1(z_1, z_2 + dz_2), h_2(z_1, z_2 + dz_2))$$

and has components

$$v_2 = \left(\frac{\partial h_1}{\partial z_2} dz_2, \frac{\partial h_2}{\partial z_2} dz_2\right).$$

Hence by (3.70) the area of the parallelogram A_x is

$$\text{Area } (A_x) = \left| \begin{vmatrix} \dfrac{\partial h_1}{\partial z_1} dz_1 & \dfrac{\partial h_2}{\partial z_1} dz_1 \\ \\ \dfrac{\partial h_1}{\partial z_2} dz_2 & \dfrac{\partial h_2}{\partial z_2} dz_2 \end{vmatrix} \right| = \left| J \begin{pmatrix} x_1 & x_2 \\ z_1 & z_2 \end{pmatrix} \right| |dz_1 \, dz_2|,$$

whence (3.66) because Area $(A_z) = |dz_1 \, dz_2|$.

If more than one point $(\mathbf{x}_1, \mathbf{x}_2)$ in the X_1X_2-plane goes into the same point $(\mathbf{z}_1, \mathbf{z}_2)$ in the Z_1Z_2-plane, the transformation in (3.64) is "many–one." Equations (3.64) then have more than one solution (3.65). The probability

$$f_{\mathbf{z}_1\mathbf{z}_2}(z_1, z_2) \cdot \text{Area } (A_z)$$

of the event "the relabeled outcome $(\mathbf{z}_1, \mathbf{z}_2)$ lies in the rectangle A_z" is now the sum of the probability measures of the infinitesimal parallelograms A_x attached to each point $(\mathbf{x}_1^{(k)}, \mathbf{x}_2^{(k)})$ that is transformed into $(\mathbf{z}_1, \mathbf{z}_2)$, and we can write

$$\begin{aligned} f_{\mathbf{z}_1\mathbf{z}_2}(z_1, z_2) &= \sum_k \frac{f_{\mathbf{x}_1\mathbf{x}_2}(x_1^{(k)}, x_2^{(k)})}{\left| J \begin{pmatrix} z_1 & z_2 \\ x_1^{(k)} & x_2^{(k)} \end{pmatrix} \right|} \\ &= \sum_k \left| J \begin{pmatrix} x_1^{(k)} & x_2^{(k)} \\ z_1 & z_2 \end{pmatrix} \right| f_{\mathbf{x}_1\mathbf{x}_2}(x_1^{(k)}, x_2^{(k)}). \end{aligned} \tag{3.71}$$

The right-hand side must be converted to a function of z_1 and z_2 by means of (3.65).

EXAMPLE 3.18. Linear Transformation

Let

$$\begin{aligned} \mathbf{z}_1 &= a_{11}\mathbf{x}_1 + a_{12}\mathbf{x}_2 \\ \mathbf{z}_2 &= a_{21}\mathbf{x}_1 + a_{22}\mathbf{x}_2 \end{aligned} \tag{3.72}$$

define our new random variables $\mathbf{z}_1$, $\mathbf{z}_2$ in terms of the original ones, $\mathbf{x}_1$, $\mathbf{x}_2$. Then to (3.65) correspond the linear equations

$$\begin{aligned} \mathbf{x}_1 &= b_{11}\mathbf{z}_1 + b_{12}\mathbf{z}_2 \\ \mathbf{x}_2 &= b_{21}\mathbf{z}_1 + b_{22}\mathbf{z}_2 \end{aligned} \tag{3.73}$$

found by solving (3.72) for $\mathbf{x}_1$ and $\mathbf{x}_2$. The matrix

$$\mathsf{b} = \begin{bmatrix} b_{11} & b_{12} \\ b_{21} & b_{22} \end{bmatrix} = \mathsf{a}^{-1}$$

is the inverse of the matrix

$$\mathsf{a} = \begin{bmatrix} a_{11} & a_{12} \\ a_{21} & a_{22} \end{bmatrix},$$

and the Jacobians in (3.67) and (3.68) are

$$J\begin{pmatrix} z_1 & z_2 \\ x_1 & x_2 \end{pmatrix} = \begin{vmatrix} a_{11} & a_{12} \\ a_{21} & a_{22} \end{vmatrix} = \det \mathsf{a},$$

$$J\begin{pmatrix} x_1 & x_2 \\ z_1 & z_2 \end{pmatrix} = \begin{vmatrix} b_{11} & b_{12} \\ b_{21} & b_{22} \end{vmatrix} = \det \mathsf{b} = (\det \mathsf{a})^{-1}.$$

Hence the joint probability density function of z_1 and z_2 is

$$f_{\mathbf{z}_1\mathbf{z}_2}(z_1, z_2) = |\det \mathsf{b}|\, f_{\mathbf{x}_1\mathbf{x}_2}(b_{11}z_1 + b_{12}z_2, b_{21}z_1 + b_{22}z_2). \tag{3.74}$$

If in particular the random variables $\mathbf{x}_1$ and $\mathbf{x}_2$ have a bivariate Gaussian density function as in Sec. 3-3—we put x_1 in place of x and x_2 in place of y in (3.40)—the new random variables $\mathbf{z}_1$ and $\mathbf{z}_2$ will also acquire a bivariate Gaussian probability density function, for upon making the substitution required by (3.74) we should find after some algebra that the probability density function $f_{\mathbf{z}_1\mathbf{z}_2}(z_1, z_2)$ is also proportional to the exponential function of a quadratic form in the new variables $z_1 - \bar{z}_1$ and $z_2 - \bar{z}_2$, where (z_1, z_2) are linked to (x_1, x_2) and $(\bar{z}_1, \bar{z}_2)$ to $(\bar{x}_1, \bar{x}_2)$ by linear equations like (3.72); $\bar{x}_k = E(\mathbf{x}_k)$, $k = 1, 2$. This result exemplifies the important general principle that linear combinations of jointly Gaussian random variables also have jointly Gaussian probability density functions. ■

EXAMPLE 3.19. Define new random variables $\mathbf{z}$ and $\mathbf{w}$ by the transformations

$$\mathbf{z} = \mathbf{xy}, \qquad \mathbf{w} = \frac{\mathbf{y}}{\mathbf{x}}.$$

Solved for $\mathbf{x}$ and $\mathbf{y}$, these yield

$$\mathbf{x} = \pm\sqrt{\frac{\mathbf{z}}{\mathbf{w}}}, \qquad \mathbf{y} = \pm\sqrt{\mathbf{zw}}.$$

Thus both the point $(\sqrt{z/w}, \sqrt{zw})$ and the point $(-\sqrt{z/w}, -\sqrt{zw})$ in the XY-plane are transformed into the same point (z, w) of the ZW-plane. Furthermore, $\mathbf{z}$ and $\mathbf{w}$ are either both positive, $\mathbf{x}$ and $\mathbf{y}$ having the same sign, or both negative, $\mathbf{x}$ and $\mathbf{y}$ having opposite signs. Hence

$$f_{\mathbf{zw}}(z, w) = 0 \begin{cases} z > 0, & w < 0, \\ z < 0, & w > 0. \end{cases}$$

The Jacobian of our transformation is, by (3.68),

$$J\begin{pmatrix} z & w \\ x & y \end{pmatrix} = \begin{vmatrix} \dfrac{\partial z}{\partial x} & \dfrac{\partial z}{\partial y} \\[2ex] \dfrac{\partial w}{\partial x} & \dfrac{\partial w}{\partial y} \end{vmatrix} = \begin{vmatrix} y & x \\ \dfrac{-y}{x^2} & \dfrac{1}{x} \end{vmatrix}.$$

$$= \frac{2y}{x} = 2w.$$

Hence by (3.71) the joint probability density function of $\mathbf{z}$ and $\mathbf{w}$ is

$$f_{\mathbf{zw}}(z, w) = \frac{1}{2|w|}\left[f_{\mathbf{xy}}\left(\sqrt{\frac{z}{w}}, \sqrt{zw}\right) + f_{\mathbf{xy}}\left(-\sqrt{\frac{z}{w}}, -\sqrt{zw}\right)\right], \tag{3.75}$$
$$z > 0, \quad w > 0; \qquad z < 0, \quad w < 0.$$

If, for instance, the random variables $\mathbf{x}$ and $\mathbf{y}$ are jointly Gaussian with means zero, variances 1, and correlation coefficient r, as in (3.41), we find that

$$f_{\mathbf{zw}}(z, w) = (2\pi|w|)^{-1}(1 - r^2)^{-1/2} \exp\left[\frac{z(w + w^{-1} - 2r)}{2(1 - r^2)}\right]U(zw). \tag{3.76}$$

Integrating over w,

$$f_{\mathbf{z}}(z) = \begin{cases} \int_0^{\infty} f_{\mathbf{zw}}(z, w)\, dw, & z > 0, \\ \int_{-\infty}^{0} f_{\mathbf{zw}}(z, w)\, dw, & z < 0, \end{cases}$$

would yield the result in (3.56). The marginal probability density function of $\mathbf{w}$, on the other hand, is, for $w > 0$,

$$f_{\mathbf{w}}(w) = \int_0^{\infty} f_{\mathbf{zw}}(z, w)\, dz$$

$$= \frac{(1 - r^2)^{1/2}}{\pi(w^2 - 2rw + 1)}$$

and the same form is found to hold for $w < 0$ as well. ■

This example illustrates a fourth method of solving the problem posed in Sec. 3-4, to find the probability density function of a function $\mathbf{z} = g_1(\mathbf{x}, \mathbf{y})$ of two random variables $\mathbf{x}$ and $\mathbf{y}$. One introduces a second, "dummy" variable $\mathbf{w} = g_2(\mathbf{x}, \mathbf{y})$ and finds the joint probability density function $f_{\mathbf{zw}}(z, w)$ by the method of this section, which requires no integrations. One then integrates out the variable w to find the marginal probability density function

$$f_{\mathbf{z}}(z) = \int_{-\infty}^{\infty} f_{\mathbf{zw}}(z, w)\, dw. \tag{3.77}$$

Sometimes it is convenient to take $\mathbf{w}$ as $\mathbf{x}$ or $\mathbf{y}$, in which case this method is equivalent to method 2 of Sec. 3-4. Indeed, with $\mathbf{w} = \mathbf{y}$,

$$J\begin{pmatrix} z & w \\ x & y \end{pmatrix} = \begin{vmatrix} \dfrac{\partial z}{\partial x} & \dfrac{\partial z}{\partial y} \\ 0 & 1 \end{vmatrix} = \frac{\partial g_1}{\partial x}, \tag{3.78}$$

and (3.69) reduces to (3.53).

B. Continuous or Discrete Random Variables: Joint Cumulative Distribution Function

If the transformation from $(\mathbf{x}_1, \mathbf{x}_2)$ to $(\mathbf{z}_1, \mathbf{z}_2)$ is not a differentiable one, it will not be possible to calculate the Jacobian needed in the method of part A. One can then resort to finding the joint cumulative distribution function of the random variables $\mathbf{z}_1$ and $\mathbf{z}_2$, that is, the probability

$$F_{\mathbf{z}_1\mathbf{z}_2}(z_1, z_2) = \Pr(\mathbf{z}_1 \le z_1, \mathbf{z}_2 \le z_2)$$

$$= \Pr\{g_1(\mathbf{x}_1, \mathbf{x}_2) \le z_1, g_2(\mathbf{x}_1, \mathbf{x}_2) \le z_2\}.$$

The inequalities $g_1(\mathbf{x}_1, \mathbf{x}_2) \le z_1$, $g_2(\mathbf{x}_1, \mathbf{x}_2) \le z_2$ define a region $R(z_1, z_2)$ in the X_1X_2-plane for each pair of values (z_1, z_2), and the cumulative distribution function is

$$F_{\mathbf{z}_1\mathbf{z}_2}(z_1, z_2) = \iint_{R(z_1,z_2)} f_{\mathbf{x}_1\mathbf{x}_2}(x_1, x_2)\, dx_1\, dx_2. \tag{3.79}$$

The region $R(z_1, z_2)$ is bounded by the curves

$$z_1 = g_1(x_1, x_2), \qquad z_2 = g_2(x_1, x_2).$$

If, on the other hand, the random variables $\mathbf{x}_1$ and $\mathbf{x}_2$ are discrete, the probability $F_{\mathbf{z}_1\mathbf{z}_2}(z_1, z_2)$ is obtained by summing the probabilities of all the mass points lying in the region $R(z_1, z_2)$; symbolically,

$$F_{\mathbf{z}_1\mathbf{z}_2}(z_1, z_2) = \sum_{(x_1,x_2)\in R(z_1,z_2)} \Pr(\mathbf{x}_1 = x_1, \mathbf{x}_2 = x_2). \tag{3.80}$$

EXAMPLE 3.20. $\mathbf{z}_1 = \max(\mathbf{x}_1, \mathbf{x}_2)$, $\mathbf{z}_2 = \min(\mathbf{x}_1, \mathbf{x}_2)$, where $\max(\mathbf{x}_1, \mathbf{x}_2)$ is the greater of the random variables $\mathbf{x}_1$ and $\mathbf{x}_2$ and $\min(\mathbf{x}_1, \mathbf{x}_2)$ is the smaller of them. For $z_1 > z_2$ the region $R(z_1, z_2)$ is illustrated in Fig. 3.22(a). Hence we find for the joint cumulative distribution function of $\mathbf{z}_1$ and $\mathbf{z}_2$,

$$\begin{aligned} F_{\mathbf{z}_1\mathbf{z}_2}(z_1, z_2) &= \Pr(\mathbf{x}_1 < z_2, \mathbf{x}_2 < z_1) + \Pr(z_2 < \mathbf{x}_1 < z_1, \mathbf{x}_2 < z_2) \\ &= F_{\mathbf{x}_1\mathbf{x}_2}(z_2, z_1) + F_{\mathbf{x}_1\mathbf{x}_2}(z_1, z_2) - F_{\mathbf{x}_1\mathbf{x}_2}(z_2, z_2). \end{aligned}$$

If $\mathbf{x}_1$ and $\mathbf{x}_2$ are continuous random variables, so are $\mathbf{z}_1$ and $\mathbf{z}_2$, and we find their joint probability density function by (3.5),

$$f_{\mathbf{z}_1\mathbf{z}_2}(z_1, z_2) = f_{\mathbf{x}_1\mathbf{x}_2}(z_1, z_2) + f_{\mathbf{x}_1\mathbf{x}_2}(z_2, z_1), \qquad z_1 > z_2. \tag{3.81}$$

For $z_1 < z_2$, on the other hand, the region $R(z_1, z_2)$, is the quadrant in Fig. 3.22(b), and

$$F_{\mathbf{z}_1\mathbf{z}_2}(z_1, z_2) = F_{\mathbf{x}_1\mathbf{x}_2}(z_1, z_1),$$

and the joint probability density function is

$$f_{\mathbf{z}_1\mathbf{z}_2}(z_1, z_2) = 0, \qquad z_1 < z_2,$$

as expected. ■

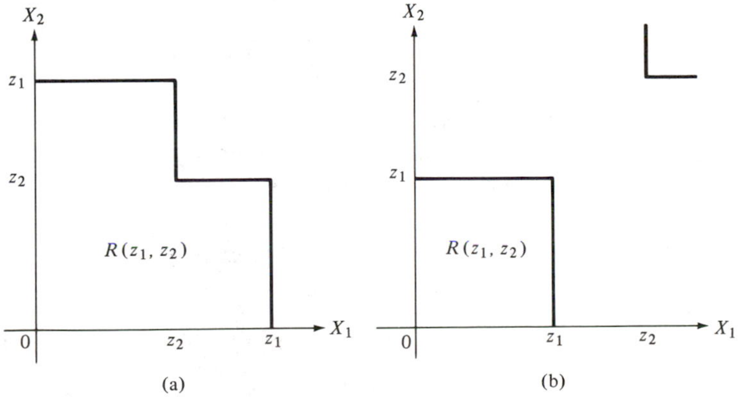

FIGURE 3.22 Defining regions for joint distribution of max $(\mathbf{x}_1, \mathbf{x}_2)$ and min $(\mathbf{x}_1, \mathbf{x}_2)$.

EXPECTED VALUES AND CORRELATION

If $\mathbf{z} = g(\mathbf{x}, \mathbf{y})$ is any function of the random variables $\mathbf{x}$ and $\mathbf{y}$, its expected value is

$$\begin{aligned} E[g(\mathbf{x}, \mathbf{y})] &= \overline{g(x, y)} \\ &= E(\mathbf{z}) = \int_{-\infty}^{\infty} z f_{\mathbf{z}}(z)\, dz \qquad (3.82) \\ &= \int_{-\infty}^{\infty} \int_{-\infty}^{\infty} g(x, y) f_{\mathbf{xy}}(x, y)\, dx\, dy. \end{aligned}$$

To demonstrate this, we write the expected value as

$$E(\mathbf{z}) = \int_{-\infty}^{\infty} \int_{-\infty}^{\infty} z f_{\mathbf{z}}(z \mid y) f_{\mathbf{y}}(y)\, dy\, dz.$$

But by (2.74),

$$\int_{-\infty}^{\infty} z f_{\mathbf{z}}(z \mid y)\, dz = \int_{-\infty}^{\infty} g(x, y) f_{\mathbf{x}}(x \mid y)\, dx,$$

because y being fixed at $\mathbf{y} = y$, $g(\mathbf{x}, \mathbf{y}) = g(\mathbf{x}, y)$ is random only through its dependence on $\mathbf{x}$, whose probability density function is now $f_{\mathbf{x}}(x \mid y)$. Hence we find that

$$E(\mathbf{z}) = \int_{-\infty}^{\infty} \int_{-\infty}^{\infty} g(x, y) f_{\mathbf{x}}(x \mid y) f_{\mathbf{y}}(y)\, dx\, dy,$$

which becomes (3.82) by virtue of (3.28). Thus if all that is wanted is the expected value of $g(\mathbf{x}, \mathbf{y})$, it is unnecessary first to find its probability density function as in Sec. 3-4; eq. (3.82) can be used and the integration carried instead over the XY-plane.

If the random variables $\mathbf{x}$ and $\mathbf{y}$ are discrete, as in (3.13) and (3.14), the expected value of a function $g(\mathbf{x}, \mathbf{y})$ of $\mathbf{x}$ and $\mathbf{y}$ is

$$E[g(\mathbf{x}, \mathbf{y})] = \overline{g(x, y)} = \sum_{k=1}^{\infty} g(x_k, y_k)\, P(x_k, y_k), \qquad (3.83)$$

where $P(x_k, y_k)$ is the joint probability that $\mathbf{x} = x_k$ and $\mathbf{y} = y_k$.

The expected value of the sum of two random variables equals the sum of their expected values, for taking $g(\mathbf{x}, \mathbf{y}) = \mathbf{x} + \mathbf{y}$, we find from (3.82)

$$\begin{aligned} E(\mathbf{x} + \mathbf{y}) &= \int_{-\infty}^{\infty} \int_{-\infty}^{\infty} (x + y) f_{\mathbf{xy}}(x, y)\, dx\, dy \\ &= \int_{-\infty}^{\infty} x f_{\mathbf{x}}(x)\, dx + \int_{-\infty}^{\infty} y f_{\mathbf{y}}(y)\, dy = E(\mathbf{x}) + E(\mathbf{y}) \qquad (3.84) \end{aligned}$$

in terms of the marginal probability density functions defined in (3.10) and (3.11).

The expected value of $(\mathbf{x} + \mathbf{y})^2$, on the other hand, is

$$E[(\mathbf{x} + \mathbf{y})^2] = \int_{-\infty}^{\infty}\int_{-\infty}^{\infty} (x^2 + 2xy + y^2) f_{\mathbf{xy}}(x, y)\, dx\, dy$$
$$= E(\mathbf{x}^2) + 2E(\mathbf{xy}) + E(\mathbf{y}^2),$$

where

$$\overline{xy} = E(\mathbf{xy}) = \int_{-\infty}^{\infty}\int_{-\infty}^{\infty} xy f_{\mathbf{xy}}(x, y)\, dx\, dy \tag{3.85}$$

is called the *correlation* of the random variables **x** and **y**.

If we now form the variance of $\mathbf{x} + \mathbf{y}$ by (2.79), we obtain

$$\begin{aligned} \text{Var}\,(\mathbf{x} + \mathbf{y}) &= \overline{x^2} + 2\overline{xy} + \overline{y^2} - (\overline{x} + \overline{y})^2 \\ &= \text{Var}\,\mathbf{x} + \text{Var}\,\mathbf{y} + 2(\overline{xy} - \overline{x}\,\overline{y}). \end{aligned} \tag{3.86}$$

The quantity

$$\text{Cov}\,(\mathbf{x}, \mathbf{y}) = \overline{xy} - \overline{x}\,\overline{y} \tag{3.87}$$

is called the *covariance* of the random variables **x** and **y**. It can be written

$$\begin{aligned} \text{Cov}\,(\mathbf{x}, \mathbf{y}) &= E[(\mathbf{x} - \overline{x})(\mathbf{y} - \overline{y})] \\ &= \int_{-\infty}^{\infty}\int_{-\infty}^{\infty} (x - \overline{x})(y - \overline{y}) f_{\mathbf{xy}}(x, y)\, dx\, dy, \end{aligned} \tag{3.88}$$

as follows by expanding the integrand and using the definition of the marginal probability density functions,

$$\begin{aligned} \text{Cov}\,(\mathbf{x}, \mathbf{y}) = &\int_{-\infty}^{\infty}\int_{-\infty}^{\infty} xy f_{\mathbf{xy}}(x, y)\, dx\, dy - \overline{x}\int_{-\infty}^{\infty}\int_{-\infty}^{\infty} y f_{\mathbf{xy}}(x, y)\, dx\, dy \\ &- \overline{y}\int_{-\infty}^{\infty}\int_{-\infty}^{\infty} x f_{\mathbf{xy}}(x, y)\, dx\, dy + \overline{x}\,\overline{y}\int_{-\infty}^{\infty}\int_{-\infty}^{\infty} f_{\mathbf{xy}}(x, y)\, dx\, dy \\ = &\ \overline{xy} - \overline{x}\,\overline{y} - \overline{y}\,\overline{x} + \overline{x}\,\overline{y} = \overline{xy} - \overline{x}\,\overline{y}. \end{aligned}$$

Thus by (3.86),

$$\text{Var}\,(\mathbf{x} + \mathbf{y}) = \text{Var}\,\mathbf{x} + 2\,\text{Cov}\,(\mathbf{x}, \mathbf{y}) + \text{Var}\,\mathbf{y}. \tag{3.89}$$

If $(\mathbf{x}, \mathbf{y})$ is a pair of discrete random variables, taking on values (x_k, y_k) at a mass point in the XY-plane with probability $P(x_k, y_k)$, as in (3.13) and (3.83), their covariance is defined by

$$\begin{aligned} \text{Cov}\,(\mathbf{x}, \mathbf{y}) &= \sum_k (x_k - \overline{x})(y_k - \overline{y}) P(x_k, y_k), \\ \overline{x} &= E(\mathbf{x}), \qquad \overline{y} = E(\mathbf{y}). \end{aligned} \tag{3.90}$$

If $\text{Cov}\,(\mathbf{x}, \mathbf{y}) = 0$, the random variables **x** and **y** are said to be uncorrelated. Statistically independent random variables are always uncorrelated—demonstrate this!—but the opposite is only exceptionally true. The variance of the sum of two or more uncorrelated random variables equals the sum of their variances,

$$\text{Var}(\mathbf{x} + \mathbf{y}) = \text{Var}\,\mathbf{x} + \text{Var}\,\mathbf{y}, \qquad \text{Cov}(\mathbf{x}, \mathbf{y}) = 0, \tag{3.91}$$

a useful relation.

The ratio

$$r = \frac{\text{Cov}(\mathbf{x}, \mathbf{y})}{\sqrt{\text{Var}\,\mathbf{x}\,\text{Var}\,\mathbf{y}}} \tag{3.92}$$

is called the *correlation coefficient* of the random variables $\mathbf{x}$ and $\mathbf{y}$. It always lies between -1 and $+1$,

$$-1 \leq r \leq 1. \tag{3.93}$$

The correlation coefficient is unaltered by shifting the means of $\mathbf{x}$ and $\mathbf{y}$, and scaling the random variables by constant factors, $\mathbf{x}' = a\mathbf{x}$, $\mathbf{y}' = b\mathbf{y}$, does not change it. It crudely measures the degree to which the random variables $\mathbf{x}$ and $\mathbf{y}$ are statistically related, and we shall see later how its value can be estimated from pairs of samples (x, y) of $\mathbf{x}$ and $\mathbf{y}$ taken in independent trials of the chance experiment.

Before proving (3.93) we derive the useful *Schwarz inequality* for expectations,

$$[E(\mathbf{xy})]^2 \leq E(\mathbf{x}^2)E(\mathbf{y}^2). \tag{3.94}$$

Since the mean-square value of a random variable can never be negative,

$$E[(\mathbf{x} - \lambda\mathbf{y})^2] \geq 0$$

for any value of λ. After expansion this becomes

$$E(\mathbf{x}^2) - 2\lambda E(\mathbf{xy}) + \lambda^2 E(\mathbf{y}^2) \geq 0.$$

For λ we choose the value for which the left-hand side of this inequality is minimum,

$$\lambda = \frac{E(\mathbf{xy})}{E(\mathbf{y}^2)},$$

obtaining the inequality

$$E(\mathbf{x}^2) - \frac{[E(\mathbf{xy})]^2}{E(\mathbf{y}^2)} \geq 0,$$

whence (3.94). Equality obtains in (3.94) if and only if the random variables $\mathbf{x}$ and $\mathbf{y}$ are proportional: $\mathbf{y} = c\mathbf{x}$ for some constant c.

Now we replace $\mathbf{x}$ by the random variable $\mathbf{x} - \bar{x}$, $\mathbf{y}$ by the random variable $\mathbf{y} - \bar{y}$ in (3.94), obtaining

$$[\text{Cov}(\mathbf{x}, \mathbf{y})]^2 \leq \text{Var}\,\mathbf{x}\,\text{Var}\,\mathbf{y}, \tag{3.95}$$

whence (3.93) by the definition of r in (3.92).

EXAMPLE 3.21. Let the random variables $\mathbf{x}$ and $\mathbf{y}$ have the joint probability density function of Example 3.2. Then

$$
\begin{aligned}
E(\mathbf{x}) &= \int_{-\infty}^{\infty} x f_{\mathbf{x}}(x)\, dx = \int_0^1 3x(1-x)^2\, dx = 3\int_0^1 u^2(1-u)\, du \\
&= \tfrac{1}{4} = E(\mathbf{y}), \\
E(\mathbf{x}^2) &= \int_{-\infty}^{\infty} x^2 f_{\mathbf{x}}(x)\, dx = 3\int_0^1 x^2(1-x)^2\, dx = \tfrac{1}{10} \\
&= E(\mathbf{y}^2), \\
E(\mathbf{xy}) &= 6 \iint_{\triangle} xy(1-x-y)\, dx\, dy \\
&= 6\int_0^1 x\, dx \int_0^{1-x} y(1-x-y)\, dy \\
&= \int_0^1 x(1-x)^3\, dx = \int_0^1 u^3(1-u)\, du = \tfrac{1}{20}.
\end{aligned}
$$

The covariance of **x** and **y** is therefore, by (3.87),

$$\text{Cov}\,(\mathbf{x}, \mathbf{y}) = \tfrac{1}{20} - \tfrac{1}{16} = -\tfrac{1}{80}.$$

The variances of **x** and **y** are

$$\text{Var}\,\mathbf{x} = \text{Var}\,\mathbf{y} = \tfrac{1}{10} - \tfrac{1}{16} = \tfrac{3}{80}.$$

and their correlation coefficient is $r = -\frac{1}{3}$. The random variables **x** and **y** are negatively correlated: when one is large, the other tends to be small, as is evident also from the form of their joint probability density function in Example 3.2. ■

EXAMPLE 3.22. Linear Transformation (continued)

Let $\mathbf{z}_1$ and $\mathbf{z}_2$ be related through the linear transformation of Example 3.18. Then their means are related by a similar linear transformation. The covariance of $\mathbf{z}_1$ and $\mathbf{z}_2$ is, from (3.72),

$$
\begin{aligned}
\text{Cov}\,(\mathbf{z}_1, \mathbf{z}_2) &= \text{Cov}\,(a_{11}\mathbf{x}_1 + a_{12}\mathbf{x}_2, a_{21}\mathbf{x}_1 + a_{22}\mathbf{x}_2) \\
&= a_{11}a_{21}\,\text{Var}\,\mathbf{x}_1 + (a_{11}a_{22} + a_{12}a_{21})\,\text{Cov}\,(\mathbf{x}_1, \mathbf{x}_2) + a_{12}a_{22}\,\text{Var}\,\mathbf{x}_2.
\end{aligned}
$$

In particular, let the matrix a represent a rotation of axes through an angle θ,

$$\mathsf{a} = \begin{bmatrix} \cos\theta & \sin\theta \\ -\sin\theta & \cos\theta \end{bmatrix}.$$

Then with $\sigma_1^2 = \text{Var}\,\mathbf{x}_1$, $\sigma_2^2 = \text{Var}\,\mathbf{x}_2$, we find that

$$
\begin{aligned}
\text{Cov}\,(\mathbf{z}_1, \mathbf{z}_2) &= \sin\theta\cos\theta\,(\sigma_2^2 - \sigma_1^2) + r\sigma_1\sigma_2(\cos^2\theta - \sin^2\theta) \\
&= \tfrac{1}{2}[(\sigma_2^2 - \sigma_1^2)\sin 2\theta + 2r\sigma_1\sigma_2\cos 2\theta]
\end{aligned}
$$

By picking the angle θ so that

$$\tan 2\theta = \frac{2r\sigma_1\sigma_2}{\sigma_1^2 - \sigma_2^2} = \frac{2\,\text{Cov}\,(\mathbf{x}_1, \mathbf{x}_2)}{\text{Var}\,\mathbf{x}_1 - \text{Var}\,\mathbf{x}_2}$$

we obtain uncorrelated random variables

$$
\begin{aligned}
\mathbf{z}_1 &= \mathbf{x}_1\cos\theta + \mathbf{x}_2\sin\theta, \\
\mathbf{z}_2 &= -\mathbf{x}_1\sin\theta + \mathbf{x}_2\cos\theta.
\end{aligned}
$$

(If $\sigma_1 = \sigma_2$, take $\theta = \pi/4$.) ■

Moments and central moments of higher order,

$$m_{kn} = E(\mathbf{x}^k\mathbf{y}^n) = \int_{-\infty}^{\infty}\int_{-\infty}^{\infty} x^k y^n f_{\mathbf{xy}}(x, y)\,dx\,dy,$$

$$\mu_{kn} = E[(\mathbf{x} - \bar{x})^k(\mathbf{y} - \bar{y})^n]$$

are occasionally useful. The moments m_{kn} can be derived from the bivariate characteristic function

$$\Phi_{\mathbf{xy}}(\omega_x, \omega_y) = E\{\exp[j(\omega_x\mathbf{x} + \omega_y\mathbf{y})]\}$$

$$= \int_{-\infty}^{\infty}\int_{-\infty}^{\infty} f_{\mathbf{xy}}(x, y)\exp[j(\omega_x x + \omega_y y)]\,dx\,dy. \quad (3.96)$$

This characteristic function is the two-dimensional Fourier transform of the bivariate joint probability density function, which can be recovered by the inverse Fourier transformation,

$$f_{\mathbf{xy}}(x, y) = \int_{-\infty}^{\infty}\int_{-\infty}^{\infty} \Phi_{\mathbf{xy}}(\omega_x, \omega_y)\exp[-j(\omega_x x + \omega_y y)]\,\frac{d\omega_x\,d\omega_y}{(2\pi)^2}. \quad (3.97)$$

If the power series of the exponential function is substituted into (3.96), we find

$$\Phi_{\mathbf{xy}}(\omega_x, \omega_y) = \sum_{k=0}\sum_{m=0}\frac{(j\omega_x)^k(j\omega_y)^m}{k!m!}E(\mathbf{x}^k\mathbf{y}^m), \quad (3.98)$$

the series terminating with a remainder term just before any term for which the moment $E(\mathbf{x}^k\mathbf{y}^m)$ happens to be infinite. Thus the joint moments of the bivariate probability density function can be found by expanding the characteristic function $\Phi_{\mathbf{xy}}(\omega_x, \omega_y)$ as a power series in the neighborhood of the origin of the (ω_x, ω_y)-plane.

It will follow from our general analysis of the multivariate Gaussian distribution in Chapter 4 that the characteristic function of two jointly Gaussian random variables with means $\bar{x}$, $\bar{y}$, variances σ_x^2, σ_y^2, and correlation coefficient r has the Gaussian form

$$\Phi_{\mathbf{xy}}(\omega_x, \omega_y) = \exp[j(\omega_x\bar{x} + \omega_y\bar{y}) - \tfrac{1}{2}(\sigma_x^2\omega_x^2 + 2r\sigma_x\sigma_y\omega_x\omega_y + \sigma_y^2\omega_y^2)]. \quad (3.99)$$

Taking $\bar{x} = \bar{y} = 0$ simply shifts the center of mass of the joint distribution of the random variables **x** and **y** to the origin. Expanding the remaining factor in (3.99), we have

$$\exp\{\tfrac{1}{2}[\sigma_x^2(j\omega_x)^2 + 2r\sigma_x\sigma_y(j\omega_x)(j\omega_y) + \sigma_y^2(j\omega_y)^2]\}$$

$$\doteq 1 + \tfrac{1}{2}[\sigma_x^2(j\omega_x)^2 + 2r\sigma_x\sigma_y(j\omega_x)(j\omega_y) + \sigma_y^2(j\omega_y)^2]$$

$$+ \tfrac{1}{8}[\sigma_x^2(j\omega_x)^2 + 2r\sigma_x\sigma_y(j\omega_x)(j\omega_y) + \sigma_y^2(j\omega_y)^2]^2 + \cdots,$$

and comparing the terms with the expansion in (3.98), we see from the term proportional to $(j\omega_x)(j\omega_y)$ that

$$E(\mathbf{xy}) = r\sigma_x\sigma_y,$$

and since $\bar{x} = \bar{y} = 0$ this is

$$\text{Cov}(\mathbf{x}, \mathbf{y}) = r\sigma_x\sigma_y,$$

confirming our remark in connection with (3.40) that the r in that equation is the correlation coefficient of **x** and **y**.

Let $\mathbf{z} = \mathbf{x} + \mathbf{y}$. Then the characteristic function of the random variable **z** in terms of the joint characteristic function of **x** and **y** is

$$\Phi_{\mathbf{z}}(\omega) = E(e^{j\omega\mathbf{z}}) = E(e^{j\omega(\mathbf{x}+\mathbf{y})}) = \Phi_{\mathbf{xy}}(\omega, \omega). \tag{3.100}$$

If the random variables **x** and **y** are statistically independent, (3.96) with (3.38) shows that their joint characteristic function factors,

$$\Phi_{\mathbf{xy}}(\omega_x, \omega_y) = \Phi_{\mathbf{x}}(\omega_x)\Phi_{\mathbf{y}}(\omega_y), \tag{3.101}$$

into the product of the characteristic functions of **x** and **y** individually. Then the characteristic function of **z** is

$$\Phi_{\mathbf{z}}(\omega) = \Phi_{\mathbf{x}}(\omega)\Phi_{\mathbf{y}}(\omega). \tag{3.102}$$

By virtue of the convolution theorem for Fourier transforms, the probability density function of $\mathbf{z} = \mathbf{x} + \mathbf{y}$ is given by the convolution

$$f_{\mathbf{z}}(z) = \int_{-\infty}^{\infty} f_{\mathbf{x}}(z - y)f_{\mathbf{y}}(y)\, dy, \tag{3.103}$$

as in (3.51).

Equation (3.101) exemplifies the general rule that if **x** and **y** are independent random variables,

$$E[g(\mathbf{x})h(\mathbf{y})] = E[g(\mathbf{x})]E[h(\mathbf{y})],$$

which we ask the reader to prove.

3-7

CONDITIONAL EXPECTED VALUES

If as in Sec. 3-2A the random variables **x** and **y** are conditioned by a nonnumerical attribute, their joint conditional probability density function will have expected values defined as in (3.82), and these will be called *conditional expected values*. Thus if $\mathcal{A}$ is the conditioning event

$$E[g(x, y) \mid \mathcal{A}] = \int_{-\infty}^{\infty}\int_{-\infty}^{\infty} g(x, y)f_{\mathbf{xy}}(x, y \mid \mathcal{A})\, dx\, dy \tag{3.104}$$

in terms of the joint conditional probability density function $f_{\mathbf{xy}}(x, y \mid \mathcal{A})$. The same definition applies if $\mathcal{A}$ is an event described by certain inequalities on **x** and **y**, or by saying that the outcome point (**x**, **y**) lies in a certain region $\mathcal{A}$ of the *XY*-plane:

$$E[g(\mathbf{x}, \mathbf{y}) \mid (\mathbf{x}, \mathbf{y}) \in \mathscr{A}]$$

$$= \iint_{\mathscr{A}} g(x, y) f_{\mathbf{xy}}(x, y \mid (\mathbf{x}, \mathbf{y}) \in \mathscr{A})\, dx\, dy \tag{3.105}$$

$$= \iint_{\mathscr{A}} g(x, y) f_{\mathbf{xy}}(x, y)\, dx\, dy \Big/ \iint_{\mathscr{A}} f_{\mathbf{xy}}(x, y)\, dx\, dy,$$

in which the integrations are carried out over the region $\mathscr{A}$; compare (3.20).

EXAMPLE 3.23. For the conditional probability density function

$$f_{\mathbf{xy}}(x, y \mid \mathbf{x}^2 + \mathbf{y}^2 < b^2)$$

defined in Example 3.7, we seek the conditional expected value

$$E(\mathbf{x}^2 \mid \mathbf{x}^2 + \mathbf{y}^2 < b^2).$$

This could be calculated from

$$E(\mathbf{x}^2 \mid \mathbf{x}^2 + \mathbf{y}^2 < b^2) = \int_{-\infty}^{\infty} x^2 f_{\mathbf{x}}(x \mid \mathbf{x}^2 + \mathbf{y}^2 < b^2)\, dx$$

by using the rather complicated conditional probability density function $f_{\mathbf{x}}(x \mid \mathbf{x}^2+\mathbf{y}^2 < b^2)$ derived there. It is simpler to note that by virtue of the circular symmetry of both the probability density function and the condition,

$$\begin{aligned} E(\mathbf{x}^2 \mid \mathbf{x}^2 + \mathbf{y}^2 < b^2) &= E(\mathbf{y}^2 \mid \mathbf{x}^2 + \mathbf{y}^2 < b^2) \\ &= \tfrac{1}{2}E(\mathbf{x}^2 + \mathbf{y}^2 \mid \mathbf{x}^2 + \mathbf{y}^2 < b^2) \\ &= \frac{\tfrac{1}{2} \displaystyle\iint_{x^2+y^2<b^2} (x^2 + y^2) \exp\left(-\frac{x^2 + y^2}{2\sigma^2}\right) dx\, dy}{\displaystyle\iint_{x^2+y^2<b^2} \exp\left(-\frac{x^2 + y^2}{2\sigma^2}\right) dx\, dy} \end{aligned}$$

The denominator was evaluated in Example 3.7; the numerator is likewise most easily evaluated by converting the integrand to polar coordinates, and we find that

$$\begin{aligned} E(\mathbf{x}^2 \mid \mathbf{x}^2 + \mathbf{y}^2 < b^2) &= \frac{1}{2} \frac{\displaystyle\int_0^b r^3 e^{-r^2/2\sigma^2}\, dr}{\displaystyle\int_0^b r e^{-r^2/2\sigma^2}\, dr} \\ &= \sigma^2 \frac{\displaystyle\int_0^\beta x e^{-x} dx}{\displaystyle\int_0^\beta e^{-x} dx} = \sigma^2 \frac{1 - (1 + \beta)e^{-\beta}}{1 - e^{-\beta}}, \end{aligned}$$

$$\beta = \frac{b^2}{2\sigma^2}.$$

This expected value of course becomes equal to σ^2 when $b \to \infty$. ■

The conditional expected value of $g(\mathbf{x}, \mathbf{y})$, given $\mathbf{y} = y$, is given by

$$E[g(\mathbf{x}, \mathbf{y}) \mid y] = \int_{-\infty}^{\infty} g(x, y) f_{\mathbf{x}}(x \mid y)\, dx \tag{3.106}$$

in terms of the conditional density function $f_{\mathbf{x}}(x \mid y)$ of $\mathbf{x}$, given $\mathbf{y} = y$, as defined in (3.28). In particular,

$$E(\mathbf{x} \mid y) = \int_{-\infty}^{\infty} x f_{\mathbf{x}}(x \mid y)\, dx \tag{3.107}$$

is the *conditional mean* of the random variable $\mathbf{x}$, given $\mathbf{y} = y$. Similarly,

$$E(\mathbf{x}^2 \mid y) = \int_{-\infty}^{\infty} x^2 f_{\mathbf{x}}(x \mid y)\, dx \tag{3.108}$$

and we define the *conditional variance* of $\mathbf{x}$, given $\mathbf{y} = y$, as in (2.79),

$$\begin{aligned} \operatorname{Var}(\mathbf{x} \mid y) &= E(\mathbf{x}^2 \mid y) - [E(\mathbf{x} \mid y)]^2 \\ &= \int_{-\infty}^{\infty} [x - E(\mathbf{x} \mid y)]^2 f_{\mathbf{x}}(x \mid y)\, dx. \end{aligned} \tag{3.109}$$

EXAMPLE 3.24. In reference to Examples 3.2, 3.9, and 3.21, we find for the conditional expected value of $\mathbf{x}$, given $\mathbf{y} = y$,

$$E(\mathbf{x} \mid y) = \int_0^{1-y} \frac{2x(1 - y - x)}{(1 - y)^2}\, dx = \tfrac{1}{3}(1 - y), \qquad 0 < y < 1,$$

$$E(\mathbf{x}^2 \mid y) = \int_0^{1-y} \frac{2x^2(1 - x - y)}{(1 - y)^2}\, dx = \tfrac{1}{6}(1 - y)^2, \qquad 0 < y < 1,$$

so that the conditional variance of $\mathbf{x}$, given $\mathbf{y} = y$, is

$$\begin{aligned} \operatorname{Var}(\mathbf{x} \mid y) &= \tfrac{1}{6}(1 - y)^2 - \tfrac{1}{9}(1 - y)^2 \\ &= \tfrac{1}{18}(1 - y)^2, \qquad 0 < y < 1. \end{aligned}$$

■

The conditional expected value $E[g(\mathbf{x}, \mathbf{y}) \mid y]$ is in general a function of y, say

$$h(y) = E[g(\mathbf{x}, \mathbf{y}) \mid y] = \int_{-\infty}^{\infty} g(x, y) f_{\mathbf{x}}(x \mid y)\, dx.$$

We can make a random variable $\mathbf{z}$ out of this by now replacing y by the random variable $\mathbf{y}$. It is written

$$\mathbf{z} = h(\mathbf{y}) = E[g(\mathbf{x}, \mathbf{y}) \mid \mathbf{y}].$$

We ask for the expected value $E(\mathbf{z})$ of $\mathbf{z}$. By the definition in (2.74) and by (3.28) it must be

$$\begin{aligned} E(\mathbf{z}) &= \int_{-\infty}^{\infty} h(y) f_{\mathbf{y}}(y)\, dy \\ &= \int_{-\infty}^{\infty} \int_{-\infty}^{\infty} g(x, y) f_{\mathbf{x}}(x \mid y) f_{\mathbf{y}}(y)\, dx\, dy \\ &= \int_{-\infty}^{\infty} \int_{-\infty}^{\infty} g(x, y) f_{\mathbf{xy}}(x, y)\, dx\, dy \\ &= E[g(\mathbf{x}, \mathbf{y})]. \end{aligned}$$

In this way we have shown that

$$E_{\mathbf{y}}\{E_{\mathbf{x}}[g(\mathbf{x}, \mathbf{y}) \mid \mathbf{y}]\} = E[g(\mathbf{x}, \mathbf{y})], \tag{3.110}$$

a useful relation. By writing $E_{\mathbf{x}}$, $E_{\mathbf{y}}$ we emphasize that the first—conditional—expectation is with respect to the random variable $\mathbf{x}$; the second—unconditional—expectation is with respect to the random variable $\mathbf{y}$. In particular,

$$E_{\mathbf{y}}\{E_{\mathbf{x}}(\mathbf{xy} \mid \mathbf{y})\} = E(\mathbf{xy}).$$

But

$$E_{\mathbf{x}}(\mathbf{xy} \mid y) = yE(\mathbf{x} \mid y)$$

since with $\mathbf{y} = y$, $\mathbf{xy} = \mathbf{x}y$, and in $\mathbf{x}y$, only $\mathbf{x}$ is random. Thus we obtain

$$\mathbf{E}(\mathbf{xy}) = E_{\mathbf{y}}[\mathbf{y}E_{\mathbf{x}}(\mathbf{x} \mid \mathbf{y})]. \tag{3.111}$$

EXAMPLE 3.24 (continued). We have shown that

$$E_{\mathbf{x}}(\mathbf{x} \mid \mathbf{y}) = \tfrac{1}{3}(1 - \mathbf{y}), \qquad 0 < \mathbf{y} < 1.$$

Hence by (3.111) and the results of Example 3.21,

$$E(\mathbf{xy}) = E_{\mathbf{y}}[\tfrac{1}{3}\,\mathbf{y}(1 - \mathbf{y})] = \tfrac{1}{3}[E(\mathbf{y}) - E(\mathbf{y}^2)] = \tfrac{1}{3}\left(\tfrac{1}{4} - \tfrac{1}{10}\right)$$
$$= \tfrac{1}{20},$$

as found in Example 3.21 by a double integration. ■

EXAMPLE 3.25. In Sec. 3-3 we showed that for jointly Gaussian random variables with means zero and variances 1,

$$f_{\mathbf{x}}(x \mid y) = [2\pi(1 - r^2)]^{-1/2} \exp\left[-\frac{(x - ry)^2}{2(1 - r^2)}\right].$$

Hence $E(\mathbf{x} \mid y) = ry$, and $E_{\mathbf{x}}(\mathbf{x} \mid \mathbf{y}) = r\mathbf{y}$. Thereupon, by (3.111),

$$E(\mathbf{xy}) = E_{\mathbf{y}}(r\mathbf{y}^2) = r,$$

and we find by (3.92) that r is the correlation coefficient of Gaussian random variables $\mathbf{x}$ and $\mathbf{y}$ whose joint probability density function is given by (3.41). Since the definition in (3.92) is invariant to scaling the random variables or shifting their means, we have verified that r is also the correlation coefficient for random variables whose joint probability density function has the general bivariate Gaussian form in (3.40). ■

PROBLEMS

3-1 The independent random variables $\mathbf{x}$ and $\mathbf{y}$ have the following probability density functions,

$$f_{\mathbf{x}}(x) = \begin{cases} 1, & 0 < x < 1, \\ 0, & x < 0, \quad x > 1, \end{cases}$$

$$f_{\mathbf{y}}(y) = \begin{cases} 1 - |y|, & -1 < y < 1, \\ 0, & |y| > 1. \end{cases}$$

Their joint probability density function is $f_{\mathbf{xy}}(x, y) = f_{\mathbf{x}}(x)f_{\mathbf{y}}(y)$. Find the probability that $\mathbf{x} > \mathbf{y}$.

3-2 Calculate $\Pr(\mathbf{x} + \mathbf{y} \le p)$ in Example 3.4 for $\nu_1 = \nu_2$. Also calculate $\Pr(\mathbf{x} \ge \mathbf{y})$ for $\nu_1 \ne \nu_2$.

3-3 The random variables $\mathbf{x}$ and $\mathbf{y}$, always positive, have the joint probability density function

$$f_{\mathbf{xy}}(x, y) = \begin{cases} C(1 - x^2 - y^2), & 0 < x^2 + y^2 < 1, \quad x > 0, \quad y > 0, \\ 0 & \text{elsewhere.} \end{cases}$$

Determine the constant C and the probability $\Pr(\mathbf{x} > \frac{1}{2})$. Note that the function $f_{\mathbf{xy}}(x, y)$ is nonzero only over a quadrant of a circle of unit radius.

3-4 Under hypothesis H_0, "noise alone is present at the input to our receiver," two independent samples $\mathbf{x}$ and $\mathbf{y}$ of the input, taken at different times, have the joint probability density function

$$f_{\mathbf{xy}}(x, y \mid H_0) = (2\pi\sigma^2)^{-1} \exp\left(-\frac{x^2 + y^2}{2\sigma^2}\right).$$

Under hypothesis H_1, "both signal and noise are present," their joint probability density function is

$$f_{\mathbf{xy}}(x, y \mid H_1) = (2\pi\sigma^2)^{-1} \exp\left[-\frac{(x - a)^2 + (y - b)^2}{2\sigma^2}\right].$$

The prior probabilities of the hypotheses are

$$p_0 = \Pr(H_0), \qquad p_1 = \Pr(H_1), \qquad p_0 + p_1 = 1.$$

Find the posterior probability $\Pr(H_1 \mid x, y)$ that a signal is present, given that particular values x and y of the samples have been observed. Suppose that the receiver decides that a signal is present for each pair (x, y) of samples for which $\Pr(H_1 \mid x, y) > \Pr(H_0 \mid x, y)$, and that otherwise it decides that no signal is present. Represent this strategy as a dichotomy of the XY-plane.

3-5 The random variables $\mathbf{x}$ and $\mathbf{y}$ have the joint probability density function

$$f_{\mathbf{xy}}(x, y) = (2\pi\sigma^2)^{-1} \exp\left[-\frac{x^2 + y^2}{2\sigma^2}\right].$$

Find the conditional probability density function

$$f_{\mathbf{xy}}(x, y \mid \mathbf{x}^2 + \mathbf{y}^2 > c^2).$$

3-6 The random variables $\mathbf{x}$ and $\mathbf{y}$ are independent and uniformly distributed over (0, 1). Find the probability density function of the absolute value of their difference,

$$\mathbf{z} = |\mathbf{x} - \mathbf{y}|.$$

3-7 The random variables $\mathbf{x}$ and $\mathbf{y}$ are independent and have Gaussian distributions with means $\bar{x}$ and $\bar{y}$, respectively, and variances σ_x^2 and σ_y^2, respectively. Calculate the probability $\Pr(\mathbf{xy} > 0)$.

3-8 Calculate the probability density function of the angle $\boldsymbol{\theta} = \arg(\mathbf{x} + j\mathbf{y})$ when $\mathbf{x}$ and $\mathbf{y}$ are independently Gaussian with means m_x and m_y, respectively, and equal variances σ^2.

3-9 Let $\mathbf{x}$ and $\mathbf{y}$ be independent random variables uniformly distributed over (0, 1). Define new random variables $\mathbf{z}$ and $\mathbf{w}$ by

$$\mathbf{z} = \sqrt{-2 \ln \mathbf{x}} \cos 2\pi\mathbf{y}, \qquad \mathbf{w} = \sqrt{-2 \ln \mathbf{x}} \sin 2\pi\mathbf{y}.$$

Find the joint probability density function of $\mathbf{z}$ and $\mathbf{w}$.

3-10 For $-\infty < z < \infty$ find the cumulative distribution function $F_{\mathbf{z}}(z) = \Pr(\mathbf{z} \le z)$ of the product $\mathbf{z} = \mathbf{xy}$ of the random variables $\mathbf{x}$ and $\mathbf{y}$ whose joint probability density function is given in Example 3.2.

3-11 Calculate the probability density functions worked out in Examples 3.11 to 3.13, but by using method 2 or method 3 of Sec. 3-4 instead.

3-12 Find the joint probability density function of the random variables

$$\mathbf{x} = \mathbf{A} \cos \boldsymbol{\psi}, \qquad \mathbf{y} = \mathbf{A} \cos (\boldsymbol{\psi} + \beta),$$

where β is constant and $\mathbf{A}$ and $\boldsymbol{\psi}$ are independent random variables, $\mathbf{A}$ having a Rayleigh distribution,

$$f_{\mathbf{A}}(A) = \left(\frac{A}{\sigma^2}\right) \exp\left(-\frac{A^2}{2\sigma^2}\right) U(A),$$

and $\boldsymbol{\psi}$ a uniform distribution over $(0, 2\pi)$,

$$f_{\boldsymbol{\psi}}(\psi) = (2\pi)^{-1}, \qquad 0 \le \psi < 2\pi.$$

3-13 For the distribution in Problem 3-3, calculate the conditional probability density function $f_{\mathbf{y}}(y \mid x)$, the conditional expected value $E(\mathbf{y} \mid x)$, and the covariance Cov $(\mathbf{x}, \mathbf{y})$ of the random variables $\mathbf{x}$ and $\mathbf{y}$.

3-14 The random variables $\mathbf{x}$ and $\mathbf{y}$ are independent and have Cauchy distributions,

$$f_{\mathbf{x}}(x) = \frac{1}{\pi(x^2 + 1)}, \qquad f_{\mathbf{y}}(y) = \frac{1}{\pi(y^2 + 1)}.$$

Find the joint probability density function of the random variables

$$\mathbf{z} = 2\mathbf{xy}, \qquad \mathbf{w} = \mathbf{x}^2 + \mathbf{y}^2.$$

Find the marginal probability density functions of $\mathbf{z}$ and $\mathbf{w}$.

3-15 Find the joint probability density function of $\mathbf{z} = \mathbf{xy}$ and $\mathbf{w} = \mathbf{y}/\mathbf{x}$ by the method of Sec. 3-5B, that is, by first finding the joint cumulative distribution function $F_{\mathbf{zw}}(z, w)$ of $\mathbf{z}$ and $\mathbf{w}$. In the XY-plane sketch the regions in which $\mathbf{xy} < z$, $\mathbf{y}/\mathbf{x} < w$, taking $z > 0$, $w > 0$, and write down the integral of $f_{\mathbf{xy}}(x, y)$ over them, being careful to use the correct limits on the double integrals. Then differentiate with respect to z and w to find an expression for the joint density function $f_{\mathbf{zw}}(z, w)$.

3-16 The random variables $\mathbf{x}$ and $\mathbf{y}$ are independent and have Cauchy distributions

$$f_{\mathbf{x}}(z) = f_{\mathbf{y}}(z) = \frac{a/\pi}{z^2 + a^2}.$$

Find the probability density function $f_{\boldsymbol{\theta}}(\theta)$ of the polar angle

$$\boldsymbol{\theta} = \arg (\mathbf{x} + j\mathbf{y}).$$

3-17 The Gaussian random variables whose joint probability density function is given in (3.41) are subjected to the transformation

$$\mathbf{u} = \tfrac{1}{2}(\mathbf{x}^2 + \mathbf{y}^2)$$
$$\mathbf{v} = \tfrac{1}{2}(\mathbf{x}^2 - \mathbf{y}^2)$$

into two new random variables $\mathbf{u}$ and $\mathbf{v}$. Find the joint probability density function of $\mathbf{u}$ and $\mathbf{v}$.

3-18 Two positive random variables $\mathbf{x}$ and $\mathbf{y}$ have the joint probability density function

$$f_{\mathbf{xy}}(x, y) = Cx^2e^{-x-xy}U(x)U(y).$$

(a) Evaluate the constant C.
(b) Determine the marginal probability density functions of both $\mathbf{x}$ and $\mathbf{y}$.
(c) Determine the conditional probability density functions $f_{\mathbf{x}}(x \mid \mathbf{y} = y)$ and $f_{\mathbf{y}}(y \mid \mathbf{x} = x)$.
(d) Calculate the conditional expected values $E(\mathbf{x} \mid \mathbf{y} = y)$ and $E(\mathbf{y} \mid \mathbf{x} = x)$.

3-19 Evaluate the covariance Cov $(\mathbf{x}, \mathbf{y})$ and the correlation coefficient r of the winnings $\mathbf{x}$ and $\mathbf{y}$ of the persons watching the toss of a die, as listed in the table at the beginning of Sec. 3-1C.

3-20 The random variables $\mathbf{x}$ and $\mathbf{y}$ are jointly Gaussian with mean values 0, variances σ_x^2 and σ_y^2, respectively, and covariance Cov $(\mathbf{x}, \mathbf{y}) = r\sigma_x\sigma_y$. Determine the probability density function of the random variable $\mathbf{z} = a\mathbf{x} + b\mathbf{y}$ by using characteristic functions.

3-21 In Problem 3-4 determine the probability density functions $f_{\mathbf{z}}(z \mid H_0)$ and $f_{\mathbf{z}}(z \mid H_1)$, where $\mathbf{z} = a\mathbf{x} + b\mathbf{y}$. Use these to calculate the probability P_e of error incurred by the decision strategy derived in Problem 3-4, averaged over all possible outcomes $(\mathbf{x}, \mathbf{y}, H_k)$, $k = 0, 1$, of the grand experiment in which a signal is either sent or not and samples $\mathbf{x}$ and $\mathbf{y}$ are observed. This probability P_e will be a function of p_0, p_1, a, b, and σ^2. Also determine probabilities of errors of the first and second kinds (false alarm and false dismissal) as defined in Problem 1-26.

3-22 The random variables $\mathbf{y}_1$ and $\mathbf{y}_2$ are defined by

$$\mathbf{y}_1 = \mathbf{x}_1 + \mathbf{x}_2, \qquad \mathbf{y}_2 = \mathbf{x}_1 - \mathbf{x}_2$$

in terms of the random variables $\mathbf{x}_1$ and $\mathbf{x}_2$. Determine their joint characteristic function $\Phi_{\mathbf{y}_1\mathbf{y}_2}(\omega_1, \omega_2)$ in terms of the joint characteristic function $\Phi_{\mathbf{x}_1\mathbf{x}_2}(\omega_1, \omega_2)$ of $\mathbf{x}_1$ and $\mathbf{x}_2$. Show that the joint characteristic function $\Phi_{\mathbf{x}_1\mathbf{x}_2}(\omega_1, \omega_2)$ is a function only of $\omega_1 + \omega_2$ if and only if the random variables $\mathbf{x}_1$ and $\mathbf{x}_2$ are equal with probability 1.

3-23 For Gaussian random variables $\mathbf{x}$ and $\mathbf{y}$ with mean zero, variance 1, and correlation coefficient r, calculate $E(\mathbf{x}^2\mathbf{y}^2)$ and $E(|\mathbf{xy}|)$.

3-24 The independent random variables $\mathbf{x}$ and $\mathbf{y}$ take on only nonnegative integral values 0, 1, 2,. . . . Find the probability generating function of the random variable $\mathbf{w} = \mathbf{x} + \mathbf{y}$ in terms of the probability generating functions $h_{\mathbf{x}}(z)$ and $h_{\mathbf{y}}(z)$ of $\mathbf{x}$ and $\mathbf{y}$ as defined in Sec. 2-4G. Apply this result to the solution of Example 3.14.

3-25 Let $\mathbf{x}$ and $\mathbf{y}$ be independent integer-valued random variables with Poisson distributions with means λ_x and λ_y, respectively. Calculate the probability distribution of $\mathbf{z} = \mathbf{x} + \mathbf{y}$. (*Hint:* Use characteristic functions or probability generating functions.)

CHAPTER 4

Multiple Random Variables

4-1

MULTIVARIATE PROBABILITY DISTRIBUTIONS

The concepts developed in Chapter 3 for handling pairs of random variables are straightforwardly extended to apply to an arbitrary number n of random variables, all defined in the same chance experiment. Its outcome ζ is now labeled with n numbers $(\mathbf{x}_1, \mathbf{x}_2, \ldots, \mathbf{x}_n)$, which it is convenient to represent as a point in an n-dimensional Cartesian space R_n. When a probability measure has been assigned to the events in the experiment, that is, to sets of outcomes, the identifying numbers $\mathbf{x}_1, \mathbf{x}_2, \ldots, \mathbf{x}_n$ become random variables.

The experiment, for instance, might consist of measuring the voltages at n points of a circuit at the same time, or the voltage at a single point of the circuit at n successive instants of time, or the voltages at a number of points at a number of instants. The three rectangular components B_x, B_y, B_z of the magnetic intensity vector B might be measured at various points of some region of space, along perhaps with the components of the electric current and the local density of charged particles. The class of images to be submitted to an image-processing technique might be described in terms of the probability distribution of light intensity at centers of a rectangular grid of picture elements. The kinetic theory of gases establishes the joint probability distribution of the components of the velocities of gas molecules and relates it to the temperature of the gas. In all such circumstances it is necessary to deal with a group of random variables defined on the same conceptual or actual chance experiment.

For any set of n values $x_1, x_2, \ldots, x_n$ the outcomes of the experiment for which $\mathbf{x}_1 \le x_1, \mathbf{x}_2 \le x_2, \ldots, \mathbf{x}_n \le x_n$ constitute an event whose probability

$$F_{\mathbf{x}}(x_1, x_2, \ldots, x_n) = \Pr(\mathbf{x}_1 \le x_1, \mathbf{x}_2 \le x_2, \ldots, \mathbf{x}_n \le x_n) \qquad (4.1)$$

is defined as the joint cumulative distribution function of the random variables $\mathbf{x}_1, \mathbf{x}_2, \ldots, \mathbf{x}_n$. The region in the n-dimensional space R_n representing this event is a rectangular polyhedral angle opening toward the point $(-\infty, -\infty, \ldots, -\infty)$, the multidimensional version of that shown in Fig. 3.1. The cumulative distribution function $F_{\mathbf{x}}$ must be such that

$$F_{\mathbf{x}}(-\infty, -\infty, \ldots, -\infty) = 0, \qquad (4.2)$$

$$F_{\mathbf{x}}(\infty, \infty, \ldots, \infty) = 1. \qquad (4.3)$$

From it can be derived the probability of the event

$$x_k^{(1)} < \mathbf{x}_k \le x_k^{(2)}, \qquad k = 1, 2, \ldots, n,$$

which is represented by an n-dimensional rectangular parallelepiped. The expression for this probability in terms of $F_{\mathbf{x}}$ is analogous to that in (3.2). For $n = 3$, for instance,

$$\begin{aligned}
&\Pr(x_1^{(1)} \le \mathbf{x}_1 \le x_1^{(2)}, x_2^{(1)} < \mathbf{x}_2 \le x_2^{(2)}, x_3^{(1)} < \mathbf{x}_3 \le x_3^{(2)}) \\
&\quad = F_{\mathbf{x}}(x_1^{(2)}, x_2^{(2)}, x_3^{(2)}) - F_{\mathbf{x}}(x_1^{(1)}, x_2^{(2)}, x_3^{(2)}) \\
&\qquad - F_{\mathbf{x}}(x_1^{(2)}, x_2^{(1)}, x_3^{(2)}) - F_{\mathbf{x}}(x_1^{(2)}, x_2^{(2)}, x_3^{(1)}) \\
&\qquad + F_{\mathbf{x}}(x_1^{(2)}, x_2^{(1)}, x_3^{(1)}) + F_{\mathbf{x}}(x_1^{(1)}, x_2^{(2)}, x_3^{(1)}) \\
&\qquad + F_{\mathbf{x}}(x_1^{(1)}, x_2^{(1)}, x_3^{(2)}) - F_{\mathbf{x}}(x_1^{(1)}, x_2^{(1)}, x_3^{(1)}).
\end{aligned} \qquad (4.4)$$

The cumulative distribution function $F_{\mathbf{x}}$ must be such that this probability is nonnegative for all sets $(x_1^{(1)}, x_2^{(1)}, \ldots, x_n^{(1)}), (x_1^{(2)}, x_2^{(2)}, \ldots, x_n^{(2)})$ for which $x_k^{(2)} \ge x_k^{(1)}, k = 1, 2, \ldots, n$.

When the cumulative distribution function is a continuous and differentiable function of its variables, we can define the joint probability density function of the n random variables $\mathbf{x}_1, \mathbf{x}_2, \ldots, \mathbf{x}_n$ by

$$f_{\mathbf{x}}(x_1, x_2, \ldots, x_n) = \frac{\partial^n}{\partial x_1\, \partial x_2 \cdots \partial x_n} F_{\mathbf{x}}(x_1, x_2, \ldots, x_n). \qquad (4.5)$$

It provides a more convenient way of describing the probability measure assigned to the outcomes of the experiment; we saw that even for $n = 2$ the expression for the joint cumulative distribution function is usually cumbersome.

The probability that the outcome point $\zeta = (\mathbf{x}_1, \mathbf{x}_2, \ldots, \mathbf{x}_n)$ lies in an infinitesimal rectangular parallelepiped of sides $dx_1, dx_2, \ldots, dx_n$ equals

$$\Pr(x_1 < \mathbf{x}_1 < x_1 + dx_1, x_2 < \mathbf{x}_2 < x_2 + dx_2, \ldots, x_n < \mathbf{x}_n < x_n + dx_n)$$
$$= \Pr(x_k < \mathbf{x}_k < x_k + dx_k, \forall k) \tag{4.6}$$
$$= f_{\mathbf{x}}(x_1, x_2, \ldots, x_n)\, dx_1\, dx_2 \cdots dx_n.$$

(We write "$\forall$ k," read "all k," for $k = 1, 2, \ldots, n$.) The probability of the event represented by a particular region $\mathscr{A}$ of the n-dimensional space R_n is, as in (3.9), given by the integral of the joint probability density function over that region,

$$\Pr((\mathbf{x}_1, \mathbf{x}_2, \ldots, \mathbf{x}_n) \in \mathscr{A}) \tag{4.7}$$
$$= \iint_{\mathscr{A}} f_{\mathbf{x}}(x_1, x_2, \ldots, x_n)\, dx_1\, dx_2 \cdots dx_n.$$

In terms of the probability density function, the joint cumulative distribution is

$$F_{\mathbf{x}}(x_1, x_2, \ldots, x_n) \tag{4.8}$$
$$= \int_{-\infty}^{x_1}\int_{-\infty}^{x_2} \cdots \int_{-\infty}^{x_n} f_{\mathbf{x}}(x_1', x_2', \ldots, x_n')\, dx_1'\, dx_2' \cdots dx_n'.$$

The joint probability density function is nonnegative everywhere,

$$f_{\mathbf{x}}(x_1, x_2, \ldots, x_n) \ge 0, \tag{4.9}$$

and by (4.3) and (4.7) its integral over the entire space R_n must equal 1,

$$\int_{-\infty}^{\infty}\int_{-\infty}^{\infty} \cdots \int_{-\infty}^{\infty} f_{\mathbf{x}}(x_1, x_2, \ldots, x_n)\, dx_1\, dx_2 \cdots dx_n = 1. \tag{4.10}$$

When our chance experiment has but a countable number of outcomes ζ and each is labeled by n numbers $\mathbf{x}_1, \mathbf{x}_2, \ldots, \mathbf{x}_n$, these numbers $\mathbf{x}_1, \mathbf{x}_2, \ldots, \mathbf{x}_n$ constitute a set of n discrete random variables. They might, for instance, be the numbers of photoelectrons emitted by n photoelectric detectors placed at n different places during a certain interval of time, or the numbers emitted by a single detector during n successive but disjoint intervals. Such outcomes can be represented by probability "masses" located at various points $(x_1^{(k)}, x_2^{(k)}, \ldots, x_n^{(k)})$ in an n-dimensional space R_n. To the kth such point a probability "mass" $\Pr(\mathbf{x}_1 = x_1^{(k)}, \mathbf{x}_2 = x_2^{(k)}, \ldots, \mathbf{x}_n = x_n^{(k)})$ is assigned, and the sum of the probabilities assigned to all such points must equal 1,

$$\sum_k \Pr(\mathbf{x}_1 = x_1^{(k)}, \mathbf{x}_2 = x_2^{(k)}, \ldots, \mathbf{x}_n = x_n^{(k)}) = 1, \tag{4.11}$$

where $(x_1^{(k)}, x_2^{(k)}, \ldots, x_n^{(k)})$ are the coordinates of the kth point to which some probability has been allocated. Equivalently, we can represent such a probability measure through a joint probability density function involving n-dimensional delta functions,

$$f_{\mathbf{x}}(x_1, x_2, \ldots, x_n) = \sum_k \Pr(\mathbf{x}_1 = x_1^{(k)}, \mathbf{x}_2 = x_2^{(k)}, \ldots, \mathbf{x}_n = x_n^{(k)})$$
$$\times\, \delta(x_1 - x_1^{(k)})\delta(x_2 - x_2^{(k)}) \cdots \delta(x_n - x_n^{(k)}). \tag{4.12}$$

The n-dimensional delta function is a product of n ordinary delta functions, in direct analogy to the two-dimensional ones defined on page 123. Now the probability of an event represented by a region $\mathscr{A}$ in the space R_n is the sum of the probabilities attached to those mass points lying within $\mathscr{A}$,

$$\begin{aligned}\Pr\,((\mathbf{x}_1, \mathbf{x}_2, \ldots, \mathbf{x}_n) \in \mathscr{A}) \\ = \sum_{\mathbf{x}^{(k)} \in \mathscr{A}} \Pr\,(\mathbf{x}_1 = x_1^{(k)}, \mathbf{x}_2 = x_2^{(k)}, \ldots, \mathbf{x}_n = x_n^{(k)}),\end{aligned} \tag{4.13}$$

where $\mathbf{x}^{(k)} \in \mathscr{A}$ means that the point with coordinates $(x_1^{(k)}, x_2^{(k)}, \ldots, x_n^{(k)})$ lies within the region $\mathscr{A}$. Equation (4.12) substituted into (4.7) of course yields (4.13), the delta functions situated within $\mathscr{A}$ each contributing the amount of probability attached to them by (4.12), those situated outside $\mathscr{A}$ contributing zero.

Marginal density functions of subsets of the n original random variables are formed by integrating over the variables to be eliminated. If, for example, we are given the four-dimensional joint probability density function $f_{\mathbf{x}}(x_1, x_2, x_3, x_4)$, we can define such marginal density functions of lower order as

$$f_{\mathbf{x}_1\mathbf{x}_2\mathbf{x}_4}(x_1, x_2, x_4) = \int_{-\infty}^{\infty} f_{\mathbf{x}}(x_1, x_2, x_3, x_4)\, dx_3, \tag{4.14}$$

$$f_{\mathbf{x}_1\mathbf{x}_4}(x_1, x_4) = \int_{-\infty}^{\infty}\int_{-\infty}^{\infty} f_{\mathbf{x}}(x_1, x_2, x_3, x_4)\, dx_2\, dx_3, \tag{4.15}$$

$$f_{\mathbf{x}_2}(x_2) = \int_{-\infty}^{\infty}\int_{-\infty}^{\infty}\int_{-\infty}^{\infty} f_{\mathbf{x}}(x_1, x_2, x_3, x_4)\, dx_1\, dx_3\, dx_4. \tag{4.16}$$

If $f_{\mathbf{x}}(x_1, x_2, x_3, x_4)$ is a proper density function, that is, satisfies (4.9) and (4.10), these marginal density functions will also possess those essential characteristics.

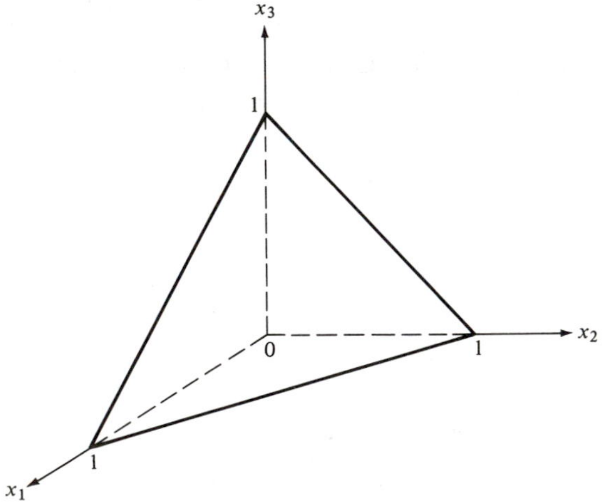

FIGURE 4.1 Defining region of density function in Example 4.1; $n = 3$.

EXAMPLE 4.1. The joint probability density function of n random variables $\mathbf{x}_1$, $\mathbf{x}_2, \ldots, \mathbf{x}_n$ is constant and equal to

$$f_{\mathbf{x}}(x_1, x_2, \ldots, x_n) \equiv C > 0$$

over a region Δ_n in R_n for which $x_k \geq 0$, $\forall\, k$, and

$$0 \leq x_1 + x_2 + \cdots x_n \leq 1.$$

This region is illustrated for $n = 3$ in Fig. 4.1. Elsewhere the joint probability density function vanishes. The constant C can be determined from (4.10), but it is simpler first to calculate all the marginal probability density functions of order less than n, after which C can be determined by normalizing, say, $f_{\mathbf{x}_1}(x_1)$.

Writing down only the nonzero portions of these probability density functions, we obtain successively

$$\begin{aligned} &f_{\mathbf{x}}(x_1, x_2, \ldots, x_{n-1}) \\ &\quad = \int_0^{1-x_1-\cdots-x_{n-1}} C\, dx_n = C(1 - x_1 - x_2 - \cdots - x_{n-1}), \end{aligned}$$

$$x_k \geq 0, \quad 1 \leq k \leq n-1, \qquad 0 \leq x_1 + x_2 + \cdots + x_{n-1} \leq 1;$$

$$\begin{aligned} &f_{\mathbf{x}}(x_1, x_2, \ldots, x_{n-2}) \\ &\quad = C\int_0^{1-x_1-\cdots-x_{n-2}} (1 - x_1 - x_2 - \cdots - x_{n-2} - x_{n-1})\, dx_{n-1} \\ &\quad = \tfrac{1}{2}C(1 - x_1 - \cdots - x_{n-2})^2, \qquad 0 \leq x_1 + \cdots + x_{n-2} \leq 1; \end{aligned}$$

and so on. By induction we find

$$f_{\mathbf{x}}(x_1, x_2, \ldots, x_{n-k}) = \frac{1}{k!} C(1 - x_1 - \cdots - x_{n-k})^k,$$

or

$$f_{\mathbf{x}}(x_1, x_2, \ldots, x_m) = \frac{1}{(n-m)!} C(1 - x_1 - \cdots - x_m)^{n-m},$$
$$0 \leq x_1 + x_2 \cdots + x_m \leq 1;$$

and finally, with $m = 1$,

$$f_{\mathbf{x}}(x_1) = \frac{C}{(n-1)!}(1 - x_1)^{n-1}, \qquad 0 \leq x_1 \leq 1.$$

Integrating this over $0 \leq x_1 \leq 1$, and setting the result equal to 1, we find that $C = n!$. Hence

$$\begin{aligned} f_{\mathbf{x}}(x_1, x_2, \ldots, x_m) &= \frac{n!}{(n-m)!}(1 - x_1 - x_2 - \cdots - x_m)^{n-m}, \\ x_k \geq 0, \quad &\forall\, k, \quad 0 \leq x_1 + x_2 + \cdots + x_m \leq 1, \quad 1 \leq m \leq n, \end{aligned} \tag{4.17}$$

is the marginal joint probability density function of the m random variables $\mathbf{x}_1, \mathbf{x}_2, \ldots, \mathbf{x}_m$. By virtue of the symmetry among our random variables in this example, other joint probability density functions can be formed by replacing $\mathbf{x}_1, \mathbf{x}_2, \ldots, \mathbf{x}_m$ by any selection of m out of the n original random variables. For instance,

$$f_{\mathbf{x}_2\mathbf{x}_3\mathbf{x}_5}(x_2, x_3, x_5) = \frac{n!}{(n-3)!}(1 - x_2 - x_3 - x_5)^{n-3},$$
$$0 \leq x_2 + x_3 + x_5 \leq 1, \quad x_2 \geq 0, \quad x_3 \geq 0, \quad x_5 \geq 0.$$

In this way we have generated a great many multivariate joint probability density functions. The one we treated in Example 3.2 is a member of this class for $n = 3$, $m = 2$. ■

EXAMPLE 4.2. Spherically Symmetrical Density Functions

A spherically symmetrical probability density function has the form

$$f_{\mathbf{x}}(x_1, x_2, \ldots, x_n) = p\left(\sqrt{x_1^2 + x_2^2 + \cdots + x_n^2}\right) \tag{4.18}$$

everywhere in R_n, where $p(r)$ is a nonnegative function of the single variable r. To subject this joint density function to the normalization condition (4.10) we must evaluate its integral over all of R_n. We can do so by utilizing the fact that the probability density function $f_{\mathbf{x}}(x_1, x_2, \ldots, x_n)$ is constant over a hyperspherical shell. The volume dV of a hyperspherical shell of radius r and thickness dr in n-dimensional space is proportional to $r^{n-1}\,dr$,

$$dV = k_n r^{n-1}\,dr. \tag{4.19}$$

Hence

$$\int_{-\infty}^{\infty}\int_{-\infty}^{\infty}\int_{-\infty}^{\infty} f_{\mathbf{x}}(x_1, x_2, \ldots, x_n)\,dx_1\,dx_2 \cdots dx_n = k_n\int_0^{\infty} r^{n-1}p(r)\,dr = 1. \tag{4.20}$$

The value of the constant k_n can be looked up in a text on the calculus,* or we can work it out for ourselves by picking a particular joint density function for which both sides of (4.20) are easily evaluated. Let us try an nth-order multivariate Gaussian density function

$$f_{\mathbf{x}}(x_1, x_2, \ldots, x_n) = (2\pi)^{-n/2}\exp\left[-\tfrac{1}{2}(x_1^2 + x_2^2 + \cdots + x_n^2)\right],$$

for which

$$p(r) = (2\pi)^{-n/2}\exp\left(-\tfrac{1}{2}r^2\right).$$

It is easy to integrate this over R_n to verify that it is properly normalized,

$$\int_{-\infty}^{\infty}\int_{-\infty}^{\infty}\cdots\int_{-\infty}^{\infty} f_{\mathbf{x}}(x_1, x_2, \ldots, x_n)\,dx_1\,dx_2 \cdots dx_n$$

$$= (2\pi)^{-n/2}\int_{-\infty}^{\infty}\exp\left(-\frac{x_1^2}{2}\right)dx_1\int_{-\infty}^{\infty}\exp\left(-\frac{x_2^2}{2}\right)dx_2\cdots\int_{-\infty}^{\infty}\exp\left(-\frac{x_n^2}{2}\right)dx_n$$

$$= 1.$$

The integral in (4.20) is then

$$\begin{aligned} 1 &= k_n(2\pi)^{-n/2}\int_0^{\infty} r^{n-1}e^{-r^2/2}\,dr \\ &= \tfrac{1}{2}k_n\pi^{-n/2}\int_0^{\infty} x^{(n-2)/2}e^{-x}\,dx \\ &= \tfrac{1}{2}k_n\pi^{-n/2}\Gamma\left(\frac{n}{2}\right) \end{aligned}$$

in terms of the gamma function defined in (2.27). Hence

$$k_n = \frac{2\pi^{n/2}}{\Gamma(n/2)}. \tag{4.21}$$

*C. H. Edwards, *Advanced Calculus of Several Variables*. New York: Academic Press, 1973. See p. 339.

For $n = 3$, for instance, this gives

$$k_3 = \frac{2\pi^{3/2}}{\Gamma(\frac{3}{2})} = \frac{2\pi^{3/2}}{\frac{1}{2}\sqrt{\pi}} = 4\pi,$$

as expected; the area of a sphere of radius r is $4\pi r^2$.

Thus if (4.18) is to be a proper density function,

$$p(r) \geq 0,$$

and $p(r)$ must be so normalized that

$$\int_0^\infty r^{n-1} p(r)\, dr = \frac{1}{k_n} = \tfrac{1}{2}\pi^{-n/2}\Gamma\left(\frac{n}{2}\right). \tag{4.22}$$

If, for instance, the density function $f_{\mathbf{x}}(x_1, x_2, \ldots, x_n)$ equals a constant C over a hypersphere of radius ρ and equals zero outside it,

$$C\int_0^\rho r^{n-1}\, dr = \frac{C}{n}\rho^n = \frac{1}{k_n},$$

whence

$$C = \tfrac{1}{2} n \pi^{-n/2}\Gamma\left(\frac{n}{2}\right)\rho^{-n}. \tag{4.23}$$

The quantity C^{-1} equals the volume of an n-dimensional hypersphere of radius ρ. ■

EXAMPLE 4.3. The n discrete random variables $\mathbf{x}_1, \mathbf{x}_2, \ldots, \mathbf{x}_n$ take on only nonnegative integral values $k_1, k_2, \ldots, k_n$ with such joint probabilities that

$$\Pr(\mathbf{x}_1 = k_1, \mathbf{x}_2 = k_2, \ldots, \mathbf{x}_n = k_n) = q(k_1 + k_2 + \cdots + k_n) \tag{4.24}$$

for some nonnegative function $q(m)$ of the integers $m = 0, 1, 2, \ldots$. The probability is concentrated at the points of an n-dimensional hypercubical lattice. In order for these probabilities to sum to 1, the function $q(m)$ must be such that

$$\sum_{m=0}^\infty K(n, m) q(m) = 1,$$

where $K(n, m)$ is the number of lattice points in the plane $x_1 + x_2 + \cdots + x_n = m$. This number can be derived by clever combinatorics, or we can evaluate it by taking the special distribution

$$\Pr(\mathbf{x}_1 = k_1, \ldots, \mathbf{x}_n = k_n) = (1 - v)^n v^{k_1} \cdots v^{k_n}, \qquad 0 < v < 1, \quad k_j \geq 0, \forall j,$$

for which

$$q(m) = (1 - v)^n v^m.$$

The reader should verify that this distribution satisfies (4.11). Then for all values of v in $0 < v < 1$,

$$(1 - v)^n \sum_{m=0}^\infty K(n, m) v^m = 1$$

or

$$\begin{aligned}\sum_{m=0}^\infty K(n, m) v^m &= (1 - v)^{-n} \\ &= 1 + nv + \tfrac{1}{2}n(n + 1)v^2 \\ &\quad + \cdots \frac{n(n + 1)\cdots(n + m - 1)}{m!} v^m + \cdots\end{aligned}$$

by Taylor's theorem. Equating the coefficients of v^m on both sides, we find that the number of lattice points in the hyperplane $x_1 + x_2 + \cdots + x_n = m$ is

$$K(n, m) = \binom{n + m - 1}{n - 1}. \tag{4.25}$$

By drawing two- and three-dimensional lattices it is easy to check that

$$K(2, m) = \binom{m + 1}{1} = m + 1,$$

$$K(3, m) = \binom{m + 2}{2} = \tfrac{1}{2}(m + 1)(m + 2) = \sum_{k=1}^{m+1} k.$$

The function $q(m)$ in (4.24) must therefore be normalized so that

$$\sum_{m=0}^{\infty} \binom{n + m - 1}{n - 1} q(m) = 1. \tag{4.26}$$

If now the n discrete random variables $\mathbf{x}_1, \mathbf{x}_2, \ldots, \mathbf{x}_n$ have equal joint probabilities C_N for all nonnegative integers such that

$$\mathbf{x}_1 + \mathbf{x}_2 + \cdots + \mathbf{x}_n \leq N,$$

where N is some nonnegative integer, and if their joint probabilities are zero elsewhere, the constant C_N must be such that

$$C_N^{-1} = \sum_{m=0}^{N} \binom{n + m - 1}{n - 1} = \binom{n + N}{n},$$

as can be shown by induction:

$$C_{N+1}^{-1} = C_N^{-1} + \binom{n + N}{n - 1} = \binom{n + N}{n} + \binom{n + N}{n - 1}$$

$$= \binom{n + N + 1}{n}.$$

(Write out the binomial coefficients in terms of factorials and combine.) Here C_N^{-1} is the number of lattice points $k_1, k_2, \ldots, k_n$ in the positive orthant of R_n such that $0 \leq k_1 + k_2 + \cdots + k_n \leq N$. ■

4-2

MULTIVARIATE CONDITIONAL DISTRIBUTIONS

A. General Conditioning Events

Multivariate joint probability distributions can be conditioned by nonnumerical attributes in the same way as bivariate distributions of random variables are so conditioned, as described in Sec. 3-2A. Nothing fundamentally new in this respect arises with sets of more than two random variables. We turn to a summary of the treatment of conditioning events defined in numerical terms.

When it is observed that the outcome $(\mathbf{x}_1, \mathbf{x}_2, \ldots, \mathbf{x}_n)$ of the experiment lies in a certain region $\mathscr{A}$ of the n-dimensional space R_n, the conditional probability density function of these random variables, given this event

$$(\mathbf{x}_1, \mathbf{x}_2, \ldots, \mathbf{x}_n) \in \mathscr{A},$$

is defined as in (3.20),

$$\begin{aligned} f_{\mathbf{x}}(x_1, x_2, \ldots, x_n | (\mathbf{x}_1, \mathbf{x}_2, \ldots, \mathbf{x}_n) \in \mathscr{A}) &= 0, \qquad (x_1, x_2, \ldots, x_n) \notin \mathscr{A}, \\ &= \frac{f_{\mathbf{x}}(x_1, x_2, \ldots, x_n)}{\Pr(\mathscr{A})}, \qquad (x_1, x_2, \ldots, x_n) \in \mathscr{A}, \end{aligned} \tag{4.27}$$

where

$$\begin{aligned} \Pr(\mathscr{A}) &= \Pr((\mathbf{x}_1, \mathbf{x}_2, \ldots, \mathbf{x}_n) \in \mathscr{A}) \\ &= \int \cdots \int_{\mathscr{A}} f_{\mathbf{x}}(x_1, x_2, \ldots, x_n)\, dx_1\, dx_2 \cdots dx_n \end{aligned} \tag{4.28}$$

is the probability of the conditioning event. If the $\mathbf{x}$'s are discrete random variables,

$$\begin{aligned} \Pr(\mathbf{x}_1 = x_1^{(k)}, \ldots, \mathbf{x}_n = x_n^{(k)} | (\mathbf{x}_1, \ldots, \mathbf{x}_n) \in \mathscr{A}) &= 0, \qquad (x_1^{(k)}, \ldots, x_n^{(k)}) \notin \mathscr{A}, \\ &= \frac{\Pr(\mathbf{x}_1 = x_1^{(k)}, \ldots, \mathbf{x}_n = x_n^{(k)})}{\Pr(\mathscr{A})}, \quad (x_1^{(k)}, \ldots, x_n^{(k)}) \in \mathscr{A}, \end{aligned} \tag{4.29}$$

where now

$$\Pr(\mathscr{A}) = \sum_{\mathbf{x}^{(k)} \in \mathscr{A}} \Pr(\mathbf{x}_1 = x_1^{(k)}, \ldots, \mathbf{x}_n = x_n^{(k)}), \tag{4.30}$$

the sum being taken over those mass points

$$\mathbf{x}^{(k)} = (x_1^{(k)}, x_2^{(k)}, \ldots, x_n^{(k)})$$

lying within the region $\mathscr{A}$.

EXAMPLE 4.4. As in Example 4.2, let the joint density function of the n random variables be

$$\begin{aligned} f_{\mathbf{x}}(x_1, x_2, \ldots, x_n) &= \tfrac{1}{2} n \pi^{-n/2} \Gamma\left(\frac{n}{2}\right) \rho^{-n}, \qquad x_1^2 + x_2^2 + \cdots + x_n^2 < \rho^2. \\ &= 0, \qquad x_1^2 + x_2^2 + \cdots + x_n^2 > \rho^2. \end{aligned}$$

It is observed in a trial of the experiment that all the random variables are positive, $\mathbf{x}_k > 0$, $\forall\, k$. The probability of this event is 2^{-n}. Then the joint conditional density function of these random variables, given that they are positive, is

$$f_{\mathbf{x}}(x_1, x_2, \ldots, x_n \mid \mathbf{x}_k > 0, \forall k) = 2^{n-1} n \pi^{-n/2} \Gamma\left(\frac{n}{2}\right) \rho^{-n},$$
$$x_1 > 0, x_2 > 0, \ldots, x_n > 0, x_1^2 + x_2^2 + \cdots + x_n^2 < \rho^2,$$

and it equals zero everywhere else in the n-dimensional space R_n. ■

EXAMPLE 4.5. The joint probability density function of the n random variables is

$$f_{\mathbf{x}}(x_1, x_2, \ldots, x_n) = (2\pi)^{-n/2} \exp\left[-\tfrac{1}{2}(x_1^2 + x_2^2 + \cdots + x_n^2)\right]$$

It is observed that

$$\mathbf{x}_1^2 + \mathbf{x}_2^2 + \cdots + \mathbf{x}_n^2 > b^2.$$

The probability of this event $\mathcal{A}$ is calculated by using the fact that the joint density function is constant at all points on a hypersphere

$$x_1^2 + x_2^2 + \cdots + x_n^2 = r^2.$$

Hence

$$\Pr(\mathcal{A}) = \int \cdots \int_{x_1^2 + \cdots + x_n^2 > b^2} f_{\mathbf{x}}(x_1, x_2, \ldots, x_n)\, dx_1\, dx_2 \cdots dx_n$$
$$= k_n \int_b^\infty r^{n-1} p(r)\, dr,$$

where k_n is given in (4.21) and

$$p(r) = (2\pi)^{-n/2} \exp\left(-\tfrac{1}{2} r^2\right).$$

We find for this probability

$$\Pr(\mathcal{A}) = k_n (2\pi)^{-n/2} \int_b^\infty r^{n-1} e^{-r^2/2}\, dr$$
$$= \tfrac{1}{2} k_n \pi^{-n/2} \int_{b^2/2}^\infty x^{(n-2)/2} e^{-x}\, dx$$
$$= \frac{1}{\Gamma(n/2)} \int_{b^2/2}^\infty x^{(n-2)/2} e^{-x}\, dx,$$

whose value can be determined from a table of the incomplete gamma function. For n even, the integral can be evaluated by integration by parts.

The joint conditional density function is now

$$f_{\mathbf{x}}(x_1, x_2, \ldots, x_n \mid \mathbf{x}_1^2 + \mathbf{x}_2^2 + \cdots + \mathbf{x}_n^2 > b^2) = 0, \quad x_1^2 + x_2^2 + \cdots + x_n^2 < b^2$$
$$= [\Pr(\mathcal{A})]^{-1} (2\pi)^{-n/2} \exp\left[-\tfrac{1}{2}(x_1^2 + x_2^2 + \cdots + x_n^2)\right],$$
$$x_1^2 + x_2^2 + \cdots + x_n^2 > b^2.$$ ■

B. Assumption of Particular Values by Certain Random Variables

Considering only continuous random variables, we look for counterparts to conditional probability density functions like $f_{\mathbf{y}}(y \mid x)$ defined in (3.27). Let us first take $\mathscr{A}$ to be the region of R_n bounded by the hyperplanes $x_m = a$, $x_m = b$, $b > a$. Then by (4.28),

$$\begin{aligned} f_{\mathbf{x}}(x_1, x_2, \ldots, x_n \mid a < \mathbf{x}_m < b) & \\ &= 0, \qquad x_m < a, \quad x_m > b, \\ &= \frac{f_{\mathbf{x}}(x_1, x_2, \ldots, x_n)}{\Pr(a < \mathbf{x}_m < b)}, \qquad a < x_m < b. \end{aligned} \tag{4.31}$$

The random variables other than $\mathbf{x}_m$ are left arbitrary. The denominator is

$$\Pr(a < \mathbf{x}_m < b) = \int_a^b f_{\mathbf{x}_m}(x)\, dx$$

in terms of the marginal density function of $\mathbf{x}_m$. Integrating now over $-\infty < x_m < \infty$, we obtain the conditional joint probability density function

$$\begin{aligned} f_{\mathbf{x}}(x_1, \ldots, x_{m-1}, x_{m+1}, \ldots, x_n \mid a < \mathbf{x}_m < b) & \\ = \frac{\int_a^b f_x(x_1, \ldots, x_{m-1}, x'_m, x_{m+1}, \ldots, x_n)\, dx'_m}{\Pr(a < \mathbf{x}_m < b)}. \end{aligned}$$

If we now put

$$a = x_m - \tfrac{1}{2}\varepsilon, \qquad b = x_m + \tfrac{1}{2}\varepsilon$$

and let $\varepsilon > 0$ decrease to zero, we find as in (3.25) to (3.27),

$$\begin{aligned} f_{\mathbf{x}}(x_1, \ldots, x_{m-1}, x_{m+1}, \ldots, x_n \mid \mathbf{x}_m = x_m) & \\ = \frac{f_{\mathbf{x}}(x_1, \ldots, x_{m-1}, x_m, x_{m+1}, \ldots, x_n)}{f_{\mathbf{x}_m}(x_m)}. \end{aligned} \tag{4.32}$$

This conditional probability density function is ordinarily written

$$f_{\mathbf{x}}(x_1, \ldots, x_{m-1}, x_{m+1}, \ldots, x_n \mid x_m).$$

Similar joint conditional probability density functions can be defined in which the members of any subset of the random variables $\mathbf{x}_1, \mathbf{x}_2, \ldots, \mathbf{x}_n$ take on particular values. For instance,

$$\begin{aligned} & f_{\mathbf{x}}(x_1, \ldots, x_{k-1}, x_{k+1}, \ldots, x_{m-1}, x_{m+1}, \ldots, x_n \mid x_k, x_m) \\ &= f_{\mathbf{x}}(x_1, \ldots, x_{k-1}, x_{k+1}, \ldots, x_{m-1}, x_{m+1}, \ldots, x_n \mid \mathbf{x}_k = x_k, \mathbf{x}_m = x_m) \\ &= \frac{f_{\mathbf{x}}(x_1, \ldots, x_{k-1}, x_k, x_{k+1}, \ldots, x_{m-1}, x_m, x_{m+1}, \ldots, x_n)}{f_{\mathbf{x}_k\mathbf{x}_m}(x_k, x_m)}. \end{aligned} \tag{4.33}$$

The rule for forming such conditional probability density functions is this: if the members of any subset of the random variables are observed to take on particular numerical values, the joint conditional density function of the remaining variables is found by dividing the joint probability density function of all the random variables $\mathbf{x}_1, \mathbf{x}_2, \ldots, \mathbf{x}_n$ by the joint probability density function of the observed (conditioning) random variables. That is, to move any random variables to the right-hand side of the conditioning bar (|), divide by their joint density function. For instance, when $\mathbf{x}_2 = x_2$, $\mathbf{x}_3 = x_3$, $\mathbf{x}_5 = x_5$, the joint conditional density function of the remaining random variables has x_2, x_3, x_5 to the right of the bar,

$$f_{\mathbf{x}}(x_1, x_2, x_3, x_4, x_5|) \rightarrow f_{\mathbf{x}}(x_1, x_4 \mid x_2, x_3, x_5)$$

and our rule prescribes that

$$f_{\mathbf{x}_1\mathbf{x}_4}(x_1, x_4 \mid x_2, x_3, x_5) = \frac{f_{\mathbf{x}}(x_1, x_2, x_3, x_4, x_5)}{f_{\mathbf{x}}(x_2, x_3, x_5)}. \tag{4.34}$$

The density function appearing in the denominator of such a definition must of course be positive and not zero.

Let us derive a special case of this rule,

$$f_{\mathbf{x}_3}(x_3 \mid x_1, x_2) = \frac{f_{\mathbf{x}_1\mathbf{x}_2\mathbf{x}_3}(x_1, x_2, x_3)}{f_{\mathbf{x}_1\mathbf{x}_2}(x_1, x_2)} \tag{4.35}$$

in order to illustrate the import of such conditional density functions. We work in three-dimensional space R_3, as in Fig. 4.2, considering first the event $(x_1 < \mathbf{x}_1 < x_1 + \varepsilon_1, x_2 < \mathbf{x}_2 < x_2 + \varepsilon_2)$ with ε_1 and ε_2 positive, but infinitesimal. It is represented by an infinitely tall, vertical cylinder standing perpendicular to the X_1X_2-plane and cutting it in the infinitesimal $\varepsilon_1 \times \varepsilon_2$ rectangle defined by

$$x_1 < \mathbf{x}_1 < x_1 + \varepsilon_1, \qquad x_2 < \mathbf{x}_2 < x_2 + \varepsilon_2,$$

as shown in Fig. 4.2. By the definition of conditional probability,

$$\begin{aligned}&\Pr(x_3 < \mathbf{x}_3 < x_3 + dx_3 \mid x_1 < \mathbf{x}_1 < x_1 + \varepsilon_1, x_2 < \mathbf{x}_2 < x_2 + \varepsilon_2)\\ &= \frac{\Pr[(x_3 < \mathbf{x}_3 < x_3 + dx_3) \cap (x_1 < \mathbf{x}_1 < x_1 + \varepsilon_1, x_2 < \mathbf{x}_2 < x_2 + \varepsilon_2)]}{\Pr(x_1 < \mathbf{x}_1 < x_1 + \varepsilon_1, x_2 < \mathbf{x}_2 < x_2 + \varepsilon_2)}.\end{aligned}$$

The event $x_3 < \mathbf{x}_3 < x_3 + dx_3$ is represented by the region between two horizontal planes at $\mathbf{x}_3 = x_3$ and $\mathbf{x}_3 = x_3 + dx_3$, respectively. The intersection of the slab of thickness dx_3 with the thin vertical rectangular cylinder

$$(x_1 < \mathbf{x}_1 < x_1 + \varepsilon_1, x_2 < \mathbf{x}_2 < x_2 + \varepsilon_2)$$

is the infinitesimal rectangular parallelepiped $\varepsilon_1 \times \varepsilon_2 \times dx_3$ whose lower left back corner is located at (x_1, x_2, x_3). The probability in the numerator of our quotient equals the probability measure

$$f_{\mathbf{x}_1\mathbf{x}_2\mathbf{x}_3}(x_1, x_2, x_3)\varepsilon_1\varepsilon_2\, dx_3$$

of that infinitesimal parallelepiped, and the probability in the denominator, according to Chapter 3, is

$$f_{\mathbf{x}_1\mathbf{x}_2}(x_1, x_2)\varepsilon_1\varepsilon_2.$$

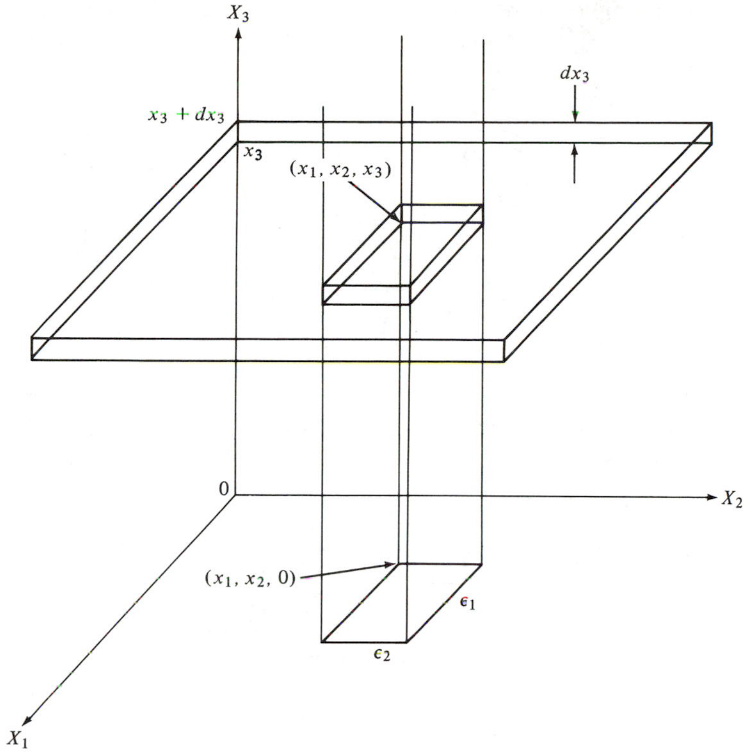

FIGURE 4.2 Regions involved in the course of defining $f_{\mathbf{x}_3}(x_3 \mid x_1, x_2)$.

Hence the conditional probability sought is

$$\Pr(x_3 < \mathbf{x}_3 < x_3 + dx_3 \mid x_1 < \mathbf{x}_1 < x_1 + \varepsilon_1, x_2 < \mathbf{x}_2 < x_2 + \varepsilon_2)$$

$$\doteq \frac{f_{\mathbf{x}_1\mathbf{x}_2\mathbf{x}_3}(x_1, x_2, x_3)\varepsilon_1\varepsilon_2\, dx_3}{f_{\mathbf{x}_1\mathbf{x}_2}(x_1, x_2)\varepsilon_1\varepsilon_2}.$$

If we now let ε_1 and ε_2 go to zero and recall the definition of a probability density function (2.12), we find that

$$f_{\mathbf{x}_3}(x_3 \mid \mathbf{x}_1 = x_1, \mathbf{x}_2 = x_2)\, dx_3 = \frac{f_{\mathbf{x}_1\mathbf{x}_2\mathbf{x}_3}(x_1, x_2, x_3)\, dx_3}{f_{\mathbf{x}_1\mathbf{x}_2}(x_1, x_2)},$$

which yields (4.35). A similar argument can be invoked to derive all other conditional density functions of this kind.

These conditional density functions are probability density functions in their own right and obey the requirements that they be nonnegative and normalized to 1. For example,

$$\int_{-\infty}^{\infty}\int_{-\infty}^{\infty} f_{\mathbf{x}_1\mathbf{x}_4}(x_1, x_4 \mid x_2, x_3, x_5)\, dx_1\, dx_4$$

$$= \frac{\displaystyle\int_{-\infty}^{\infty}\int_{-\infty}^{\infty} f_{\mathbf{x}}(x_1, x_2, x_3, x_4, x_5)\, dx_1\, dx_4}{f_{\mathbf{x}}(x_2, x_3, x_5)}$$

$$= \frac{f_{\mathbf{x}}(x_2, x_3, x_5)}{f_{\mathbf{x}}(x_2, x_3, x_5)} = 1$$

by (4.34).

If, for instance, n random variables $\mathbf{x}_1, \mathbf{x}_2, \ldots, \mathbf{x}_n$, $n > 5$, have the constant joint probability density function $C = n!$ over the polyhedron Δ_n described in Example 4.1, then

$$f_{\mathbf{x}_1\mathbf{x}_4}(x_1, x_4 \mid x_2, x_3, x_5)$$

$$= \frac{n!}{(n-5)!}\,\frac{(n-3)!}{n!}\,\frac{(1 - x_1 - x_2 - x_3 - x_4 - x_5)^{n-5}}{(1 - x_2 - x_3 - x_5)^{n-3}},$$

$$x_1 \geq 0,\ x_4 \geq 0, \quad 0 \leq x_2 + x_3 + x_5 \leq 1,$$

$$0 \leq x_1 + x_4 \leq 1 - x_2 - x_3 - x_5.$$

This joint conditional density function vanishes elsewhere in the X_1X_4-plane, and it is meaningless if the point (x_2, x_3, x_5) lies outside the polyhedron specified by

$$x_2 \geq 0, \qquad x_3 \geq 0, \qquad x_5 \geq 0, \qquad 0 \leq x_2 + x_3 + x_5 \leq 1$$

in the three-dimensional space of these variables.

The rule for forming conditional probability density functions holds even when we start with conditional density functions. For instance,

$$\frac{f_{\mathbf{x}_1\mathbf{x}_3\mathbf{x}_4}(x_1, x_3, x_4 \mid x_2, x_5)}{f_{\mathbf{x}_3}(x_3 \mid x_2, x_5)} = f_{\mathbf{x}_1\mathbf{x}_4}(x_1, x_4 \mid x_2, x_3, x_5).$$

Moving x_3 to the right of the bar in

$$f_{\mathbf{x}_1\mathbf{x}_3\mathbf{x}_4}(x_1, x_3, x_4 \mid x_2, x_5)$$

entails dividing it by $f_{\mathbf{x}_3}(x_3 \mid x_2, x_5)$. This equation should be checked by substituting for each of the conditional density functions its definition in terms of unconditional joint probability density functions.

C. Statistical Independence

The n random variables $\mathbf{x}_1, \mathbf{x}_2, \ldots, \mathbf{x}_n$ are statistically independent if and only if their joint probability density function factors as the product of their marginal probability density functions,

$$f_{\mathbf{x}}(x_1, x_2, \ldots, x_n) = f_{\mathbf{x}_1}(x_1) f_{\mathbf{x}_2}(x_2) \cdots f_{\mathbf{x}_n}(x_n). \tag{4.36}$$

Conditional joint density functions of the kind defined in part B then do not depend on the conditioning variables:

$$f_{\mathbf{x}}(x_1, \ldots, x_{m-1}, x_{m+1}, \ldots x_n \mid x_m) = f_{\mathbf{x}}(x_1, \ldots, x_{m-1}, x_{m+1}, \ldots x_n),$$

$$f_{\mathbf{x}}(x_1, \ldots, x_{k-1}, x_{k+1}, \ldots, x_{m-1}, x_{m+1}, \ldots x_n \mid x_k, x_m)$$
$$= f_{\mathbf{x}}(x_1, \ldots, x_{k-1}, x_{k+1}, \ldots, x_{m-1}, x_{m+1}, \ldots x_n),$$

$$f_{\mathbf{x}_1\mathbf{x}_4}(x_1, x_4 \mid x_2, x_3, x_5) = f_{\mathbf{x}_1\mathbf{x}_4}(x_1, x_4),$$

and so on. Observing that some of the random variables have taken on particular values gives us no information about the rest. The random variables of Example 4.1 are not independent. If, on the other hand, the joint probability density function has the form

$$f_{\mathbf{x}}(x_1, x_2, \ldots, x_n)$$
$$= (2\pi)^{-n/2}(\sigma_1\sigma_2 \cdots \sigma_n)^{-1} \exp\left[-\frac{(x_1 - \bar{x}_1)^2}{2\sigma_1^2} - \frac{(x_2 - \bar{x}_2)^2}{2\sigma_2^2} - \cdots - \frac{(x_n - \bar{x}_n)^2}{2\sigma_n^2}\right] \tag{4.37}$$
$$= \prod_{k=1}^{n} (2\pi\sigma_k^2)^{-1/2} \exp\left[-\frac{(x_k - \bar{x}_k)^2}{2\sigma_k^2}\right],$$

the random variables are statistically independent. This is a special case of the nth-order multivariate Gaussian density function, whose general form will be treated in Sec. 4-5.

4-3

DISTRIBUTIONS OF FUNCTIONS OF MULTIPLE RANDOM VARIABLES

A. Single Function of *n* Random Variables

The methods developed in Sec. 3-4 for determining the probability density function of a function $g(\mathbf{x}, \mathbf{y})$ of two random variables can be extended to apply to calculating the probability density function of a function

$$\mathbf{z} = g(\mathbf{x}_1, \mathbf{x}_2, \ldots, \mathbf{x}_n) \tag{4.38}$$

on n random variables $\mathbf{x}_1, \mathbf{x}_2, \ldots, \mathbf{x}_n$. The manipulations are more complicated, and fewer problems can be solved by simple calculations, but the basic principles are the same.

METHOD 1

The event

$$\mathbf{z} = g(\mathbf{x}_1, \mathbf{x}_2, \ldots, \mathbf{x}_n) \le z$$

corresponds to a certain region $R(z)$ in the n-dimensional space R_n. The cumulative distribution function $F_{\mathbf{z}}(z)$ of the random variable $\mathbf{z}$ is found by integrating the joint density function $f_{\mathbf{x}}(x_1, x_2, \ldots, x_n)$ over this region, which is bounded by the surface

$$z = g(x_1, x_2, \ldots, x_n);$$

according to (4.7) it is

$$\begin{aligned} F_{\mathbf{z}}(z) &= \Pr\,(g(\mathbf{x}_1, \mathbf{x}_2, \ldots, \mathbf{x}_n) \le z) \\ &= \int \cdots \int_{R(z)} f_{\mathbf{x}}(x_1, x_2, \ldots, x_n)\, dx_1\, dx_2 \cdots dx_n. \end{aligned} \tag{4.39}$$

The probability density function $f_{\mathbf{z}}(z)$ of $\mathbf{z}$ can then be determined by differentiating the cumulative distribution function $F_{\mathbf{z}}(z)$ with respect to z.

EXAMPLE 4.6. $\mathbf{z} = \max\,(\mathbf{x}_1, \mathbf{x}_2, \ldots, \mathbf{x}_n)$

Now the random variable $\mathbf{z}$ is the greatest of the random variables $\mathbf{x}_1, \mathbf{x}_2, \ldots, \mathbf{x}_n$. The region R_z is specified by the inequalities $\mathbf{x}_k \le z,\ \forall\, k$, and is a rectangular polyhedral angle with vertex at the point $\mathbf{x}_k \equiv z,\ \forall\, k$; Fig. 3.13 illustrates $R(z)$ for $n = 2$. Then in terms of the joint cumulative distribution function of our n random variables,

$$F_{\mathbf{z}}(z) = \Pr\,(\mathbf{x}_k \le z,\ \forall\, k) = F_{\mathbf{x}}(z, z, \ldots, z). \tag{4.40}$$

Differentiating and using (4.8) we find the probability density function of $\mathbf{z}$ to be

$$\begin{aligned} f_{\mathbf{z}}(z) = &\int_{-\infty}^{z} dx_2 \cdots \int_{-\infty}^{z} dx_n\, f_{\mathbf{x}}(z, x_2, \ldots, x_n) \\ &+ \int_{-\infty}^{z} dx_1 \int_{-\infty}^{z} dx_3 \cdots \int_{-\infty}^{z} dx_n\, f_{\mathbf{x}}(x_1, z, x_3, \ldots, x_n) + \cdots \\ &+ \int_{-\infty}^{z} dx_1 \int_{-\infty}^{z} dx_2 \cdots \int_{-\infty}^{z} dx_{n-1}\, f_{\mathbf{x}}(x_1, x_2, \ldots, x_{n-1}, z). \end{aligned} \tag{4.41}$$

If the random variables are independent and identically distributed,

$$f_{\mathbf{x}}(x_1, x_2, \ldots, x_n) = \prod_{k=1}^{n} f_{\mathbf{x}}(x_k), \tag{4.42}$$

all terms in (4.41) contribute equally, and we find

$$f_{\mathbf{z}}(z) = n f_{\mathbf{x}}(z)[F_{\mathbf{x}}(z)]^{n-1}, \tag{4.43}$$

where $F_{\mathbf{x}}(z)$ is the common cumulative distribution function

$$F_{\mathbf{x}}(z) = \Pr\,(\mathbf{x} \le z) = \int_{-\infty}^{z} f_{\mathbf{x}}(x)\, dx \tag{4.44}$$

of the individual random variables.

The formula in (4.43) can be directly understood. If the largest of the random variables lies between z and $z + dz$, all the others must be less than z. Suppose that $\mathbf{x}_1$ is the largest. Then the probability that the largest lies between z and $z + dz$ is

$$\Pr(z < \mathbf{x}_1 < z + dz) \Pr(\mathbf{x}_2 < z, \mathbf{x}_3 < z, \ldots, \mathbf{x}_n < z)$$

$$= f_{\mathbf{x}_1}(z)\, dz \prod_{k=2}^{n} F_{\mathbf{x}_k}(z).$$

The largest of the random variables might, however, have been any of the n variables, $\mathbf{x}_1$ or $\mathbf{x}_2$ or $\cdots$ or $\mathbf{x}_n$, and these possibilities are mutually exclusive. Hence the probability density function of the maximum is

$$f_{\mathbf{z}}(z) = \sum_{m=1}^{n} f_{\mathbf{x}_m}(z) \prod_{k \neq m} F_{\mathbf{x}_k}(z). \tag{4.45}$$

and when the random variables are identically distributed as in (4.42), (4.45) reduces to (4.43). ■

METHOD 2

All but one of the random variables, say the first, are fixed at particular values: $\mathbf{x}_k \equiv x_k$, $k = 2, 3, \ldots, n$. Then the function $g(\mathbf{x}_1, x_2, \ldots, x_n)$ is random only through its dependence on $\mathbf{x}_1$, and its conditional probability density function

$$f_{\mathbf{z}}(z \mid \mathbf{x}_2 = x_2, \mathbf{x}_3 = x_3, \ldots, \mathbf{x}_n = x_n)$$

can be derived from the conditional probability density function $f_{\mathbf{x}_1}(x_1 \mid x_2, x_3, \ldots, x_n)$ by the methods of Sec. 2-3. If, for instance, $g(x_1, x_2, \ldots, x_n)$ is a monotone function of x_1,

$$f_{\mathbf{z}}(z \mid \mathbf{x}_2 = x_2, \ldots, \mathbf{x}_n = x_n) = \frac{f_{\mathbf{x}_1}(x_1 \mid x_2, \ldots, x_n)}{|\partial g / \partial x_1|}, \tag{4.46}$$

and into the right-hand side one must substitute

$$x_1 = g^{-1}(z, x_2, \ldots, x_n),$$

which is the solution of

$$z = g(x_1, x_2, \ldots, x_n)$$

for x_1 as a function of z and other x's. Now by the definition of a conditional probability density function

$$f_{\mathbf{z}\mathbf{x}_2 \ldots \mathbf{x}_n}(z, x_2, \ldots, x_n)$$
$$= f_{\mathbf{z}}(z \mid \mathbf{x}_2 = x_2, \ldots, \mathbf{x}_n = x_n) f_{\mathbf{x}}(x_2, \ldots, x_n)$$

and we find the probability density function $f_{\mathbf{z}}(z)$ by integrating over the entire ranges of $x_2, x_3, \ldots, x_n$,

$$f_{\mathbf{z}}(z) = \int_{-\infty}^{\infty} dx_2 \cdots \int_{-\infty}^{\infty} dx_n\, f_{\mathbf{z}}(z \mid \mathbf{x}_2 = x_2, \ldots, \mathbf{x}_n = x_n) \times f_{\mathbf{x}}(x_2, \ldots, x_n), \tag{4.47}$$

after substituting from (4.46).

EXAMPLE 4.7. $\mathbf{z} = \mathbf{x}_1 + \mathbf{x}_2 + \cdots + \mathbf{x}_n$

When we fix $\mathbf{x}_2 = x_2, \ldots, \mathbf{x}_n = x_n$,

$$\mathbf{x}_1 = \mathbf{z} - x_2 - x_3 - \cdots - x_n,$$

and the conditional probability density function of $\mathbf{z}$ is simply

$$\begin{aligned} f_{\mathbf{z}}(z \mid \mathbf{x}_2 = x_2, \ldots, \mathbf{x}_n = x_n) \\ &= f_{\mathbf{x}_1}(z - x_2 - \cdots - x_n \mid x_2, \ldots, x_n) \\ &= \frac{f_{\mathbf{x}}(z - x_2 - \cdots - x_n, x_2, \ldots, x_n)}{f_{\mathbf{x}}(x_2, \ldots, x_n)}. \end{aligned}$$

Substituting into (4.47), we find that

$$f_{\mathbf{z}}(z) = \int_{-\infty}^{\infty} dx_2 \cdots \int_{-\infty}^{\infty} dx_n\, f_{\mathbf{x}}(z - x_2 - \cdots - x_n, x_2, x_3, \ldots, x_n). \tag{4.48}$$

When the random variables $\mathbf{x}_1, \mathbf{x}_2, \ldots, \mathbf{x}_n$ are statistically independent, as in (4.36), this reduces to

$$f_{\mathbf{z}}(z) = \int_{-\infty}^{\infty} dx_2 \cdots \int_{-\infty}^{\infty} dx_n\, f_{\mathbf{x}_1}(z - x_2 - \cdots - x_n) f_{\mathbf{x}_2}(x_2) \cdots f_{\mathbf{x}_n}(x_n). \tag{4.49}$$

Let us introduce the new integration variables

$$\begin{aligned} u_2 &= x_2 + x_3 + \cdots + x_n, \\ u_3 &= \phantom{x_2 + {}} x_3 + \cdots + x_n, \\ &\ \vdots \\ u_{n-1} &= \phantom{x_2 + x_3 + \cdots + {}} x_{n-1} + x_n, \\ u_n &= x_n, \end{aligned}$$

whence

$$x_2 = u_2 - u_3, \quad x_3 = u_3 - u_4, \quad \ldots, \quad x_{n-1} = u_{n-1} - u_n.$$

The Jacobian of this transformation is

$$\frac{dx_2 dx_3 \cdots dx_n}{du_2 du_3 \cdots du_n} = J\begin{pmatrix} x_2 & x_3 & \cdots & x_n \\ u_2 & u_3 & \cdots & u_n \end{pmatrix} = \begin{vmatrix} 1 & -1 & 0 & \cdots & 0 & 0 \\ 0 & 1 & -1 & \cdots & 0 & 0 \\ \cdot & \cdot\cdot & \cdot\cdot & \cdots & \cdot & \cdot\cdot \\ 0 & 0 & 0 & \cdots & 1 & -1 \\ 0 & 0 & 0 & \cdots & 0 & 1 \end{vmatrix} = 1.$$

Then we can write (4.49) as an $(n - 1)$-fold convolution,

$$\begin{aligned} f_{\mathbf{z}}(z) = \int_{-\infty}^{\infty} du_2 \cdots \int_{-\infty}^{\infty} du_n\, & f_{\mathbf{x}_1}(z - u_2) f_{\mathbf{x}_2}(u_2 - u_3) \cdots \\ & \times f_{\mathbf{x}_{n-1}}(u_{n-1} - u_n) f_{\mathbf{x}_n}(u_n). \end{aligned} \tag{4.50}$$

Often this is easiest to evaluate in the Fourier domain by using the characteristic functions

$$\Phi_{\mathbf{x}_k}(\omega) = E[\exp(j\omega\mathbf{x}_k)] = \int_{-\infty}^{\infty} f_{\mathbf{x}_k}(x) e^{j\omega x}\, dx, \tag{4.51}$$

whereupon the characteristic function of the sum $\mathbf{z}$ is the product

$$\Phi_{\mathbf{z}}(\omega) = E(e^{j\omega\mathbf{z}}) = \prod_{k=1}^{n} \Phi_{\mathbf{x}_k}(\omega) \tag{4.52}$$

of the characteristic functions of the independent random variables $\mathbf{x}_1, \mathbf{x}_2, \ldots, \mathbf{x}_n$. The probability density function of their sum

$$\mathbf{z} = \mathbf{x}_1 + \mathbf{x}_2 + \cdots + \mathbf{x}_n$$

is then obtained by the inverse Fourier transform

$$f_{\mathbf{z}}(z) = \int_{-\infty}^{\infty} \Phi_{\mathbf{z}}(\omega) e^{-j\omega z} \frac{d\omega}{2\pi}. \tag{4.53}$$

If, for instance, the random variables all have the same exponential distribution,

$$f_{\mathbf{x}}(x) = ae^{-ax}U(x), \tag{4.54}$$

their common characteristic function is

$$\Phi_{\mathbf{x}}(\omega) = a(a - j\omega)^{-1} \tag{4.55}$$

and the characteristic function of their sum $\mathbf{z}$ is

$$\Phi_{\mathbf{z}}(\omega) = a^n(a - j\omega)^{-n} = \frac{(-1)^{n-1}a^n}{(n-1)!} \frac{d^{n-1}}{da^{n-1}} (a - j\omega)^{-1}, \tag{4.56}$$

whose inverse Fourier transform yields the probability density function of the sum $\mathbf{z}$,

$$f_{\mathbf{z}}(z) = \frac{(-1)^{n-1}a^n}{(n-1)!} \frac{d^{n-1}}{da^{n-1}} (e^{-az})U(z) = \frac{a^n z^{n-1} e^{-az}}{(n-1)!} U(z). \tag{4.57}$$

■

METHOD 3

The probability density function $f_{\mathbf{z}}(z)$ of

$$\mathbf{z} = g(\mathbf{x}_1, \mathbf{x}_2, \ldots, \mathbf{x}_n)$$

can sometimes be derived by evaluating the probability

$$f_{\mathbf{z}}(z)\, dz = \Pr\,(z < g(\mathbf{x}_1, \mathbf{x}_2, \ldots, \mathbf{x}_n) < z + dz),$$

the event in question corresponding to a thin shell bounded by the hypersurfaces

$$z = g(x_1, x_2, \ldots, x_n), \qquad z + dz = g(x_1, x_2, \ldots, x_n) \tag{4.58}$$

in the n-dimensional space R_n.

EXAMPLE 4.8. Given that the joint probability density function $f_{\mathbf{x}}(x_1, x_2, \ldots, x_n)$ is a function only of $(x_1^2 + \cdots + x_n^2)^{1/2}$,

$$f_{\mathbf{x}}(x_1, x_2, \ldots, x_n) = p(\sqrt{x_1^2 + x_2^2 + \cdots + x_n^2})$$

as in (4.18), we seek the probability density function of

$$\mathbf{z} = (\mathbf{x}_1^2 + \mathbf{x}_2^2 + \cdots + \mathbf{x}_n^2)^{1/2} \tag{4.59}$$

Now the region between the hypersurfaces in (4.58) is a thin hyperspherical shell of volume $k_n z^{n-1} dz$, with k_n given in (4.21), and our joint probability density function is the more nearly constant within that shell, the smaller dz. Hence

$$f_{\mathbf{z}}(z)\, dz = p(z)U(z)k_n z^{n-1}\, dz \tag{4.60}$$

and

$$f_{\mathbf{z}}(z) = \frac{2\pi^{n/2} z^{n-1} p(z) U(z)}{\Gamma(n/2)}. \tag{4.61}$$

Because the function $p(z)$ must satisfy (4.22), the probability density function $f_{\mathbf{z}}(z)$ will be properly normalized,

$$\int_0^\infty f_{\mathbf{z}}(z)\, dz = 1.$$

If, for instance, the random variables $\mathbf{x}_1, \mathbf{x}_2, \ldots, \mathbf{x}_n$ are independently Gaussian with means zero and variances 1 as in Example 4.2, we find

$$f_{\mathbf{z}}(z) = 2^{-(n/2)+1}[\Gamma(n/2)]^{-1} z^{n-1} e^{-z^2/2} U(z) \tag{4.62}$$

as the probability density function of the random variable $\mathbf{z}$ defined in (4.59). Introducing the variable

$$\mathbf{w} = \mathbf{z}^2 = \mathbf{x}_1^2 + \mathbf{x}_2^2 + \cdots + \mathbf{x}_n^2 \tag{4.63}$$

we find by the methods of Sec. 2-3 that the probability density function of $\mathbf{w}$ is

$$f_{\mathbf{w}}(w) = 2^{-n/2}[\Gamma(n/2)]^{-1} w^{(n-2)/2} e^{-w/2} U(w). \tag{4.64}$$

The random variable $\mathbf{w}$ is often called *chi-squared* (χ^2), and (4.64) is known as the *chi-squared distribution*. By comparison with (2.26) we see that $\mathbf{w}/2$ has a gamma distribution with parameter $c = n/2$. ■

B. *n* Functions of *n* Random Variables

The outcome ζ of the chance experiment, labeled by the n random variables $(\mathbf{x}_1, \mathbf{x}_2, \ldots, \mathbf{x}_n)$, could as well be labeled by n different random variables $(\mathbf{z}_1, \mathbf{z}_2, \ldots, \mathbf{z}_n)$ related to the original ones through n single-valued transformation functions

$$\begin{aligned} \mathbf{z}_1 &= g_1(\mathbf{x}_1, \mathbf{x}_2, \ldots, \mathbf{x}_n), \\ \mathbf{z}_2 &= g_2(\mathbf{x}_1, \mathbf{x}_2, \ldots, \mathbf{x}_n), \\ \mathbf{z}_n &= g_n(\mathbf{x}_1, \mathbf{x}_2, \ldots, \mathbf{x}_n). \end{aligned} \tag{4.65}$$

These functions must be algebraically independent; that is, there must exist no function $H(\mathbf{z}_1, \mathbf{z}_2, \ldots, \mathbf{z}_n)$ of the $\mathbf{z}$'s that is identically zero. If these equations can be solved uniquely to yield the $\mathbf{x}$'s in terms of the $\mathbf{z}$'s,

$$\begin{aligned} \mathbf{x}_1 &= h_1(\mathbf{z}_1, \mathbf{z}_2, \ldots, \mathbf{z}_n), \\ \mathbf{x}_2 &= h_2(\mathbf{z}_1, \mathbf{z}_2, \ldots, \mathbf{z}_n), \\ \mathbf{x}_n &= h_n(\mathbf{z}_1, \mathbf{z}_2, \ldots, \mathbf{z}_n), \end{aligned} \tag{4.66}$$

the transformation is one-to-one; to each point in the space with coordinates $(\mathbf{x}_1, \mathbf{x}_2, \ldots, \mathbf{x}_n)$ there corresponds a unique point in the space with coordinates $(\mathbf{z}_1, \mathbf{z}_2, \ldots, \mathbf{z}_n)$, and vice versa.

All outcomes ζ lying in an infinitesimal rectangular parallelepiped in the space of the $\mathbf{z}$'s,

$$z_k < \mathbf{z}_k < z_k + dz_k, \quad \forall\, k, \tag{4.67}$$

which we call V_z, must be in a corresponding infinitesimal parallelepiped V_x in the space of the **x**'s. It is bounded by pairs of surfaces

$$g_k(x_1, x_2, \ldots, x_n) = z_k, \; g_k(x_1, x_2, \ldots, x_n) = z_k + dz_k, \quad k = 1, 2, \ldots, n.$$

The events $(\mathbf{z}_1, \mathbf{z}_2, \ldots, \mathbf{z}_n) \in V_z$, $(\mathbf{x}_1, \mathbf{x}_2, \ldots, \mathbf{x}_n) \in V_x$, comprising the same outcomes ζ of the chance experiment, must have equal probability

$$f_{\mathbf{x}}(x_1, x_2, \ldots, x_n) \cdot \text{Volume } (V_x) = f_{\mathbf{z}}(z_1, z_2, \ldots, z_n) \cdot \text{Volume } (V_z)$$

in terms of the joint probability density functions of the random variables. Hence the ratio of the probability density functions equals the inverse ratio of the volumes of the infinitesimal parallelepipeds V_x and V_z. This ratio is in turn given as in (3.66) by Jacobian determinants, defined as the n-dimensional versions of (3.67) and (3.68). The joint probability density function of the random variables $\mathbf{z}_1, \mathbf{z}_2, \ldots, \mathbf{z}_n$ defined by the transformation equations (4.65) is therefore

$$f_{\mathbf{z}}(z_1, z_2, \ldots, z_n) = \left| J\begin{pmatrix} x_1 & x_2 & \cdots & x_n \\ z_1 & z_2 & \cdots & z_n \end{pmatrix} \right| f_{\mathbf{x}}(x_1, x_2, \ldots, x_n) = \frac{f_{\mathbf{x}}(x_1, x_2, \ldots, x_n)}{\left| J\begin{pmatrix} z_1 & z_2 & \cdots & z_n \\ x_1 & x_2 & \cdots & x_n \end{pmatrix} \right|}, \tag{4.68}$$

where

$$J\begin{pmatrix} x_1 & x_2 & \cdots & x_n \\ z_1 & z_2 & \cdots & z_n \end{pmatrix} = \det \left\| \frac{\partial x_i}{\partial z_j} \right\| = \begin{vmatrix} \dfrac{\partial x_1}{\partial z_1} & \dfrac{\partial x_2}{\partial z_1} & \cdots & \dfrac{\partial x_n}{\partial z_1} \\ \dfrac{\partial x_1}{\partial z_2} & \dfrac{\partial x_2}{\partial z_2} & \cdots & \dfrac{\partial x_n}{\partial z_2} \\ & & \cdots & \\ \dfrac{\partial x_1}{\partial z_n} & \dfrac{\partial x_2}{\partial z_n} & \cdots & \dfrac{\partial x_n}{\partial z_n} \end{vmatrix} \tag{4.69}$$

and

$$J\begin{pmatrix} z_1 & z_2 & \cdots & z_n \\ x_1 & x_2 & \cdots & x_n \end{pmatrix} = \det \left\| \frac{\partial z_i}{\partial x_j} \right\| \tag{4.70}$$

has a similar appearance, the x's and z's being interchanged. If the latter Jacobian is identically zero, the functions $g_1, g_2, \ldots, g_n$ are not algebraically independent as required. In (4.68) we must express the variables $x_1, x_2, \ldots, x_n$ in terms of $z_1, z_2, \ldots, z_n$ by means of (4.66).

That the ratio Volume (V_x) : Volume (V_z) is indeed given by the Jacobian determinant as required for (4.68) is proved for n dimensions much as for two. The details are found in the appendix of this chapter.

If the transformation in (4.65) is many–one, a number of points in the space of the **x**'s go into each point in the space of the **z**'s, and solving the equations (4.65) for the **x**'s yields several solutions of the form of (4.66). The joint probability density function $f_{\mathbf{z}}(z_1, z_2, \ldots, z_n)$ must then be calculated by summing terms of the form (4.68), one for each point $(x_1, x_2, \ldots, x_n)$ that is transformed into the point $(z_1, z_2, \ldots, z_n)$ under consideration.

EXAMPLE 4.9. Linear Transformation of Random Variables

Let

$$\mathbf{z}_k = \sum_{j=1}^{n} a_{kj}\mathbf{x}_j, \qquad 1 \le k \le n, \tag{4.71}$$

where the coefficients a_{kj} are constants. Provided that the matrix $\mathsf{a} = \|a_{kj}\|$ is not singular, that is,

$$\det \mathsf{a} = \det\|a_{kj}\| \neq 0,$$

we can solve (4.71) for the **x**'s in terms of the **z**'s,

$$\mathbf{x}_m = \sum_{k=1}^{n} b_{mk}\mathbf{z}_k, \qquad 1 \le m \le n, \tag{4.72}$$

where the matrix $\mathsf{b} = \|b_{mk}\|$ is the inverse, $\mathsf{b} = \mathsf{a}^{-1}$, of the matrix a. The joint probability density function of the new random variables is now, by (4.68),

$$f_{\mathbf{z}}(z_1, z_2, \ldots, z_n) = |\det \mathsf{b}| f_{\mathbf{x}}\left(\sum_{k=1}^{n} b_{1k}z_k, \sum_{k=1}^{n} b_{2k}z_k, \ldots, \sum_{k=1}^{n} b_{nk}z_k\right), \tag{4.73}$$

with $|\det \mathsf{b}| = |\det \mathsf{a}|^{-1}$.

In particular, let

$$\begin{array}{rclcrcl}
\mathbf{z}_1 &=& \mathbf{x}_1 + \mathbf{x}_2 + \cdots + \mathbf{x}_{n-1} + \mathbf{x}_n & \qquad & \mathbf{x}_1 &=& \mathbf{z}_1 - \mathbf{z}_2 \\
\mathbf{z}_2 &=& \mathbf{x}_2 + \cdots + \mathbf{x}_{n-1} + \mathbf{x}_n & & \mathbf{x}_2 &=& \mathbf{z}_2 - \mathbf{z}_3 \\
\vdots & & & & \vdots & & \\
\mathbf{z}_k &=& \mathbf{x}_k + \cdots + \mathbf{x}_n & & \mathbf{x}_k &=& \mathbf{z}_k - \mathbf{z}_{k+1} \\
\vdots & & & & \vdots & & \\
\mathbf{z}_{n-1} &=& \mathbf{x}_{n-1} + \mathbf{x}_n & & \mathbf{x}_{n-1} &=& \mathbf{z}_{n-1} - \mathbf{z}_n \\
\mathbf{z}_n &=& \mathbf{x}_n & & \mathbf{x}_n &=& \mathbf{z}_n.
\end{array} \tag{4.74}$$

Then it is easy to show that $\det \mathsf{b} = 1$, and

$$f_{\mathbf{z}}(z_1, z_2, \ldots, z_n) = f_{\mathbf{x}}(z_1 - z_2, z_2 - z_3, \ldots, z_{n-1} - z_n, z_n). \tag{4.75}$$

If we want the probability density function of

$$\mathbf{z}_1 = \mathbf{x}_1 + \mathbf{x}_2 + \cdots + \mathbf{x}_n$$

alone, we integrate over $-\infty < z_k < \infty$, $k = 2, 3, \ldots, n$, and for independent random variables we obtain (4.50). ■

This artifice exemplifies the possibility of obtaining the joint probability density function of fewer than n functions of $\mathbf{x}_1, \mathbf{x}_2, \ldots, \mathbf{x}_n$ by introducing enough dummy random variables to make up a set of n random variables $\mathbf{z}_1, \mathbf{z}_2, \ldots, \mathbf{z}_n$ as in (4.65). One uses the Jacobian method to find their joint density function as in (4.68) and then integrates over the entire range of each dummy variable.

EXAMPLE 4.10. Thermal noise that has passed through a narrowband pass filter can be represented by a phasor $x(t) + jy(t)$ whose x- and y-components change significantly only during intervals of the order of the reciprocal of the bandwidth W of the filter. The output of the filter, whose pass frequency is $\Omega/2\pi$, is

$$\text{Re}\,\{[x(t) + jy(t)]e^{j\Omega t}\}.$$

If we sample this filtered noise at two distinct times t_1 and t_2, we obtain two phasors

$$x(t_1) + jy(t_1), \qquad x(t_2) + jy(t_2)$$

whose components $\mathbf{x}_1 = x(t_1)$, $\mathbf{y}_1 = y(t_1)$, $\mathbf{x}_2 = x(t_2)$, $\mathbf{y}_2 = y(t_2)$ constitute four Gaussian random variables. It can be shown that if the transfer function of the filter is symmetrical about angular frequency Ω, the joint probability density function of these four random variables is the product of two bivariate Gaussian density functions,

$$f_{\mathbf{x}_1\mathbf{y}_1\mathbf{x}_2\mathbf{y}_2}(x_1, y_1, x_2, y_2) = (2\pi\sigma^2)^{-2}(1 - r^2)^{-1} \exp\left[-\frac{x_1^2 - 2rx_1x_2 + x_2^2 + y_1^2 - 2ry_1y_2 + y_2^2}{2\sigma^2(1 - r^2)}\right]. \tag{4.76}$$

The correlation coefficient r depends on the interval $|t_2 - t_1|$ and on the transfer function of the filter and is the smaller, the larger $W|t_2 - t_1|$; σ^2 measures the average power in the noise output.

We seek the joint probability density function of the amplitudes

$$\mathbf{p}_1 = (\mathbf{x}_1^2 + \mathbf{y}_1^2)^{1/2}, \qquad \mathbf{p}_2 = (\mathbf{x}_2^2 + \mathbf{y}_2^2)^{1/2}$$

and of the phase difference

$$\boldsymbol{\theta} = \boldsymbol{\phi}_2 - \boldsymbol{\phi}_1,$$

where as shown in Fig. 4.3

$$\boldsymbol{\phi}_1 = \arg\,(\mathbf{x}_1 + j\mathbf{y}_1), \qquad \boldsymbol{\phi}_2 = \arg\,(\mathbf{x}_2 + j\mathbf{y}_2).$$

We first find the joint density function of $\mathbf{p}_1$, $\mathbf{p}_2$, $\boldsymbol{\phi}_1$, and $\boldsymbol{\phi}_2$. Our transformation equations, as in Fig. 4.3, yield

$$\begin{aligned} \mathbf{x}_1 &= \mathbf{p}_1 \cos \boldsymbol{\phi}_1, \qquad \mathbf{y}_1 = \mathbf{p}_1 \sin \boldsymbol{\phi}_1, \\ \mathbf{x}_2 &= \mathbf{p}_2 \cos \boldsymbol{\phi}_2, \qquad \mathbf{y}_2 = \mathbf{p}_2 \sin \boldsymbol{\phi}_2, \end{aligned} \tag{4.77}$$

and the Jacobian we need is

$$J\begin{pmatrix} x_1 & y_1 & x_2 & y_2 \\ p_1 & \phi_1 & p_2 & \phi_2 \end{pmatrix} = p_1p_2.$$

Hence we find from (4.76), (4.77), and (4.68)

$$\begin{aligned} f_{\mathbf{p}_1\boldsymbol{\phi}_1\mathbf{p}_2\boldsymbol{\phi}_2}(p_1, \phi_1, p_2, \phi_2) &= \frac{p_1p_2}{(2\pi\sigma^2)^2(1 - r^2)} \\ &\times \exp\left[-\frac{p_1^2 + p_2^2 - 2rp_1p_2 \cos(\phi_2 - \phi_1)}{2\sigma^2(1 - r^2)}\right] U(p_1)U(p_2). \end{aligned} \tag{4.78}$$

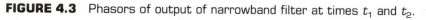

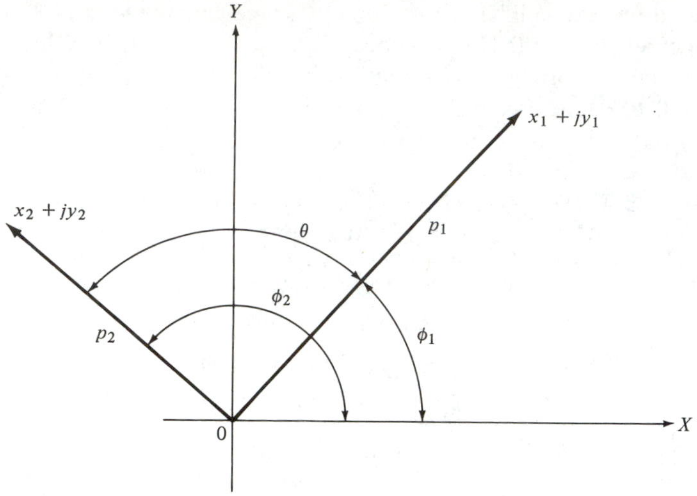

FIGURE 4.3 Phasors of output of narrowband filter at times t_1 and t_2.

If we now introduce the new random variables

$$\boldsymbol{\theta} = \boldsymbol{\phi}_2 - \boldsymbol{\phi}_1$$

$$\boldsymbol{\psi} = \boldsymbol{\phi}_2,$$

whose Jacobian is

$$J\begin{pmatrix} \theta & \psi \\ \phi_1 & \phi_2 \end{pmatrix} = \begin{vmatrix} -1 & 1 \\ 0 & 1 \end{vmatrix} = -1,$$

we find the joint density function of $\mathbf{p}_1$, $\mathbf{p}_2$, $\boldsymbol{\theta}$, and $\boldsymbol{\psi}$ to be

$$f_{\mathbf{p}_1\mathbf{p}_2\boldsymbol{\theta}\boldsymbol{\psi}}(p_1, p_2, \theta, \psi)$$
$$= \frac{p_1 p_2}{(2\pi\sigma^2)^2(1 - r^2)} \exp\left[-\frac{p_1^2 + p_2^2 - 2rp_1p_2\cos\theta}{2\sigma^2(1 - r^2)}\right] U(p_1)U(p_2).$$

Integrating over $0 \le \psi < 2\pi$, we obtain the joint density function we were looking for,

$$f_{\mathbf{p}_1\mathbf{p}_2\boldsymbol{\theta}}(p_1, p_2, \theta) \tag{4.79}$$
$$= \frac{p_1 p_2}{2\pi\sigma^4(1 - r^2)} \exp\left[-\frac{p_1^2 + p_2^2 - 2rp_1p_2\cos\theta}{2\sigma^2(1 - r^2)}\right] U(p_1)U(p_2).$$

The joint probability density function of the amplitudes $\mathbf{p}_1$ and $\mathbf{p}_2$ is found by integrating over $0 \le \theta < 2\pi$,

$$f_{\mathbf{p}_1\mathbf{p}_2}(p_1, p_2) \tag{4.80}$$
$$= \frac{p_1 p_2}{\sigma^4(1 - r^2)} \exp\left[-\frac{p_1^2 + p_2^2}{2\sigma^2(1 - r^2)}\right] I_0\left(\frac{rp_1p_2}{\sigma^2(1 - r^2)}\right) U(p_1)U(p_2)$$

in terms of the modified Bessel function defined in (3.62). These amplitudes could be observed by sampling at times t_1 and t_2 the output of a linear full-wave rectifier attached to the output of the narrowband filter. For $W|t_2 - t_1| >> 1$, $r \doteq 0$, and $\mathbf{p}_1$ and $\mathbf{p}_2$ are independent and have Rayleigh distributions.

The probability density function of the difference $\boldsymbol{\theta} = \boldsymbol{\phi}_2 - \boldsymbol{\phi}_1$ of the phases of the noise phasors at times t_2 and t_1 is calculated by integrating (4.79) over the amplitudes, $0 < p_1 < \infty$, $0 < p_2 < \infty$. This integration is most easily carried out by changing to polar coordinates in the first quadrant of the (p_1, p_2)-plane,

$$p_1 = P \cos \phi, \qquad p_2 = P \sin \phi,$$

$$0 < P < \infty, \qquad 0 < \phi < \frac{\pi}{2}.$$

The element of area in this plane is

$$dp_1 \, dp_2 = P \, dP \, d\phi,$$

and from (4.79) the probability density function of $\boldsymbol{\theta}$ is

$$\begin{aligned} f_{\boldsymbol{\theta}}(\theta) &= [2\pi\sigma^4(1 - r^2)]^{-1} \int_0^{\pi/2} d\phi \int_0^\infty P^3 \sin \phi \cos \phi \\ &\quad \times \exp\left[-\frac{P^2(1 - 2r \cos \theta \sin \phi \cos \phi)}{2\sigma^2(1 - r^2)} \right] dP \\ &= \frac{1 - r^2}{\pi} \int_0^{\pi/2} \frac{\sin \phi \cos \phi \, d\phi}{(1 - 2r \cos \theta \sin \phi \cos \phi)^2} \\ &= \frac{1 - r^2}{2\pi} \int_0^{\pi/2} \frac{\sin 2\phi \, d\phi}{(1 - r \cos \theta \sin 2\phi)^2} = \frac{1 - r^2}{2\pi} \frac{d}{da} \int_0^{\pi/2} \frac{d\phi}{1 - a \sin 2\phi} \Bigg|_{a = r \cos \theta} \\ &= \frac{1 - r^2}{2\pi} (1 - r^2 \cos^2 \theta)^{-3/2} \\ &\quad \times \left[(1 - r^2 \cos^2 \theta)^{1/2} + r \cos \theta \left(\frac{\pi}{2} + \sin^{-1}(r \cos \theta) \right) \right]. \end{aligned} \tag{4.81}$$

It is plotted for $0 < \theta < \pi$ in Fig. 4.4 for various values of the correlation coefficient r; $f_{\boldsymbol{\theta}}(-\theta) = f_{\boldsymbol{\theta}}(\theta)$. ■

EXAMPLE 4.11. The three random variables $\mathbf{x}$, $\mathbf{y}$, and $\mathbf{z}$ have the joint probability density function

$$f_{\mathbf{xyz}}(x, y, z) = (2\pi\sigma^2)^{-3/2} \exp\left(-\frac{x^2 + y^2 + z^2}{2\sigma^2} \right).$$

We seek the joint probability density function of the random variables

$$\mathbf{u} = \frac{\mathbf{x}}{\mathbf{z}}, \qquad \mathbf{v} = \frac{\mathbf{y}}{\mathbf{z}}.$$

We introduce the dummy variable $\mathbf{w} = \mathbf{z}$. Then $\mathbf{x} = \mathbf{wu}$, $\mathbf{y} = \mathbf{wv}$, $\mathbf{z} = \mathbf{w}$, and the Jacobian of this transformation is

$$J\begin{pmatrix} x & y & z \\ u & v & w \end{pmatrix} = \begin{vmatrix} w & 0 & u \\ 0 & w & v \\ 0 & 0 & 1 \end{vmatrix} = w^2.$$

From (4.68) the joint probability density function of $\mathbf{u}$, $\mathbf{v}$, and $\mathbf{w}$ is

$$f_{\mathbf{uvw}}(u, v, w) = (2\pi\sigma^2)^{-3/2} w^2 \exp\left(-\frac{w^2u^2 + w^2v^2 + w^2}{2\sigma^2} \right),$$

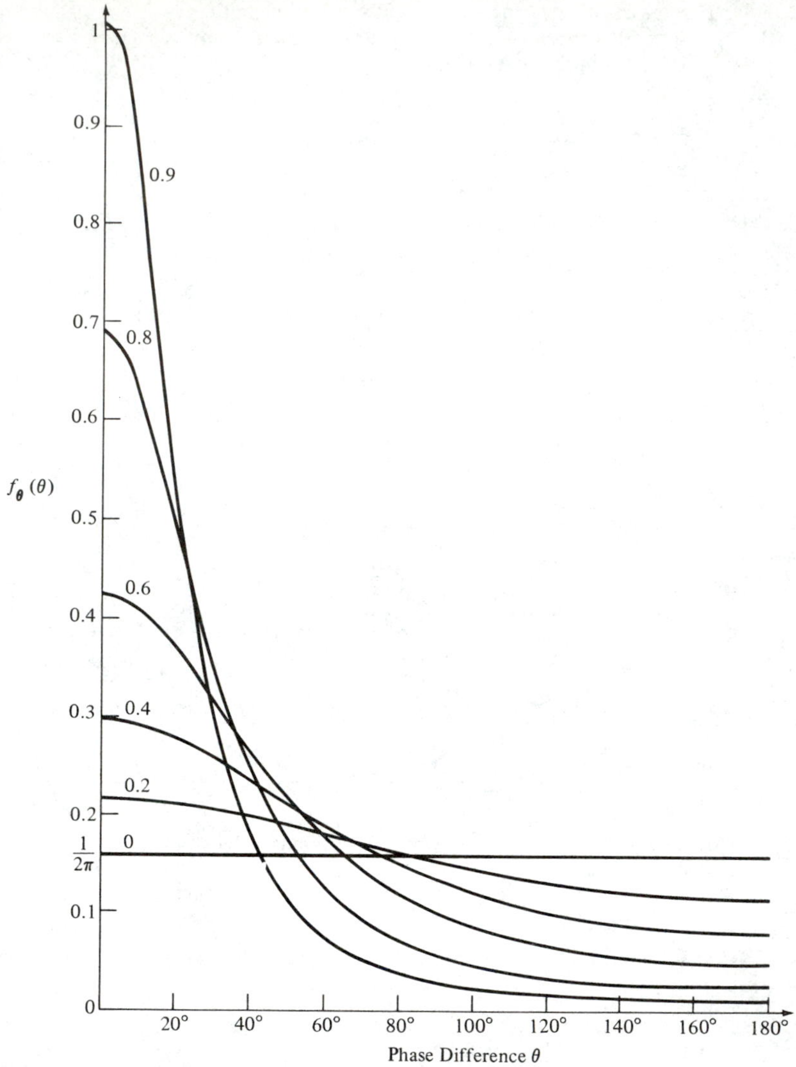

FIGURE 4.4 Probability density function: difference in phases of narrowband Gaussian random noise at two sampling times. Curves are indexed with the correlation coefficient $r(t_2 - t_1)$.

and the density function sought is

$$f_{\mathbf{uv}}(u, v) = (2\pi\sigma^2)^{-3/2} \int_{-\infty}^{\infty} w^2 \exp\left[-\frac{w^2(u^2 + v^2 + 1)}{2\sigma^2}\right] dw$$
$$= (2\pi)^{-1} (1 + u^2 + v^2)^{-3/2}. \qquad \blacksquare$$

EXPECTED VALUES

A. Means, Variances, and Covariances

If $\mathbf{z} = g(\mathbf{x}_1, \mathbf{x}_2, \ldots, \mathbf{x}_n)$ is a function of the n random variables $\mathbf{x}_1, \mathbf{x}_2, \ldots,$ $\mathbf{x}_n$, its expected value, in direct analogy to (3.82), is

$$\begin{aligned} E[g(\mathbf{x}_1, \mathbf{x}_2, \ldots, \mathbf{x}_n)] &= \overline{g(x_1, x_2, \ldots, x_n)} \\ &= E(\mathbf{z}) = \int_{-\infty}^{\infty} z f_{\mathbf{z}}(z)\, dz \\ &= \int_{-\infty}^{\infty} \cdots \int_{-\infty}^{\infty} g(x_1, x_2, \ldots, x_n) \\ &\quad \times f_{\mathbf{x}}(x_1, x_2, \ldots, x_n)\, dx_1\, dx_2 \cdots dx_n, \end{aligned} \tag{4.82}$$

where $f_{\mathbf{x}}(x_1, x_2, \ldots, x_n)$ is the joint density function of $\mathbf{x}_1, \mathbf{x}_2, \ldots, \mathbf{x}_n$. The proof of this follows the same lines as that in Sec. 3-6 and will not be repeated. If the random variables are discrete, as in (4.12), the expected value of this function is

$$\begin{aligned} E[g(\mathbf{x}_1, \mathbf{x}_2, \ldots, \mathbf{x}_n)] &= \overline{g(x_1, x_2, \ldots, x_n)} \\ &= \sum_k g(x_1^{(k)}, x_2^{(k)}, \ldots ; x_n^{(k)}) \\ &\quad \times \Pr(\mathbf{x}_1 = x_1^{(k)}, \mathbf{x}_2 = x_2^{(k)}, \ldots, \mathbf{x}_n = x_n^{(k)}). \end{aligned} \tag{4.83}$$

The mean or expected value of one of the random variables, say $\mathbf{x}_k$, is determined from its marginal density function $f_{\mathbf{x}_k}(x)$ as in (2.72) or (2.73). Indeed, in accordance with (4.82),

$$\begin{aligned} \bar{x}_k = E(\mathbf{x}_k) &= \int_{-\infty}^{\infty} \cdots \int_{-\infty}^{\infty} x_k f_{\mathbf{x}}(x_1, x_2, \ldots, x_n)\, dx_1\, dx_2 \cdots dx_n \\ &= \int_{-\infty}^{\infty} x_k f_{\mathbf{x}_k}(x_k)\, dx_k \end{aligned} \tag{4.84}$$

after integrating over $-\infty < x_m < \infty,\ \forall\ m \neq k$.

The expected value of the sum of n random variables equals the sum of their expected values,

$$E(\mathbf{x}_1 + \mathbf{x}_2 + \cdots + \mathbf{x}_n) = E(\mathbf{x}_1) + E(\mathbf{x}_2) + \cdots + E(\mathbf{x}_n) \tag{4.85}$$

as follows immediately from (4.82) and (4.84).

The variance Var $\mathbf{x}_k$ is also found from the marginal density function $f_{\mathbf{x}_k}(x)$ of $\mathbf{x}_k$, as in (2.77). The correlation of any two of the random variables, say $\mathbf{x}_k$ and $\mathbf{x}_m$, is calculated from their joint bivariate density function as in (3.85),

$$E(\mathbf{x}_k \mathbf{x}_m) = \int_{-\infty}^{\infty} x_k x_m f_{\mathbf{x}}(x_1, x_2, \ldots, x_n)\, dx_1\, dx_2 \cdots dx_n$$
$$= \int_{-\infty}^{\infty}\int_{-\infty}^{\infty} x_k x_m f_{\mathbf{x}_k\mathbf{x}_m}(x_k, x_m)\, dx_k\, dx_m \tag{4.86}$$

after integrating over all the x's except x_k and x_m. The covariance of any such pair is then defined as in (3.87) and (3.88),

$$\begin{aligned} \text{Cov}\,(\mathbf{x}_k, \mathbf{x}_m) &= E(\mathbf{x}_k \mathbf{x}_m) - \bar{x}_k \bar{x}_m \\ &= E[(\mathbf{x}_k - \bar{x}_k)(\mathbf{x}_m - \bar{x}_m)]. \end{aligned} \tag{4.87}$$

There are n variances Var $\mathbf{x}_k$ and $\frac{1}{2}n(n-1)$ such covariances, which can be combined into the $n \times n$ *covariance matrix* c, whose elements are

$$c_{ij} = \text{Cov}\,(\mathbf{x}_i, \mathbf{x}_j), \qquad 1 \le (i, j) \le n, \tag{4.88}$$

for $i \neq j$; its diagonal elements

$$c_{ii} = \text{Var}\ \mathbf{x}_i, \qquad 1 \le i \le n,$$

are the variances of the random variables $\mathbf{x}_1, \mathbf{x}_2, \ldots, \mathbf{x}_n$. The covariance matrix is symmetrical,

$$c_{ij} = c_{ji}.$$

The variance of the linear combination

$$\mathbf{z} = a_1\mathbf{x}_1 + a_2\mathbf{x}_2 + \cdots + a_n\mathbf{x}_n = \sum_{k=1}^{n} a_k \mathbf{x}_k \tag{4.89}$$

of our random variables is

$$\begin{aligned} \text{Var}\ \mathbf{z} &= \sum_{k=1}^{n}\sum_{m=1}^{n} a_k a_m E[(\mathbf{x}_k - \bar{x}_k)(\mathbf{x}_m - \bar{x}_m)] \\ &= \sum_{k=1}^{n}\sum_{m=1}^{n} a_k c_{km} a_m = \mathsf{a}^T \mathsf{c} \mathsf{a}, \end{aligned} \tag{4.90}$$

where a is the n-element column vector of coefficients a_k, $1 \le k \le n$, and

$$\mathsf{a}^T = (a_1 \quad a_2 \quad \cdots \quad a_n)$$

is its transposed row vector. We shall find this matrix notation most useful hereafter.

If the n random variables $\mathbf{x}_k$ are uncorrelated,

$$\text{Cov}\,(\mathbf{x}_k, \mathbf{x}_m) = \begin{cases} \text{Var}\ \mathbf{x}_k, & k = m \\ 0, & k \neq m, \end{cases} \tag{4.91}$$

the covariance matrix c is diagonal, and

$$\text{Var}\ \mathbf{z} = \sum_{k=1}^{n} a_k^2\ \text{Var}\ \mathbf{x}_k. \tag{4.92}$$

In particular, the variance of the sum of n uncorrelated random variables equals the sum of their variances,

$$\operatorname{Var}(\mathbf{x}_1 + \mathbf{x}_2 + \cdots + \mathbf{x}_n) = \operatorname{Var}\mathbf{x}_1 + \operatorname{Var}\mathbf{x}_2 + \cdots + \operatorname{Var}\mathbf{x}_n, \tag{4.93}$$

$$\operatorname{Cov}(\mathbf{x}_i, \mathbf{x}_j) \equiv 0, \qquad i \neq j,$$

a useful relation.

EXAMPLE 4.12. The covariance matrix of $\mathbf{x}_1$, $\mathbf{x}_2$, and $\mathbf{x}_3$ is

$$\mathsf{c} = \begin{bmatrix} 10 & 3 & 2 \\ 3 & 7 & -1 \\ 2 & -1 & 5 \end{bmatrix}.$$

Then the variance of

$$\mathbf{z} = 3\mathbf{x}_1 - \mathbf{x}_2 + 2\mathbf{x}_3$$

is, in matrix notation,

$$\operatorname{Var}\mathbf{z} = (3 \quad -1 \quad 2)\begin{bmatrix} 10 & 3 & 2 \\ 3 & 7 & -1 \\ 2 & -1 & 5 \end{bmatrix}\begin{pmatrix} 3 \\ -1 \\ 2 \end{pmatrix}$$

$$= (31 \quad 0 \quad 17)\begin{pmatrix} 3 \\ -1 \\ 2 \end{pmatrix} = 127.$$

■

B. Characteristic Functions

The joint characteristic function of the random variables $\mathbf{x}_1, \mathbf{x}_2, \ldots, \mathbf{x}_n$ is the n-dimensional Fourier transform of their joint probability density function,

$$\Phi_{\mathbf{x}}(\omega_1, \omega_2, \ldots, \omega_n) = E\{\exp[j(\omega_1\mathbf{x}_1 + \omega_2\mathbf{x}_2 + \cdots + \omega_n\mathbf{x}_n)]\}$$

$$= \int_{-\infty}^{\infty} \cdots \int_{-\infty}^{\infty} \exp[j(\omega_1 x_1 + \omega_2 x_2 + \cdots + \omega_n x_n)]$$

$$\times f_{\mathbf{x}}(x_1, x_2, \ldots, x_n)\, dx_1\, dx_2 \cdots dx_n, \tag{4.94}$$

and the joint density function can be recovered by the inverse Fourier transform

$$f_{\mathbf{x}}(x_1, x_2, \ldots, x_n) = \int_{-\infty}^{\infty} \cdots \int_{-\infty}^{\infty} \exp[-j(\omega_1 x_1 + \omega_2 x_2 + \cdots + \omega_n x_n)]$$

$$\times \Phi_{\mathbf{x}}(\omega_1, \omega_2, \ldots, \omega_n)\, \frac{d\omega_1\, d\omega_2 \cdots d\omega_n}{(2\pi)^n}. \tag{4.95}$$

Joint moments of the random variables $\mathbf{x}_1, \mathbf{x}_2, \ldots, \mathbf{x}_n$ can be formed in analogy with those in Sec. 3-6,

$$\overline{x_1^{k_1}x_2^{k_2}\cdots x_n^{k_n}} = E\left(\mathbf{x}_1^{k_1}\mathbf{x}_2^{k_2}\cdots\mathbf{x}_n^{k_n}\right)$$
$$= \int_{-\infty}^{\infty}\cdots\int_{-\infty}^{\infty} x_1^{k_1}x_2^{k_2}\cdots x_n^{k_n} \quad (4.96)$$
$$\times\, f_{\mathbf{x}}(x_1, x_2, \ldots, x_n)\, dx_1\, dx_2 \cdots dx_n,$$

for all positive integers $k_1, k_2, \ldots, k_n$. Depending on the density function, these moments may be infinite for large enough exponents k_i, $1 \le i \le n$. When they are finite, they can be derived by differentiating the characteristic function at the origin of the Fourier domain:

$$E\left(\mathbf{x}_1^{k_1}\mathbf{x}_2^{k_2}\cdots\mathbf{x}_n^{k_n}\right) \quad (4.97)$$
$$= \left.\frac{\partial^{k_1+k_2+\cdots+k_n}\Phi_{\mathbf{x}}(\omega_1, \omega_2, \ldots, \omega_n)}{\partial(j\omega_1)^{k_1}\partial(j\omega_2)^{k_2}\cdots\partial(j\omega_n)^{k_n}}\right|_{\omega_1=\omega_2=\cdots\omega_n=0},$$

for each time we differentiate

$$\exp\,(j\omega_1\mathbf{x}_1 + j\omega_2\mathbf{x}_2 + \cdots + j\omega_n\mathbf{x}_n)$$

with respect to $(j\omega_i)$, we bring down a factor $\mathbf{x}_i$, and setting the ω's equal to zero after the differentiation replaces the exponential function by 1.

The joint characteristic function of a subset of the random variables is the Fourier transform of their marginal density function and is obtained simply by setting to zero the ω's corresponding to the discarded random variables. Thus if $\Phi_{\mathbf{x}}(\omega_1, \omega_2, \omega_3, \omega_4)$ is the characteristic function of $\mathbf{x}_1$, $\mathbf{x}_2$, $\mathbf{x}_3$, and $\mathbf{x}_4$,

$$\Phi_{\mathbf{x}_1\mathbf{x}_2\mathbf{x}_4}(\omega_1, \omega_2, \omega_4) = \Phi_{\mathbf{x}}(\omega_1, \omega_2, 0, \omega_4),$$
$$\Phi_{\mathbf{x}_1\mathbf{x}_4}(\omega_1, \omega_4) = \Phi_{\mathbf{x}}(\omega_1, 0, 0, \omega_4),$$
$$\Phi_{\mathbf{x}_2}(\omega_2) = \Phi_{\mathbf{x}}(0, \omega_2, 0, 0).$$

The reason for this rule is easily perceived from the example

$$\Phi_{\mathbf{x}_1\mathbf{x}_4}(\omega_1, \omega_4) = E\{\exp\,[j(\omega_1\mathbf{x}_1 + \omega_4\mathbf{x}_4)]\}$$
$$= E\{\exp\,[j(\omega_1\mathbf{x}_1 + \omega_2\mathbf{x}_2 + \omega_3\mathbf{x}_3 + \omega_4\mathbf{x}_4)]\}\Big|_{\omega_2=\omega_3=0}$$
$$= \Phi(\omega_1, 0, 0, \omega_4).$$

If the random variables $\mathbf{x}_1, \mathbf{x}_2, \ldots, \mathbf{x}_n$ are statistically independent, the expected value of the product $h_1(\mathbf{x}_1)h_2(\mathbf{x}_2)\cdots h_n(\mathbf{x}_n)$ equals the product of the expected values of its factors,

$$
\begin{aligned}
E[h_1(\mathbf{x}_1)h_2(\mathbf{x}_2) \cdots h_n(\mathbf{x}_n)]
&= \int_{-\infty}^{\infty} \cdots \int_{-\infty}^{\infty} h_1(x_1)h_2(x_2) \cdots h_n(x_n) \\
&\quad \times f_{\mathbf{x}_1}(x_1)f_{\mathbf{x}_2}(x_2) \cdots f_{\mathbf{x}_n}(x_n)\, dx_1\, dx_2 \cdots dx_n \\
&= E[h_1(\mathbf{x}_1)]E[h_2(\mathbf{x}_2)] \cdots E[h_n(\mathbf{x}_n)].
\end{aligned} \tag{4.98}
$$

In particular, their joint characteristic function factors into the product of the characteristic functions of the individual random variables,

$$
\Phi_{\mathbf{x}}(\omega_1, \omega_2, \ldots, \omega_n) = \prod_{k=1}^{n} \Phi_{\mathbf{x}_k}(\omega_k). \tag{4.99}
$$

The characteristic function of their sum

$$
\mathbf{z} = \mathbf{x}_1 + \mathbf{x}_2 + \cdots + \mathbf{x}_n
$$

is then given as in (4.52),

$$
\begin{aligned}
\Phi_{\mathbf{z}}(\omega) &= E[\exp(j\omega\mathbf{z})] \\
&= E\{\exp[j\omega(\mathbf{x}_1 + \mathbf{x}_2 + \cdots + \mathbf{x}_n)]\} \\
&= \prod_{k=1}^{n} E[\exp(j\omega\mathbf{x}_k)] = \prod_{k=1}^{n} \Phi_{\mathbf{x}_k}(\omega),
\end{aligned}
$$

and its probability density function can be calculated by taking the inverse Fourier transform of $\Phi_{\mathbf{z}}(\omega)$.

EXAMPLE 4.13. Sum of Poisson Random Variables

The n random variables $\mathbf{x}_k$ are independent and have Poisson distributions,

$$
\Pr(\mathbf{x}_k = m) = \frac{\lambda_k^m \exp(-\lambda_k)}{m!}. \tag{4.100}
$$

Their joint characteristic function is, by (2.102) with $z = e^{j\omega}$,

$$
\Phi_{\mathbf{x}}(\omega_1, \omega_2, \ldots, \omega_n) = \prod_{k=1}^{n} \exp[\lambda_k(e^{j\omega_k} - 1)]. \tag{4.101}
$$

The characteristic function of their sum is then, by (4.52),

$$
\Phi_{\mathbf{z}}(\omega) = \exp\left[\sum_{k=1}^{n} \lambda_k(e^{j\omega} - 1)\right]. \tag{4.102}
$$

The sum

$$
\mathbf{z} = \mathbf{x}_1 + \mathbf{x}_2 + \cdots + \mathbf{x}_n
$$

of n independent Poisson random variables therefore also has a Poisson distribution,

$$
\Pr(\mathbf{z} = m) = \frac{\lambda^m e^{-\lambda}}{m!} \tag{4.103}
$$

with mean

$$\lambda = \sum_{k=1}^{n} \lambda_k. \tag{4.104}$$

■

C. Transformation to Uncorrelated Random Variables

Matrix notation much simplifies the study of linear transformations of random variables, and we shall also find it useful in treating jointly Gaussian random variables in the next section. Denote by $\mathbf{x}$ a column vector made up of the random variables $\mathbf{x}_1, \mathbf{x}_2, \ldots, \mathbf{x}_n$, and denote by m the column vector of their means:

$$\mathbf{x} = \begin{bmatrix} \mathbf{x}_1 \\ \mathbf{x}_2 \\ \cdot \\ \cdot \\ \cdot \\ \mathbf{x}_n \end{bmatrix}, \qquad \mathsf{m} = E(\mathbf{x}) = \begin{bmatrix} m_1 \\ m_2 \\ \cdot \\ \cdot \\ \cdot \\ m_n \end{bmatrix}, \qquad m_k = E(\mathbf{x}_k). \tag{4.105}$$

The corresponding row vectors are

$$\mathbf{x}^T = (\mathbf{x}_1 \quad \mathbf{x}_2 \quad \cdots \quad \mathbf{x}_n), \qquad \mathsf{m}^T = (m_1 \quad m_2 \quad \cdots \quad m_n).$$

Remembering that the product of a column vector by a row vector is a matrix, we can write the $n \times n$ covariance matrix defined in (4.88) as

$$\mathsf{c} = E[(\mathbf{x} - \mathsf{m})(\mathbf{x}^T - \mathsf{m}^T)]. \tag{4.106}$$

The expected value of a matrix of random variables is defined as the matrix of the expected values of its elements.

We introduce n new random variables $\mathbf{z}_1, \mathbf{z}_2, \ldots, \mathbf{z}_n$ as linear combinations of $\mathbf{x}_1, \mathbf{x}_2, \ldots, \mathbf{x}_n$,

$$\mathbf{z}_k = \sum_{i=1}^{n} U_{ki}\mathbf{x}_i, \tag{4.107}$$

or in terms of the $n \times n$ transformation matrix U,

$$\mathbf{z} = \mathsf{U}\mathbf{x}. \tag{4.108}$$

The mean values of the new random variables form a column vector m_z given by

$$\mathsf{m}_z = \mathsf{U}\mathsf{m}, \tag{4.109}$$

as ensues when one takes the expected value of both sides of (4.107), invoking (4.85). The $n \times n$ covariance matrix of $\mathbf{z}_1, \mathbf{z}_2, \ldots, \mathbf{z}_n$ is, by (4.106), (4.108), with $\mathbf{z}^T = \mathbf{x}^T\mathsf{U}^T$,

$$\begin{aligned} \mathsf{c}_z &= E[(\mathbf{z} - \mathsf{m}_z)(\mathbf{z}^T - \mathsf{m}_z^T)] \\ &= E[\mathsf{U}(\mathbf{x} - \mathsf{m})(\mathbf{x}^T - \mathsf{m}^T)\mathsf{U}^T] \qquad (4.110) \\ &= \mathsf{U}\mathsf{c}\mathsf{U}^T, \end{aligned}$$

U^T being the transpose of the matrix U,

$$(\mathsf{U}^T)_{ij} = U_{ji}; \tag{4.111}$$

the transpose of a matrix product is the product of the transposes of the factors, but in reverse order. The reader should write out the covariance Cov $(\mathbf{z}_k, \mathbf{z}_m)$ in terms of the covariances

$$c_{ij} = \text{Cov}\ (\mathbf{x}_i, \mathbf{x}_j),$$

as in Example 3.22, in order to absorb the meaning of (4.110).

We seek a transformation matrix U for which the new random variables $\mathbf{z}_1$, $\mathbf{z}_2, \ldots, \mathbf{z}_n$ are uncorrelated, that is, for which their covariance matrix c_z is diagonal,

$$(\mathsf{c}_z)_{ij} = \lambda_i \delta_{ij} = \begin{cases} \lambda_i, & i = j, \\ 0, & i \neq j. \end{cases} \tag{4.112}$$

It is shown in courses on matrix algebra that an $n \times n$ orthogonal matrix U can be found that accomplishes this transformation. By orthogonal it is meant that

$$\mathsf{U}\mathsf{U}^T = \mathsf{U}^T\mathsf{U} = \mathsf{I}, \tag{4.113}$$

where I is the $n \times n$ identity matrix. That is,

$$\sum_{j=1}^{n} U_{ij} U_{kj} = \delta_{ik} = \begin{cases} 1, & i = k, \\ 0, & i \neq k. \end{cases} \tag{4.114}$$

The rows of the matrix U form a set of n orthogonal vectors of unit length, and so do the columns. Multiplying (4.110) on the left by U^T and using (4.113), we obtain

$$\mathsf{c}\mathsf{U}^T = \mathsf{U}^T\mathsf{c}_z,$$

which when written out with the help of (4.112) yields n sets of n linear simultaneous equations

$$\sum_{j=1}^{n} c_{ij} U_{kj} = \lambda_k U_{ki}, \qquad (i, k) = 1, 2, \ldots, n. \tag{4.115}$$

The n numbers λ_k are called the *eigenvalues* (or *characteristic values* or *latent roots*) of the matrix c, and the column vector θ_k whose ith element is $U_{ki} = (U^T)_{ik}$ is called the kth *eigenvector* (or *characteristic vector*) of c. The eigenvalues λ are the n roots of the equation

$$\det\ (\mathsf{c} - \lambda\mathsf{I}) = 0, \tag{4.116}$$

which is required in order that for each k the n equations (4.115) have solutions for U_{kj} that are not all zero. When this determinant is expanded, one obtains an algebraic equation of nth degree, whose roots λ are the n eigenvalues λ_1, $\lambda_2, \ldots, \lambda_n$. We shall assume that these are all distinct.

A covariance matrix c is positive definite, for if a is any column vector of n coefficients a_i, $1 \le i \le n$, the variance of the random variable

$$\mathbf{w} = \mathsf{a}^T\mathbf{x} = \sum_{i=1}^{n} a_i \mathbf{x}_i$$

must be positive:

$$\begin{aligned} \text{Var } \mathbf{w} &= E[\mathsf{a}^T(\mathbf{x} - \mathsf{m})(\mathbf{x}^T - \mathsf{m}^T)\mathsf{a}] \\ &= \mathsf{a}^T\mathsf{c}\mathsf{a} > 0, \end{aligned} \tag{4.117}$$

and this is the condition for positive definiteness. We assume that for no vector a is Var $\mathbf{w}$ identically zero, for that would imply a linear relationship among the $\mathbf{x}$'s,

$$\sum_{i=1}^{n} a_i \mathbf{x}_i = \sum_{j=1}^{n} a_j m_j = \text{constant}.$$

By the Chebyshev inequality (2.83), a random variable of zero variance is equal to its expected value with probability 1. It is shown in texts on matrix algebra that the eigenvalues of a positive-definite matrix are all positive: $\lambda_k > 0$, $\forall\, k$, as is indeed necessary here because $\lambda_k = \text{Var } \mathbf{z}_k$.

The eigenvalues λ_k having been found, one substitutes each in turn into (4.115) and solves for U_{kj}, $1 \le j \le n$. This can be done by taking one term of each, say U_{kn}, to the right-hand side, putting all others on the left, and solving the first $n - 1$ of the resulting linear simultaneous equations for U_{kj}, $1 \le j \le n - 1$, in terms of U_{kn}:

$$U_{kj} = b_j U_{kn}, \qquad j = 1, 2, \ldots, n - 1.$$

In order for (4.113) to be satisfied, we must pick U_{kn} so that

$$\sum_{j=1}^{n} U_{kj}^2 = \left(\sum_{j=1}^{n-1} b_j^2 + 1\right) U_{kn}^2 = 1,$$

after which the remaining U_{kj} can be determined. Equation (4.113) requires

$$\det \mathsf{U} \det \mathsf{U}^T = (\det \mathsf{U})^2 = 1,$$

and by changing the signs of at most one row or column of U one can arrange that

$$\det \mathsf{U} = 1.$$

We assume that this has been done.

When the matrix elements U_{kj} so calculated are substituted into (4.107), the random variables $\mathbf{z}_1, \mathbf{z}_2, \ldots, \mathbf{z}_n$ become uncorrelated. Statisticians call these the *principal components* of the random variables $\mathbf{x}_1, \mathbf{x}_2, \ldots, \mathbf{x}_n$, particularly when the matrix c is an empirical covariance matrix computed, as will be shown in Sec. 4-6, from a large number of samples of the n random variables $\mathbf{x}_1, \mathbf{x}_2, \ldots, \mathbf{x}_n$, each set the outcome of an independent trial of a chance experiment.* An eigenvalue, say λ_n, that is much smaller than all the rest might be taken to be "really" zero and thus to indicate a linear relation

*M. G. Kendall and A. Stuart, *The Advanced Theory of Statistics*, Vol. 3. London: Griffin, 1966, Chap. 43, pp. 285–313.

$$\sum_{i=1}^{n} U_{ni}\mathbf{x}_i = \sum_{i=1}^{n} U_{ni}m_i = \text{constant}$$

among the random variables. If one eigenvalue, say λ_1, is much greater than the rest, the random variable

$$\mathbf{z}_1 = \sum_{i=1}^{n} U_{1i}\mathbf{x}_i$$

embodies most of the variability of the $\mathbf{x}$'s and might serve as an "index," as in quoting levels of commodity and stock prices.

EXAMPLE 4.14. The covariance matrix of the random variables $\mathbf{x}_1$, $\mathbf{x}_2$, $\mathbf{x}_3$ is

$$\mathsf{C} = \begin{bmatrix} 12 & -11 & -1 \\ -11 & 33 & 3 \\ -1 & 3 & 27 \end{bmatrix}.$$

The determinantal equation (4.116) is then

$$\begin{vmatrix} 12 - \lambda & -11 & -1 \\ -11 & 33 - \lambda & 3 \\ -1 & 3 & 27 - \lambda \end{vmatrix} = 7350 - 1480\lambda + 72\lambda^2 - \lambda^3 = 0$$

and its roots are

$$\lambda_1 = 7.2896, \qquad \lambda_2 = 26.1446, \qquad \lambda_3 = 38.5658.$$

In order to calculate U_{1j}, $1 \le j \le 3$, one substitutes λ_1 into (4.115), obtaining the linear equations

$$(12 - \lambda_1)U_{11} - 11U_{12} - U_{13} = 0$$
$$-11U_{11} + (33 - \lambda_1)U_{12} + 3U_{13} = 0$$
$$-U_{11} + 3U_{12} + (27 - \lambda_1)U_{13} = 0,$$

of which the first two are rewritten as

$$4.7104\, U_{11} - 11U_{12} = U_{13}$$
$$-11U_{11} + 25.7104\, U_{12} = -3U_{13}.$$

Solved for U_{11} and U_{12}, these yield

$$U_{11} = -68.2842U_{13}, \qquad U_{12} = -29.3315U_{13}.$$

Since the vector (U_{11}, U_{12}, U_{13}) must have length 1,

$$[(68.2842)^2 + (29.3315)^2 + 1]U_{13}^2 = 1,$$

whence $U_{13} = -0.013455$, and thereupon

$$U_{11} = 0.91874, \qquad U_{12} = 0.39464.$$

This provides the first row of the matrix U and the first column of U^T. The rest are obtained in like manner from the values of λ_2 and λ_3. The matrix U so calculated is

$$\mathsf{U} = \begin{bmatrix} 0.91874 & 0.39464 & -0.013455 \\ 0.11572 & -0.23650 & 0.96472 \\ -0.37754 & 0.88788 & 0.26294 \end{bmatrix},$$

and the principal components are

$$\mathbf{z}_1 = 0.9187\mathbf{x}_1 + 0.3946\mathbf{x}_2 - 0.01345\mathbf{x}_3,$$

$$\mathbf{z}_2 = 0.1157\mathbf{x}_1 - 0.2365\mathbf{x}_2 + 0.9647\mathbf{x}_3,$$

$$\mathbf{z}_3 = -0.3775\mathbf{x}_1 + 0.8879\mathbf{x}_2 + 0.2629\mathbf{x}_3. \qquad \blacksquare$$

Computer programs are available for calculating the eigenvalues and eigenvectors of symmetric matrices such as c.

EXAMPLE 4.15. Multivariate Gaussian Integral

For future reference we utilize the results of this part to show that

$$\int_{-\infty}^{\infty} \cdots \int_{-\infty}^{\infty} \exp\left[-\tfrac{1}{2}\sum_{i=1}^{n}\sum_{j=1}^{n} C_{ij}x_i x_j + \sum_{i=1}^{n} a_i x_i\right] dx_1\, dx_2 \cdots dx_n$$

$$= \int_{-\infty}^{\infty} \cdots \int_{-\infty}^{\infty} \exp\left(-\tfrac{1}{2}\mathsf{x}^T\mathsf{C}\mathsf{x} + \mathsf{a}^T\mathsf{x}\right) dx_1\, dx_2 \cdots dx_n \tag{4.118}$$

$$= (2\pi)^{n/2} (\det \mathsf{C})^{-1/2} \exp\left(\tfrac{1}{2}\mathsf{a}^T\mathsf{C}^{-1}\mathsf{a}\right)$$

with

$$\mathsf{a}^T\mathsf{C}^{-1}\mathsf{a} = \sum_{i=1}^{n}\sum_{j=1}^{n} c_{ij}a_i a_j \tag{4.119}$$

in terms of the elements c_{ij} of the $n \times n$ matrix $\mathsf{c} = \mathsf{C}^{-1}$. It is assumed that the matrix C is positive definite, so that the quadratic form $\mathsf{x}^T\mathsf{C}\mathsf{x}$ in the exponent is always positive and the absolute value of the integrand drops off to zero at infinity. The coefficients a_i may be real or complex.

The integral is evaluated by changing to new integration variables $z_1, z_2, \ldots, z_n$ by a linear transformation $\mathsf{z} = \mathsf{U}\mathsf{x}$ like that in (4.108); the matrix U is chosen so that

$$\mathsf{C}' = \mathsf{U}\mathsf{C}\mathsf{U}^T \tag{4.120}$$

is diagonal, with diagonal elements

$$C'_{ii} = \lambda_i^{-1}, \qquad i = 1, 2, \ldots, n, \tag{4.121}$$

the λ_i being as before the eigenvalues of $\mathsf{c} = \mathsf{C}^{-1}$. Thereupon the quadratic form in the exponent becomes, with $\mathsf{x} = \mathsf{U}^T\mathsf{z}$,

$$\mathsf{x}^T\mathsf{C}\mathsf{x} = \mathsf{z}^T\mathsf{U}\mathsf{C}\mathsf{U}^T\mathsf{z} = \mathsf{z}^T\mathsf{C}'\mathsf{z}$$
$$= \sum_{k=1}^{n} \frac{z_k^2}{\lambda_k}, \tag{4.122}$$

and the linear form in the exponent becomes

$$\mathsf{a}^T\mathsf{x} = \mathsf{a}^T\mathsf{U}^T\mathsf{z} = \mathsf{b}^T\mathsf{z} = \sum_{k=1}^{n} b_k z_k,$$

where

$$\mathsf{b} = \mathsf{U}\mathsf{a} \qquad \text{or} \qquad b_k = \sum_{i=1}^{n} U_{ki}a_i. \tag{4.123}$$

The Jacobian of the transformation relates the infinitesimal volume elements of integration,

$$\frac{dx_1\, dx_2 \cdots dx_n}{dz_1\, dz_2 \ldots dz_n} = J\begin{pmatrix} x_1 & x_2 & \cdots & x_n \\ z_1 & z_2 & \cdots & z_n \end{pmatrix} = \det \mathsf{U}^T = 1. \tag{4.124}$$

Hence our integral becomes

$$\int_{-\infty}^{\infty} \cdots \int_{-\infty}^{\infty} \exp\left[-\sum_{i=1}^{n}\left(\frac{z_k^2}{2\lambda_k} - b_k z_k\right)\right] dz_1 \cdots dz_n$$

$$= \prod_{k=1}^{n} \int_{-\infty}^{\infty} \exp\left(-\frac{z_k^2}{2\lambda_k} + b_k z_k\right) dz_k \tag{4.125}$$

$$= \prod_{k=1}^{n} [\sqrt{2\pi\lambda_k} \exp(\tfrac{1}{2}\lambda_k b_k^2)]$$

$$= (2\pi)^{n/2}|\det \mathsf{C}'|^{-1/2} \exp(\tfrac{1}{2}\mathsf{b}^T\mathsf{C}'^{-1}\mathsf{b}).$$

The individual integrals over $-\infty < z_k < \infty$ are evaluated by completing the square in the exponent and using the normalization integral for the Gaussian distribution of (2.19). Furthermore, since C' is diagonal, with diagonal values λ_k^{-1},

$$\prod_{k=1}^{n} \lambda_k^{-1} = \det \mathsf{C}'$$

and

$$\det \mathsf{C}' = \det(\mathsf{UCU}^T) = \det \mathsf{U} \det \mathsf{C} \det \mathsf{U}^T$$
$$= \det \mathsf{C} = |\det \mathsf{c}|^{-1}.$$

In addition, by (4.123),

$$\mathsf{b}^T\mathsf{C}'^{-1}\mathsf{b} = \mathsf{a}^T\mathsf{U}^T\mathsf{C}'^{-1}\mathsf{U}\mathsf{a} = \mathsf{a}^T\mathsf{C}^{-1}\mathsf{a}$$
$$= \mathsf{a}^T\mathsf{c}\mathsf{a}.$$

Putting all this together with (4.125), we obtain (4.118). ■

D. Conditional Expected Values

If $\mathcal{A}$ is a conditioning event and $f_{\mathbf{x}}(x_1, x_2, \ldots, x_n \mid \mathcal{A})$ is the joint probability density function of the random variables $\mathbf{x}_1, \mathbf{x}_2, \ldots, \mathbf{x}_n$, given that the event $\mathcal{A}$ has occurred, then the conditional expected value of any function $g(\mathbf{x}_1, \mathbf{x}_2, \ldots, \mathbf{x}_n)$ of these random variables is defined by

$$E[g(\mathbf{x}_1, \mathbf{x}_2, \ldots, \mathbf{x}_n) \mid \mathcal{A}]$$
$$= \int_{-\infty}^{\infty} \cdots \int_{-\infty}^{\infty} g(\mathbf{x}_1, \mathbf{x}_2, \ldots, \mathbf{x}_n) \tag{4.126}$$
$$\times f_{\mathbf{x}}(x_1, x_2, \ldots, x_n \mid \mathcal{A})\, dx_1\, dx_2 \cdots dx_n.$$

The event $\mathcal{A}$ may represent a subset of certain nonnumerical attributes, or it may mean that the random variables $\mathbf{x}_1, \mathbf{x}_2, \ldots, \mathbf{x}_n$ were observed to lie in a certain region $\mathcal{A}$ of their n-dimensional space R_n.

In particular, the conditional expected value, or *conditional mean,* of the random variable $\mathbf{x}_j$ is

$$E(\mathbf{x}_j \mid \mathcal{A}) = \int_{-\infty}^{\infty} \cdots \int_{-\infty}^{\infty} x_j f_{\mathbf{x}}(x_1, x_2, \ldots, x_n \mid \mathcal{A})\, dx_1\, dx_2 \cdots dx_n \tag{4.127}$$
$$= \int_{-\infty}^{\infty} x f_{\mathbf{x}_j}(x \mid \mathcal{A})\, dx$$

in terms of the marginal conditional probability density function $f_{\mathbf{x}_j}(x \mid \mathcal{A})$ of $\mathbf{x}_j$, given the event $\mathcal{A}$. The conditional covariance Cov $(\mathbf{x}_j, \mathbf{x}_k \mid \mathcal{A})$ is similarly

$$\begin{aligned}\text{Cov}\,(\mathbf{x}_j, \mathbf{x}_k \mid \mathcal{A}) = \int_{-\infty}^{\infty}\int_{-\infty}^{\infty} x_j x_k f_{\mathbf{x}_j\mathbf{x}_k}(x_j, x_k \mid \mathcal{A})\, dx_j\, dx_k \\ - E(\mathbf{x}_j \mid \mathcal{A})E(\mathbf{x}_k \mid \mathcal{A})\end{aligned} \tag{4.128}$$

in terms of the marginal conditional density function of $\mathbf{x}_j$ and $\mathbf{x}_k$, given $\mathcal{A}$.

The event $\mathcal{A}$ may be the assumption of particular values by certain of the random variables, whereupon the conditional density functions involved are of the type given in (4.32)–(4.35). For example, the conditional mean of $\mathbf{x}_1$, given that $\mathbf{x}_2, \mathbf{x}_3, \ldots, \mathbf{x}_p$ have taken on particular values $x_2, x_3, \ldots, x_p$, respectively, is

$$\begin{aligned}E(\mathbf{x}_1 \mid \mathbf{x}_2 = x_2, \mathbf{x}_3 = x_3, \ldots, \mathbf{x}_p = x_p) \\ = E(\mathbf{x}_1 \mid x_2, x_3, \ldots, x_p) \\ = \int_{-\infty}^{\infty} x_1 f_{\mathbf{x}_1}(x_1 \mid x_2, x_3, \ldots, x_p)\, dx_1 \\ = \frac{\int_{-\infty}^{\infty} x_1 f_{\mathbf{x}_1}(x_1, x_2, \ldots, x_p)\, dx_1}{f_{\mathbf{x}}(x_2, x_3, \ldots, x_p)}.\end{aligned} \tag{4.129}$$

For such conditional expected values a rule generalizing (3.110) holds:

$$\begin{aligned}E_{\mathbf{x}_{r+1}\cdots\mathbf{x}_n}\{E[g(\mathbf{x}_1, \mathbf{x}_2, \ldots, \mathbf{x}_n) \mid \mathbf{x}_{r+1}, \mathbf{x}_{r+2}, \ldots, \mathbf{x}_n]\} \\ = E[g(\mathbf{x}_1, \mathbf{x}_2, \ldots, \mathbf{x}_n)].\end{aligned} \tag{4.130}$$

Here the random variable

$$E[g(\mathbf{x}_1, \mathbf{x}_2, \ldots, \mathbf{x}_n) \mid \mathbf{x}_{r+1}, \mathbf{x}_{r+2}, \ldots, \mathbf{x}_n]$$

is obtained by replacing the particular values $x_{r+1}, x_{r+2}, \ldots, x_n$ in the function

$$\begin{aligned}h(x_{r+1}, x_{r+2}, \ldots, x_n) \\ = E[g(\mathbf{x}_1, \mathbf{x}_2, \ldots, \mathbf{x}_n) \mid x_{r+1}, x_{r+2}, \ldots, x_n] \\ = \int_{-\infty}^{\infty} dx_1 \cdots \int_{-\infty}^{\infty} dx_r\, g(x_1, x_2, \ldots, x_n) \\ \times f_{\mathbf{x}}(x_1, x_2, \ldots, x_r \mid x_{r+1}, x_{r+2}, \ldots, x_n)\end{aligned} \tag{4.131}$$

by the random variables $\mathbf{x}_{r+1}, \mathbf{x}_{r+2}, \ldots, \mathbf{x}_n$, respectively, thus converting the function $h(x_{r+1}, x_{r+2}, \ldots, x_n)$ to a random variable $h(\mathbf{x}_{r+1}, \mathbf{x}_{r+2}, \ldots, \mathbf{x}_n)$. The symbol $E_{\mathbf{x}_{r+1}\cdots\mathbf{x}_n}$ denotes taking the expected value of this random variable $h(\mathbf{x}_{r+1}, \mathbf{x}_{r+2}, \ldots, \mathbf{x}_n)$. Thus the left-hand side of (4.130) is

$$
\begin{aligned}
&E[h(\mathbf{x}_{r+1}, \mathbf{x}_{r+2}, \ldots, \mathbf{x}_n)] \\
&\quad = \int_{-\infty}^{\infty} dx_{r+1} \cdots \int_{-\infty}^{\infty} dx_n\, h(x_{r+1}, x_{r+2}, \ldots, x_n) f_{\mathbf{x}}(x_{r+1}, x_{r+2}, \ldots, x_n) \\
&\quad = \int_{-\infty}^{\infty} dx_{r+1} \cdots \int_{-\infty}^{\infty} dx_n\, f_{\mathbf{x}}(x_{r+1}, x_{r+2}, \ldots, x_n) \\
&\qquad \times \int_{-\infty}^{\infty} dx_1 \cdots \int_{-\infty}^{\infty} dx_r\, g(x_1, x_2, \ldots, x_n) \\
&\qquad \times f_{\mathbf{x}}(x_1, x_2, \ldots, x_r \mid x_{r+1}, x_{r+2}, \ldots, x_n) \\
&\quad = \int_{-\infty}^{\infty} dx_1 \cdots \int_{-\infty}^{\infty} dx_r \int_{-\infty}^{\infty} dx_{r+1} \cdots \int_{-\infty}^{\infty} dx_n\, g(x_1, x_2, \ldots, x_n) \\
&\qquad \times f_{\mathbf{x}}(x_1, x_2, \ldots, x_n) = E[g(\mathbf{x}_1, \mathbf{x}_2, \ldots, \mathbf{x}_n)]
\end{aligned}
$$

because the conditional density function involved is

$$
\begin{aligned}
&f_{\mathbf{x}}(x_1, x_2, \ldots, x_r \mid x_{r+1}, x_{r+2}, \ldots, x_n) \\
&\quad = \frac{f_{\mathbf{x}}(x_1, x_2, \ldots, x_r, x_{r+1}, \ldots, x_n)}{f_{\mathbf{x}}(x_{r+1}, x_{r+2}, \ldots, x_n)}.
\end{aligned}
$$

Examples of expected values of this kind are conditional correlations like

$$
\begin{aligned}
&E(\mathbf{x}_1\mathbf{x}_2 \mid x_3, x_4, \ldots, x_n) \\
&\quad = E(\mathbf{x}_1\mathbf{x}_2 \mid \mathbf{x}_3 = x_3, \mathbf{x}_4 = x_4, \ldots, \mathbf{x}_n = x_n) \\
&\quad = \int_{-\infty}^{\infty} dx_1 \int_{-\infty}^{\infty} dx_2\, x_1 x_2 f_{\mathbf{x}_1\mathbf{x}_2}(x_1, x_2 \mid x_3, x_4, \ldots, x_n) \qquad (4.132) \\
&\quad = \frac{\int_{-\infty}^{\infty} dx_1 \int_{-\infty}^{\infty} dx_2\, x_1 x_2\, f_{\mathbf{x}}(x_1, x_2, x_3, x_4, \ldots, x_n)}{f_{\mathbf{x}}(x_3, x_4, \ldots, x_n)},
\end{aligned}
$$

and from these, conditional covariances can be formed, such as

$$
\begin{aligned}
&\operatorname{Cov}(\mathbf{x}_1, \mathbf{x}_2 \mid x_3, x_4, \ldots, x_n) \\
&\quad = E(\mathbf{x}_1\mathbf{x}_2 \mid x_3, x_4, \ldots, x_n) \qquad (4.133) \\
&\qquad - E(\mathbf{x}_1 \mid x_3, x_4, \ldots, x_n) E(\mathbf{x}_2 \mid x_3, x_4, \ldots, x_n),
\end{aligned}
$$

which are of importance in statistical estimation and signal processing.

THE MULTIVARIATE GAUSSIAN DISTRIBUTION

A. Density Function

The n random variables $\mathbf{x}_1, \mathbf{x}_2, \ldots, \mathbf{x}_n$ are said to be jointly Gaussian (or "normal") if their joint probability density function has the multivariate Gaussian form

$$f_{\mathbf{x}}(x_1, x_2, \ldots, x_n)$$
$$= (2\pi)^{-n/2} \,|\det \mathsf{C}\,|^{1/2} \exp\left[-\tfrac{1}{2}\sum_{i=1}^{n}\sum_{j=1}^{n} C_{ij}\,(x_i - m_i)(x_j - m_j)\right], \tag{4.134}$$

where the C_{ij} are elements of an $n \times n$ positive-definite matrix C that will be shown to be the inverse, $\mathsf{C} = \mathsf{c}^{-1}$, of the covariance matrix c of the random variables, as defined in (4.87) and (4.106). The quantities m_i will be shown to be the expected values of the random variables, and

$$\det \mathsf{C} = \det\,(C_{ij}) \tag{4.135}$$

is the determinant of the matrix C. The density function can be more concisely written in the matrix notation introduced in Sec. 4-4,

$$f_{\mathbf{x}}(x_1, x_2, \ldots, x_n)$$
$$= (2\pi)^{-n/2}\,|\det \mathsf{c}\,|^{-1/2} \exp\left[-\tfrac{1}{2}(\mathsf{x}^T - \mathsf{m}^T)\mathsf{c}^{-1}(\mathsf{x} - \mathsf{m})\right], \tag{4.136}$$

where x is a column vector of the x's, and

$$\mathsf{x}^T = (x_1 \quad x_2 \quad \cdots \quad x_n)$$

is a row vector; m is the column vector and m^T the row vector of expected values m_i, $1 \le i \le n$.

Because the matrices c and $\mathsf{C} = \mathsf{c}^{-1}$ are assumed positive definite, the exponents in (4.134) and (4.136) are always negative, and the density function, always positive, drops off to zero at infinity in R_n. It is properly normalized, as can be shown by substituting it into (4.10) and utilizing (4.118) with $a_i \equiv 0$, $\forall\, i$, to evaluate the integral over R_n after changing variables,

$$x_k - m_k \to x_k', \qquad \forall\, k.$$

If when $n = 2$ we take

$$\mathsf{c} = \begin{bmatrix} \sigma_1^2 & r\sigma_1\sigma_2 \\ r\sigma_1\sigma_2 & \sigma_2^2 \end{bmatrix}, \tag{4.137}$$

the inverse matrix in (4.136) is

$$\mathsf{c}^{-1} = \mathsf{C} = \frac{1}{\det \mathsf{c}}\begin{bmatrix} \sigma_2^2 & -r\sigma_1\sigma_2 \\ -r\sigma_1\sigma_2 & \sigma_1^2 \end{bmatrix}, \tag{4.138}$$

with

$$\det \mathsf{c} = \sigma_1^2\sigma_2^2(1 - r^2). \tag{4.139}$$

(A scalar in front of a matrix multiplies each of its elements.) The reader should convince himself that for $n = 2$ the forms in (4.134) and (4.136) lead to the bivariate Gaussian density function given in (3.40).

B. Gaussian Characteristic Function

When the random variables $\mathbf{x}_1, \mathbf{x}_2, \ldots, \mathbf{x}_n$ have the Gaussian density function in (4.134), (4.136), their joint characteristic function is

$$\begin{aligned}\Phi_{\mathbf{x}}(\omega_1, \omega_2, \ldots, \omega_n) &= E\{\exp[j(\omega_1\mathbf{x}_1 + \omega_2\mathbf{x}_2 + \cdots + \omega_n\mathbf{x}_n)]\}\\ &= \exp\left(j\sum_{i=1}^{n} m_i\omega_i - \tfrac{1}{2}\sum_{i=1}^{n}\sum_{k=1}^{n} c_{ik}\omega_i\omega_k\right)\\ &= \exp(j\mathsf{m}^T\mathsf{w} - \tfrac{1}{2}\,\mathsf{w}^T\mathsf{c}\mathsf{w}),\end{aligned} \tag{4.140}$$

where w is the column vector of Fourier-domain variables ω_i, $1 \le i \le n$, and

$$\mathsf{w}^T = (\omega_1 \quad \omega_2 \quad \ldots \quad \omega_n).$$

The matrix c of the quadratic form in (4.140) is the inverse $\mathsf{c} = \mathsf{C}^{-1}$ of the matrix $\mathsf{C} = \|C_{ij}\|$ of the quadratic form in the probability density function (4.134).

To demonstrate (4.140) one takes the n-dimensional Fourier transform of (4.134) in accordance with the definition (4.94). After changing variables,

$$x_i - m_i = x_i', \qquad 1 \le i \le n,$$

one can utilize (4.118) to evaluate the integral over R_n. Using matrix notation for conciseness, we write, with $\mathsf{C} = \mathsf{c}^{-1}$,

$$\begin{aligned}&\Phi_{\mathbf{x}}(\omega_1, \omega_2, \ldots, \omega_n)\\ &\quad= (2\pi)^{-n/2}\,|\det \mathsf{C}|^{1/2}\\ &\qquad\times \int\cdots\int \exp[j\mathsf{w}^T\mathsf{x} - \tfrac{1}{2}(\mathsf{x}^T - \mathsf{m}^T)\mathsf{C}(\mathsf{x} - \mathsf{m})]\,dx_1\,dx_2\cdots dx_n\\ &\quad= (2\pi)^{-n/2}\,|\det \mathsf{C}|^{1/2}\\ &\qquad\times \int\cdots\int \exp[j\mathsf{w}^T(\mathsf{x}' + \mathsf{m}) - \tfrac{1}{2}\mathsf{x}'^T\mathsf{C}\mathsf{x}']\,dx_1'\,dx_2'\cdots dx_n'\\ &\quad= \exp(j\mathsf{w}^T\mathsf{m})\exp(\tfrac{1}{2}j^2\mathsf{w}^T\mathsf{C}^{-1}\mathsf{w}),\end{aligned} \tag{4.141}$$

which yields (4.140) because $\mathsf{w}^T\mathsf{m} = \mathsf{m}^T\mathsf{w}$.

Now we can prove that the m_i are the expected values of the random variables $\mathbf{x}_i$ and that the matrix c is their covariance matrix. As in (4.97),

$$E(\mathbf{x}_i) = \frac{\partial}{\partial(j\omega_i)}\Phi_{\mathbf{x}}(\omega_1, \omega_2, \ldots, \omega_n)\bigg|_{\mathsf{w}=0} = m_i \tag{4.142}$$

by differentiating the characteristic function in (4.140) with respect to $j\omega_i$. Similarly,

$$E(\mathbf{x}_m\mathbf{x}_p) = -\frac{\partial^2}{\partial\omega_m\,\partial\omega_p}\Phi_{\mathbf{x}}(\omega_1, \omega_2, \ldots, \omega_n)\bigg|_{\mathsf{w}=0}$$

$$= c_{mp} + m_m m_p,$$

when we remember that the quadratic form in the exponent of (4.140) contains ω_m and ω_p combined as

$$c_{mp}\omega_m\omega_p + c_{pm}\omega_p\omega_m = 2c_{mp}\omega_m\omega_p.$$

Hence by the definition of covariance

$$\mathrm{Cov}\,(\mathbf{x}_m, \mathbf{x}_p) = E(\mathbf{x}_m\mathbf{x}_p) - m_m m_p = c_{mp}, \tag{4.143}$$

and we have verified that the matrix c figuring in the joint Gaussian density function (4.136) and in the Gaussian characteristic function (4.140) is indeed the covariance matrix of the random variables $\mathbf{x}_1, \mathbf{x}_2, \ldots, \mathbf{x}_n$.

If the random variables $\mathbf{x}_i$ are uncorrelated, their covariance matrix c is diagonal,

$$c_{ij} = \sigma_i^2\delta_{ij} = \begin{cases}\sigma_i^2, & i = j,\\ 0, & i \neq j,\end{cases} \tag{4.144}$$

where $\sigma_i^2 = \mathrm{Var}\,\mathbf{x}_i$. The inverse matrix $\mathsf{C} = \mathsf{c}^{-1}$ in (4.134), (4.136) is then also diagonal,

$$C_{ij} = \sigma_i^{-2}\delta_{ij}, \tag{4.145}$$

and the joint probability density function takes the form

$$f_{\mathbf{x}}(x_1, x_2, \ldots, x_n) = (2\pi)^{-n/2}\prod_{i=1}^{n}\sigma_i^{-1}\exp\left[-\tfrac{1}{2}\sum_{k=1}^{n}\frac{(x_k - m_k)^2}{\sigma_k^2}\right]$$

$$= \prod_{k=1}^{n}(2\pi\sigma_k^2)^{-1/2}\exp\left[-\frac{(x_k - m_k)^2}{2\sigma_k^2}\right]. \tag{4.146}$$

As we saw in (4.37), the random variables $\mathbf{x}_i$ are therefore statistically independent. Uncorrelated Gaussian random variables are statistically independent. Uncorrelated random variables governed by other probability distributions are *not* in general independent.

For any four Gaussian random variables $\mathbf{x}_1, \mathbf{x}_2, \mathbf{x}_3, \mathbf{x}_4$ with mean zero,

$$E(\mathbf{x}_1\mathbf{x}_2\mathbf{x}_3\mathbf{x}_4) = E(\mathbf{x}_1\mathbf{x}_2)E(\mathbf{x}_3\mathbf{x}_4) + E(\mathbf{x}_1\mathbf{x}_3)E(\mathbf{x}_2\mathbf{x}_4) + E(\mathbf{x}_1\mathbf{x}_4)E(\mathbf{x}_2\mathbf{x}_3)$$

$$= c_{12}c_{34} + c_{13}c_{24} + c_{14}c_{23}. \tag{4.147}$$

We derive this useful rule from the power-series expansion of the characteristic function

$$\Phi_{\mathbf{x}}(\omega_1, \omega_2, \omega_3, \omega_4) = 1 - \tfrac{1}{2}\mathsf{w}^T\mathsf{c}\mathsf{w} + \tfrac{1}{8}(\mathsf{w}^T\mathsf{c}\mathsf{w})^2 + O(\omega^6). \tag{4.148}$$

By (4.97) the moment we are looking for is

$$E(\mathbf{x}_1\mathbf{x}_2\mathbf{x}_3\mathbf{x}_4) = \left.\frac{\partial^4\Phi_{\mathbf{x}}(\omega_1, \omega_2, \omega_3, \omega_4)}{\partial\omega_1\,\partial\omega_2\,\partial\omega_3\,\partial\omega_4}\right|_{\mathsf{w}=0} \tag{4.149}$$

and this fourth derivative at the origin $\mathsf{w} = 0$ can be calculated from the third term on the right of (4.148), all the others vanishing either upon differentiation four times or after we set the ω's equal to zero. Writing out that third term, we find

$$\begin{aligned}&\tfrac{1}{8}(c_{11}\omega_1^2 + c_{22}\omega_2^2 + c_{33}\omega_3^2 + c_{44}\omega_4^2\\&\quad + 2c_{12}\omega_1\omega_2 + 2c_{13}\omega_1\omega_3 + 2c_{14}\omega_1\omega_4\\&\qquad + 2c_{23}\omega_2\omega_3 + 2c_{24}\omega_2\omega_4 + 2c_{34}\omega_3\omega_4)^2.\end{aligned}$$

When we square the quadratic form, the only terms proportional to $\omega_1\omega_2\omega_3\omega_4$ will be

$$\begin{aligned}&\tfrac{1}{8}(2c_{12}\omega_1\omega_2 \cdot 2c_{34}\omega_3\omega_4 + 2c_{13}\omega_1\omega_3 \cdot 2c_{24}\omega_2\omega_4\\&\quad + 2c_{14}\omega_1\omega_4 \cdot 2c_{23}\omega_2\omega_3 + 2c_{23}\omega_2\omega_3 \cdot 2c_{14}\omega_1\omega_4\\&\quad + 2c_{24}\omega_2\omega_4 \cdot 2c_{13}\omega_1\omega_3 + 2c_{34}\omega_3\omega_4 \cdot 2c_{12}\omega_1\omega_2)\\&= (c_{12}c_{34} + c_{13}c_{24} + c_{14}c_{23})\omega_1\omega_2\omega_3\omega_4.\end{aligned}$$

When we evaluate the partial derivative in (4.149), only the coefficient of $\omega_1\omega_2\omega_3\omega_4$ will remain, and we obtain (4.147).

The expected value of the product of any even number of Gaussian random variables with mean values zero is formed as in (4.147) by pairing them in all possible ways, taking the expected value of each pair, and summing over all the ways in which the variables have been paired.

C. Linear Transformations of Gaussian Random Variables

Define n new random variables $\mathbf{z}_1, \mathbf{z}_2, \ldots, \mathbf{z}_n$ by a linear transformation of $\mathbf{x}_1, \mathbf{x}_2, \ldots, \mathbf{x}_n$ of the kind given in (4.71). In our matrix notation that can be written

$$\mathbf{z} = \mathsf{a}\mathbf{x} \tag{4.150}$$

and the inverse transformation is

$$\mathbf{x} = \mathsf{a}^{-1}\mathbf{z} = \mathsf{b}\mathbf{z}. \tag{4.151}$$

By applying (4.73) to the Gaussian density function in (4.136) we find that the joint probability density function of $\mathbf{z}_1, \mathbf{z}_2, \ldots, \mathbf{z}_n$ also has the Gaussian form

$$f_{\mathbf{z}}(z_1, z_2, \ldots, z_n)$$
$$= (2\pi)^{-n/2} \,|\det \mathsf{a}^{-1}| \cdot |\det \mathsf{c}|^{-1/2}$$
$$\times \exp\left[-\tfrac{1}{2}(\mathsf{z}^T \mathsf{a}^{T^{-1}} - \mathsf{m}^T)\mathsf{c}^{-1}(\mathsf{a}^{-1}\mathsf{z} - \mathsf{m})\right] \qquad (4.152)$$
$$= (2\pi)^{-n/2} \,|\det(\mathsf{aca}^T)|^{-1/2}$$
$$\times \exp\left[-\tfrac{1}{2}(\mathsf{z}^T - \mathsf{m}^T\mathsf{a}^T)(\mathsf{aca}^T)^{-1}(\mathsf{z} - \mathsf{am})\right].$$

Recall that

$$(\mathsf{aca}^T)^{-1} = \mathsf{a}^{T^{-1}}\mathsf{c}^{-1}\mathsf{a}^{-1}$$

and that

$$\det(\mathsf{aca}^T) = \det \mathsf{a} \det \mathsf{c} \det \mathsf{a}^T = (\det \mathsf{a})^2 \det \mathsf{c}.$$

The column vector of mean values of the $\mathbf{z}$'s is

$$\mathsf{m}_z = \mathsf{am} \qquad (4.153)$$

and their covariance matrix is

$$\mathsf{c}_z = \mathsf{aca}^T. \qquad (4.154)$$

These can be verified by direct calculation and are apparent from (4.152).

All the marginal probability density functions of Gaussian random variables, such as the $\mathbf{x}$'s and the $\mathbf{z}$'s, also have the Gaussian form. That this is so can be verified by considering the characteristic function in (4.140). The joint characteristic function of a subset of the random variables $\mathbf{x}_1, \mathbf{x}_2, \ldots, \mathbf{x}_n$ is obtained by setting those ω's equal to zero that correspond to the random variables being discarded, as we learned in Sec. 4-4B. When this is done to the characteristic function in (4.140), it retains its Gaussian form, and the inverse Fourier transform with respect to the remaining variables will lead to Gaussian marginal probability density functions.

That linear transformations of Gaussian random variables create new random variables that also have Gaussian distributions will be important when we later consider the filtering of Gaussian stochastic processes such as thermal noise in linear circuits. The joint density function of the new variables is completely determined by their means and covariances, and those can be calculated as in (4.153) and (4.154).

EXAMPLE 4.16. The random variables $\mathbf{x}_1$, $\mathbf{x}_2$, and $\mathbf{x}_3$ are jointly Gaussian with mean values $m_1 = 2$, $m_2 = -1$, $m_3 = 3$, respectively, and covariance matrix

$$\mathsf{c} = \begin{bmatrix} 2 & 1 & -1 \\ 1 & 5 & 2 \\ -1 & 2 & 7 \end{bmatrix}.$$

The inverse of this covariance matrix is

$$\mathsf{c}^{-1} = \mathsf{C} = \frac{1}{46}\begin{bmatrix} 31 & -9 & 7 \\ -9 & 13 & -5 \\ 7 & -5 & 9 \end{bmatrix}$$

and the joint density function of $\mathbf{x}_1$, $\mathbf{x}_2$, $\mathbf{x}_3$ is

$$f_{\mathbf{x}}(x_1, x_2, x_3) = (2\pi)^{-3/2}(46)^{-1/2}$$

$$\times \exp\left\{-\frac{1}{92}[31(x_1 - 2)^2 - 18(x_1 - 2)(x_2 + 1) + 14(x_1 - 2)(x_3 - 3) + 13(x_2 + 1)^2 - 10(x_2 + 1)(x_3 - 3) + 9(x_3 - 3)^2]\right\}.$$

The new random variables

$$\mathbf{z}_1 = 3\mathbf{x}_1 - \mathbf{x}_2 - \mathbf{x}_3$$

$$\mathbf{z}_2 = \mathbf{x}_1 + \mathbf{x}_2 + 2\mathbf{x}_3$$

are derived from the original ones by the transformation matrix

$$\mathsf{a} = \begin{pmatrix} 3 & -1 & -1 \\ 1 & 1 & 2 \end{pmatrix}.$$

Their mean values are, by (4.153),

$$E(\mathbf{z}_1) = 4, \qquad E(\mathbf{z}_2) = 7$$

and their covariance matrix is, by (4.154),

$$\mathsf{C}_z = \begin{pmatrix} 3 & -1 & -1 \\ 1 & 1 & 2 \end{pmatrix}\begin{pmatrix} 2 & 1 & -1 \\ 1 & 5 & 2 \\ -1 & 2 & 7 \end{pmatrix}\begin{pmatrix} 3 & 1 \\ -1 & 1 \\ -1 & 2 \end{pmatrix}$$

$$= \begin{pmatrix} 6 & -4 & -12 \\ 1 & 10 & 15 \end{pmatrix}\begin{pmatrix} 3 & 1 \\ -1 & 1 \\ -1 & 2 \end{pmatrix} = \begin{pmatrix} 34 & -22 \\ -22 & 41 \end{pmatrix}.$$

Its inverse is

$$\mathsf{C}_z = \frac{1}{910}\begin{pmatrix} 41 & 22 \\ 22 & 34 \end{pmatrix}$$

and the joint density function of $\mathbf{z}_1$ and $\mathbf{z}_2$ is

$$f_{\mathbf{z}}(z_1, z_2) = (2\pi)^{-1}(910)^{-1/2}$$

$$\times \exp\left\{-\frac{1}{1820}[41(z_1 - 4)^2 + 44(z_1 - 4)(z_2 - 7) + 34(z_2 - 7)^2]\right\}.$$

■

D. Conditional Distributions and Expected Values

Suppose that a number $n + m$ of random variables have a jointly Gaussian probability density function of the kind we have been discussing, and suppose that m of them have been observed to take on particular values in an experiment. We seek the joint probability density function of the rest, which for convenience we label $\mathbf{x}_1, \mathbf{x}_2, \ldots, \mathbf{x}_n$; the ones whose values have been observed are labeled

$x_{n+1}, x_{n+2}, \ldots, x_{n+m}$. We gather the first n into a column vector $\mathbf{x}_1$, the rest into a column vector $\mathbf{x}_2$; their transposed vectors are the rows

$$\mathbf{x}_1^T = (x_1 \quad x_2 \quad \cdots \quad x_n),$$

$$\mathbf{x}_2^T = (x_{n+1} \quad x_{n+2} \quad \cdots \quad x_{n+m}).$$

The column vector for all $n + m$ random variables can then be written

$$\mathbf{x} = \begin{bmatrix} \mathbf{x}_1 \\ \mathbf{x}_2 \end{bmatrix}$$

and the transposed vector is $\mathbf{x}^T = (\mathbf{x}_1^T \quad \mathbf{x}_2^T)$. We assume for simplicity that all the random variables have mean values zero. The changes necessary to account for nonzero means are easy to make, but uninstructive.

The $(n + m) \times (n + m)$ covariance matrix of these $n + m$ random variables is divided into corresponding blocks and written as

$$\mathsf{c} = \begin{bmatrix} \mathsf{c}_{11} & \mathsf{c}_{12} \\ \mathsf{c}_{21} & \mathsf{c}_{22} \end{bmatrix}, \tag{4.155}$$

in which c_{11} is $n \times n$, c_{12} is $n \times m$, $\mathsf{c}_{21} = \mathsf{c}_{12}^T$ is $m \times n$, and c_{22} is $m \times m$. The inverse covariance matrix c^{-1} appearing in the $(n + m)$-variate joint Gaussian density function of these variables is likewise decomposed,

$$\mathsf{c}^{-1} = \begin{bmatrix} \mathsf{C}_{11} & \mathsf{C}_{12} \\ \mathsf{C}_{21} & \mathsf{C}_{22} \end{bmatrix}. \tag{4.156}$$

It is shown in texts on matrix algebra that

$$\mathsf{C}_{11} = (\mathsf{c}_{11} - \mathsf{c}_{12}\mathsf{c}_{22}^{-1}\mathsf{c}_{21})^{-1}, \tag{4.157}$$

$$\mathsf{C}_{12} = -\mathsf{C}_{11}\mathsf{c}_{12}\mathsf{c}_{22}^{-1}, \tag{4.158}$$

$$\mathsf{C}_{21} = -\mathsf{C}_{22}\mathsf{c}_{21}\mathsf{c}_{11}^{-1}, \tag{4.159}$$

$$\mathsf{C}_{22} = (\mathsf{c}_{22} - \mathsf{c}_{21}\mathsf{c}_{11}^{-1}\mathsf{c}_{12})^{-1}. \tag{4.160}$$

You can verify these by multiplying the matrices c^{-1} and c in block form and showing that the result is the $(n + m) \times (n + m)$ identity matrix

$$\mathsf{I} = \begin{bmatrix} \mathsf{I}_n & 0 \\ 0 & \mathsf{I}_m \end{bmatrix},$$

where I_n and I_m are the $n \times n$ and $m \times m$ identity matrices, and 0 indicates a matrix filled with zeros.

The joint probability density function of the $(n + m)$ random variables is, as in (4.136),

$$f_{\mathbf{x}}(x_1, x_2, \ldots, x_{n+m}) = (2\pi)^{-(n+m)/2} \,|\det \mathsf{c}\,|^{-1/2}$$
$$\times \exp\left\{-\tfrac{1}{2}(\mathsf{x}_1^T\ \mathsf{x}_2^T)\begin{bmatrix} \mathsf{C}_{11} & \mathsf{C}_{12} \\ \mathsf{C}_{21} & \mathsf{C}_{22}\end{bmatrix}\begin{bmatrix}\mathsf{x}_1 \\ \mathsf{x}_2\end{bmatrix}\right\}$$
$$= (2\pi)^{-(n+m)/2} \,|\det \mathsf{c}\,|^{-1/2} \tag{4.161}$$
$$\times \exp\,[-\tfrac{1}{2}(\mathsf{x}_1^T\mathsf{C}_{11}\mathsf{x}_1 + \mathsf{x}_1^T\mathsf{C}_{12}\mathsf{x}_2 + \mathsf{x}_2^T\mathsf{C}_{21}\mathsf{x}_1 + \mathsf{x}_2^T\mathsf{C}_{22}\mathsf{x}_2)].$$

The joint probability density function of the m conditioning variables $\mathbf{x}_{n+1}, \mathbf{x}_{n+2}, \ldots, \mathbf{x}_{n+m}$ is

$$f_{\mathsf{x}_2}(x_{n+1}, x_{n+2}, \ldots, x_{n+m}) = (2\pi)^{-m/2} \,|\det \mathsf{c}_{22}|^{-1/2} \exp\,(-\tfrac{1}{2}\mathsf{x}_2^T\mathsf{c}_{22}^{-1}\mathsf{x}_2). \tag{4.162}$$

The joint conditional probability density function of $\mathbf{x}_1, \mathbf{x}_2, \ldots, \mathbf{x}_n$, given the observed outcome $\mathbf{x}_{n+1} = x_{n+1}, \mathbf{x}_{n+2} = x_{n+2}, \ldots, \mathbf{x}_{n+m} = x_{n+m}$, is then, as in Sec. 4-2B, the quotient of these density functions,

$$f_{\mathsf{x}_1}(x_1, x_2, \ldots, x_n \mid x_{n+1}, x_{n+2}, \ldots, x_{n+m})$$
$$= (2\pi)^{-n/2} \,|\det \mathsf{c}\,|^{-1/2} \,|\det \mathsf{c}_{22}|^{1/2} \tag{4.163}$$
$$\times \exp\,[-\tfrac{1}{2}(\mathsf{x}_1^T\mathsf{C}_{11}\mathsf{x}_1 + \mathsf{x}_1^T\mathsf{C}_{12}\mathsf{x}_2 + \mathsf{x}_2^T\mathsf{C}_{21}\mathsf{x}_1 + \mathsf{x}_2^T(\mathsf{C}_{22} - \mathsf{c}_{22}^{-1})\mathsf{x}_2)].$$

Now by (4.160) and (4.159),

$$\mathsf{C}_{22} - \mathsf{c}_{22}^{-1} = (\mathsf{c}_{22} - \mathsf{c}_{21}\mathsf{c}_{11}^{-1}\mathsf{c}_{12})^{-1} - \mathsf{c}_{22}^{-1}$$
$$= \mathsf{C}_{22}[\mathsf{c}_{22} - (\mathsf{c}_{22} - \mathsf{c}_{21}\mathsf{c}_{11}^{-1}\mathsf{c}_{12})]\mathsf{c}_{22}^{-1}$$
$$= \mathsf{C}_{22}\mathsf{c}_{21}\mathsf{c}_{11}^{-1}\mathsf{c}_{12}\mathsf{c}_{22}^{-1} = -\mathsf{C}_{21}\mathsf{c}_{12}\mathsf{c}_{22}^{-1}.$$

Putting this into (4.163) we find after some algebra that the quadratic form in the exponent can be written

$$(\mathsf{x}_1^T - \mathsf{x}_2^T\mathsf{c}_{22}^{-1}\mathsf{c}_{21})\mathsf{C}_{11}(\mathsf{x}_1 - \mathsf{c}_{12}\mathsf{c}_{22}^{-1}\mathsf{x}_2).$$

Since the conditional density function in (4.163) must yield 1 when integrated over $-\infty < x_k < \infty$, $k = 1, 2, \ldots, n$, the normalizing factor in front of it must be $|\det \mathsf{C}_{11}|^{1/2}$. Hence we obtain for that conditional density function

$$f_{\mathsf{x}_1}(x_1, x_2, \ldots, x_n \mid x_{n+1}, x_{n+2}, \ldots, x_{n+m})$$
$$= (2\pi)^{-n/2} \,|\det \mathsf{c}_c|^{-1/2} \tag{4.164}$$
$$\times \exp\,[-\tfrac{1}{2}(\mathsf{x}_1^T - \mathsf{x}_2^T\mathsf{c}_{22}^{-1}\mathsf{c}_{21})\mathsf{c}_c^{-1}(\mathsf{x}_1 - \mathsf{c}_{12}\mathsf{c}_{22}^{-1}\mathsf{x}_2)],$$

where

$$\mathsf{c}_c = \mathsf{c}_{11} - \mathsf{c}_{12}\mathsf{c}_{22}^{-1}\mathsf{c}_{21} \tag{4.165}$$

is the $n \times n$ conditional covariance matrix of the random variables $\mathbf{x}_1, \mathbf{x}_2, \ldots, \mathbf{x}_n$, given that $\mathbf{x}_{n+1} = x_{n+1}, \mathbf{x}_{n+2} = x_{n+2}, \ldots, \mathbf{x}_{n+m} = x_{n+m}$. That is,

$$(\mathsf{C}_c)_{ij} = \text{Cov}\,(\mathbf{x}_i, \mathbf{x}_j \mid x_{n+1}, x_{n+2}, \ldots, x_{n+m}) \tag{4.166}$$

in the sense of (4.133). These conditional covariances do not depend on how the observation of $\mathbf{x}_{n+1}, \mathbf{x}_{n+2}, \ldots, \mathbf{x}_{n+m}$ actually turned out.

From (4.164) we see that the conditional mean of any of the random variables $\mathbf{x}_j$, $1 \le j \le n$, is the jth element of the column vector $\mathsf{C}_{12}\mathsf{C}_{22}^{-1}\mathsf{X}_2$; that is,

$$E(\mathbf{x}_j \mid x_{n+1}, x_{n+2}, \ldots, x_{n+m}) \tag{4.167}$$

$$= \sum_{k=n+1}^{n+m} \sum_{p=n+1}^{n+m} c_{jk}(\mathsf{C}_{22}^{-1})_{kp}x_p, \qquad 1 \le j \le n.$$

Here C_{22} is the lower right-hand block of the covariance matrix C, and C_{22}^{-1} is the inverse of that $m \times m$ block, with its rows and columns labeled from $n + 1$ through $n + m$.

EXAMPLE 4.17. The Gaussian random variables $\mathbf{x}_1$, $\mathbf{x}_2$, $\mathbf{x}_3$, $\mathbf{x}_4$ have zero means and the covariance matrix

$$\mathsf{C} = \begin{bmatrix} 16 & 5 & 3 & -1 \\ 5 & 7 & 1 & -3 \\ 3 & 1 & 7 & -2 \\ -1 & -3 & -2 & 4 \end{bmatrix}.$$

It is observed that $\mathbf{x}_3$ and $\mathbf{x}_4$ have taken on particular values x_3 and x_4, respectively. Find the conditional means of $\mathbf{x}_1$ and $\mathbf{x}_2$ and their conditional covariance matrix.

Here

$$\mathsf{C}_{22} = \begin{bmatrix} 7 & -2 \\ -2 & 4 \end{bmatrix}, \qquad \mathsf{C}_{22}^{-1} = \frac{1}{24}\begin{bmatrix} 4 & 2 \\ 2 & 7 \end{bmatrix}$$

$$\mathsf{C}_{12}\mathsf{C}_{22}^{-1} = \frac{1}{24}\begin{bmatrix} 3 & -1 \\ 1 & -3 \end{bmatrix}\begin{bmatrix} 4 & 2 \\ 2 & 7 \end{bmatrix} = \frac{1}{24}\begin{bmatrix} 10 & -1 \\ -2 & -19 \end{bmatrix}$$

$$\mathsf{C}_{12}\mathsf{C}_{22}^{-1}\begin{bmatrix} x_3 \\ x_4 \end{bmatrix} = \frac{1}{24}\begin{bmatrix} 10x_3 - x_4 \\ -2x_3 - 19x_4 \end{bmatrix}.$$

Hence

$$E(\mathbf{x}_1 \mid x_3, x_4) = \tfrac{5}{12}\,x_3 - \tfrac{1}{24}\,x_4$$

$$E(\mathbf{x}_2 \mid x_3, x_4) = -\tfrac{1}{12}\,x_3 - \tfrac{19}{24}\,x_4.$$

The conditional covariance matrix of $\mathbf{x}_1$ and $\mathbf{x}_2$ is, by (4.165),

$$\mathsf{C}_c = \begin{bmatrix} 16 & 5 \\ 5 & 7 \end{bmatrix} - \frac{1}{24}\begin{bmatrix} 10 & -1 \\ -2 & -19 \end{bmatrix}\begin{bmatrix} 3 & 1 \\ -1 & -3 \end{bmatrix} = \frac{1}{24}\begin{bmatrix} 353 & 107 \\ 107 & 113 \end{bmatrix}.$$

Thus

$$\text{Var}\,(\mathbf{x}_1 \mid x_3, x_4) = \frac{353}{24} < \text{Var}\,\mathbf{x}_1,$$

$$\text{Var}\,(\mathbf{x}_2 \mid x_3, x_4) = \frac{113}{24} < \text{Var}\,\mathbf{x}_2,$$

$$\text{Cov}\,(\mathbf{x}_1, \mathbf{x}_2 \mid x_3, x_4) = \frac{107}{24}. \qquad \blacksquare$$

ESTIMATING PARAMETERS OF A DISTRIBUTION

A. Estimators

Numbers such as the mean and the variance that characterize a probability density function are called *parameters*. The mean, as we have seen in Sec. 2-4, tells us where the density function of a random variable $\mathbf{x}$ is located along the x-axis, and the standard deviation—the square root of the variance—tells us over how broad a range of values about the mean the random variable is dispersed. The covariance Cov $(\mathbf{x}, \mathbf{y})$, on the other hand, is a parameter of the bivariate joint distribution of the random variables $\mathbf{x}$ and $\mathbf{y}$, and it roughly indicates the degree to which $\mathbf{x}$ and $\mathbf{y}$ are statistically related.

If the probability distribution is unknown, it is often useful to estimate the values of its parameters by making a number of independent trials of the chance experiment E and somehow processing the outcomes, which we denote by $\mathbf{x}_1$, $\mathbf{x}_2, \ldots, \mathbf{x}_n$; these are sometimes called the *data*. They collectively identify the outcome ζ of the n-fold compound experiment

$$\mathsf{E} \times \mathsf{E} \times \cdots \times \mathsf{E} = \mathsf{E}_n,$$

within which they are independent and identically distributed random variables. By some reasoning one may have concluded that a certain function

$$\hat{\boldsymbol{\theta}} = g(\mathbf{x}_1, \mathbf{x}_2, \ldots, \mathbf{x}_n)$$

may be a useful estimate of a particular parameter θ of the distribution of the random variable $\mathbf{x}$ of the chance experiment E. Such a function $g(\mathbf{x}_1, \mathbf{x}_2, \ldots, \mathbf{x}_n)$ is called an *estimator* of the parameter θ. The value that the estimator produces in a given series E_n of trials is called the *estimate* of θ.

As an estimator $\hat{\mathbf{m}}$ of the mean $E(\mathbf{x})$, for example, one usually adopts the average, or *sample mean*

$$\hat{\mathbf{m}} = \frac{\mathbf{x}_1 + \mathbf{x}_2 + \cdots + \mathbf{x}_n}{n} = n^{-1} \sum_{k=1}^{n} \mathbf{x}_k \tag{4.168}$$

of the data $\mathbf{x}_1, \mathbf{x}_2, \ldots, \mathbf{x}_n$. For the variance one can use the estimator

$$\hat{\mathbf{v}} = (n - 1)^{-1} \sum_{k=1}^{n} (\mathbf{x}_k - \hat{\mathbf{m}})^2, \tag{4.169}$$

into which one must substitute the sample mean $\hat{\mathbf{m}}$.

One would like the estimate $\hat{\boldsymbol{\theta}}$ of the parameter θ to be close to the true value of θ in some sense, at least "on the average." If one used the estimator a great many times, conducting a great many trials of the compound experiment E_n, one would hope that the errors in one's estimates would average to zero. Equivalently, the expected value $E(\hat{\boldsymbol{\theta}}) = \bar{\hat{\theta}}$ should equal the true value θ of the parameter being estimated. The discrepancy

$$b(\theta) = E(\hat{\boldsymbol{\theta}}) - \theta = \bar{\hat{\theta}} - \theta \tag{4.170}$$

is called the *bias* of the estimator, and if it is zero, the estimator is called *unbiased*. The estimates should also cluster closely about their mean; that is, the variance Var $\hat{\boldsymbol{\theta}}$ of the estimator should be small.

The mean-square error, defined as $E[(\hat{\boldsymbol{\theta}} - \theta)^2]$, is often used to express the average size of the discrepancy between estimate and true value θ. It can be written

$$\begin{aligned} E[(\hat{\boldsymbol{\theta}} - \theta)^2] &= E[\hat{\boldsymbol{\theta}} - \bar{\hat{\theta}} + \bar{\hat{\theta}} - \theta)^2] \\ &= E(\hat{\boldsymbol{\theta}} - \bar{\hat{\theta}})^2 + (\bar{\hat{\theta}} - \theta)^2 = \text{Var}\,\hat{\boldsymbol{\theta}} + [b(\theta)]^2 \end{aligned} \qquad (4.171)$$

because $E[(\hat{\boldsymbol{\theta}} - \bar{\hat{\theta}})(\bar{\hat{\theta}} - \theta)] = 0$. Thus the mean-square error incurred by an estimator is composed of its variance and the square of its bias, and the mean-square error is small if both of these are small. If the estimator is unbiased, the mean-square error and the variance Var $\hat{\boldsymbol{\theta}}$ are identical.

When an estimator $\hat{\boldsymbol{\theta}} = g(\mathbf{x}_1, \mathbf{x}_2, \ldots, \mathbf{x}_n)$ of some parameter is proposed, it is customary to assess its efficacy by calculating its bias and its variance. Both of these should be the smaller, the greater the number n of trials on which they are based. One often also works out the probability density function of the estimator under the assumption that the random variable $\mathbf{x}$ has a particular distribution in order to see how it would behave if $\mathbf{x}$ really had that distribution, and when possible this should be done for a number of different hypothetical distributions of $\mathbf{x}$.

B. Estimating the Mean

We illustrate these concepts first for the estimator $\hat{\mathbf{m}}$ of the mean, given in (4.168). Its expected value is

$$E(\hat{\mathbf{m}}) = n^{-1} \sum_{k=1}^{n} E(\mathbf{x}_k) = m, \qquad (4.172)$$

where $m = E(\mathbf{x})$ is the true mean of the random variable $\mathbf{x}$. Thus the estimator $\hat{\mathbf{m}}$ of (4.168) is unbiased.

Its variance is

$$\text{Var}\,\hat{\mathbf{m}} = n^{-2} \sum_{k=1}^{n} \text{Var}\,\mathbf{x}_k = \frac{\sigma^2}{n}, \qquad (4.173)$$

where $\sigma^2 = \text{Var}\,\mathbf{x}$ is the variance of $\mathbf{x}$. (We used the fact that the variance of the sum of independent and hence uncorrelated random variables equals the sum of their variances.) We find that the larger n, the more closely the values of the estimate cluster about the true mean m. Indeed, by the Chebyshev inequality (2.83) the probability that the estimator $\hat{\mathbf{m}}$ deviates by more than ε from the true mean m is bounded by

$$\Pr(|\hat{\mathbf{m}} - m| > \varepsilon) \leq \frac{\sigma^2}{n\varepsilon^2} \qquad (4.174)$$

and hence goes to zero as n goes to infinity.

If the random variable $\mathbf{x}$ is Gaussian, the estimator $\hat{\mathbf{m}}$ is also Gaussian, and the probability in (4.174) goes to zero with increasing ε much more rapidly than (4.174) indicates. We shall see in the next section that for a large class of distributions, the probability density function of the sample mean $\hat{\mathbf{m}}$ is closely approximated by a Gaussian distribution when the number n of independent trials of E is large.

C. Estimating the Variance

The most commonly used estimator of the variance of a random variable $\mathbf{x}$ on the basis of data $\mathbf{x}_1, \mathbf{x}_2, \ldots, \mathbf{x}_n$ from n independent trials of E is that given in (4.169), which is called the *sample variance*. Into it one substitutes the estimate $\hat{\mathbf{m}}$ of the mean, and

$$\begin{aligned}\hat{\mathbf{v}} &= (n-1)^{-1}\sum_{k=1}^{n}(\mathbf{x}_k-\hat{\mathbf{m}})^2 = (n-1)^{-1}\sum_{k=1}^{n}(\mathbf{x}_k^2-2\hat{\mathbf{m}}\mathbf{x}_k+\hat{\mathbf{m}}^2)\\ &= (n-1)^{-1}\left(\sum_{k=1}^{n}\mathbf{x}_k^2-2\hat{\mathbf{m}}\sum_{k=1}^{n}\mathbf{x}_k+n\hat{\mathbf{m}}^2\right)\\ &= (n-1)^{-1}\left(\sum_{k=1}^{n}\mathbf{x}_k^2-n\hat{\mathbf{m}}^2\right).\end{aligned} \tag{4.175}$$

This is the form most convenient for calculating the sample variance.

We observe first that if we define a new random variable $\mathbf{x}' = \mathbf{x} - m$, whose true mean is zero, and hence new data $\mathbf{x}'_k = \mathbf{x}_k - m$, then

$$\mathbf{x}_k - \hat{\mathbf{m}} = \mathbf{x}'_k - \hat{\mathbf{m}}'$$

because

$$\hat{\mathbf{m}}' = n^{-1}\sum_{k=1}^{n}\mathbf{x}'_k = \hat{\mathbf{m}} - m.$$

The estimator $\hat{\mathbf{v}}$ of the variance (4.169) is unchanged when written in terms of the new data $\mathbf{x}'_k$. In analyzing the estimator $\hat{\mathbf{v}}$ in (4.169) and (4.175) we can therefore assume that the true mean of $\mathbf{x}$ is zero without any loss of generality, but with some simplification in our equations.

The expected value of the estimator in (4.169) is

$$E(\hat{\mathbf{v}}) = (n-1)^{-1}\sum_{k=1}^{n}E(\mathbf{x}_k - \hat{\mathbf{m}})^2$$

because the expected value of a sum of random variables equals the sum of their expected values. Because all the terms in this sum are equal,

$$E(\hat{\mathbf{v}}) = \frac{n}{n-1}E(\mathbf{x}_1 - \hat{\mathbf{m}})^2.$$

Now $E(\mathbf{x}_1 - \hat{\mathbf{m}}) = m - m = 0$, whence

$$E(\hat{\mathbf{v}}) = \frac{n}{n-1}\operatorname{Var}(\mathbf{x}_1 - \hat{\mathbf{m}}).$$

By the definition (4.168) of the sample mean $\hat{\mathbf{m}}$,

$$\mathbf{x}_1 - \hat{\mathbf{m}} = \mathbf{x}_1 - n^{-1}(\mathbf{x}_1 + \mathbf{x}_2 + \cdots + \mathbf{x}_n)$$
$$= (1 - n^{-1})\mathbf{x}_1 - n^{-1}\mathbf{x}_2 - \cdots - n^{-1}\mathbf{x}_n.$$

The terms in this expression are statistically independent, and therefore uncorrelated, so that the variance of this sum equals the sum of the variances of its terms. Hence as in (4.92)

$$E(\hat{\mathbf{v}}) = \frac{n}{n-1}[(1 - n^{-1})^2 \operatorname{Var} \mathbf{x}_1 + n^{-2} \operatorname{Var} \mathbf{x}_2 + \cdots + n^{-2} \operatorname{Var} \mathbf{x}_n]$$
$$= \frac{n}{n-1}\left[\frac{(n-1)^2}{n^2} + \frac{n-1}{n^2}\right]\sigma^2 = \sigma^2. \tag{4.176}$$

Thus the expected value of the estimator $\hat{\mathbf{v}}$ of the variance of the random variable $\mathbf{x}$ equals its true variance σ^2, and the estimator $\hat{\mathbf{v}}$ in (4.169) is unbiased.

Determining the variance Var $\hat{\mathbf{v}}$ of this estimator requires some assumption about the fourth moment of the random variable $\mathbf{x}$, and we shall for simplicity assume that $\mathbf{x}$ has a Gaussian distribution. That assumption suffices to determine the probability density function of the estimator $\hat{\mathbf{v}}$ as well, and we shall work it out here as an illustration of the power of the method of characteristic functions.

We introduce the abbreviations

$$\mathbf{S} = \sum_{k=1}^{n} \mathbf{x}_k^2,$$

$$\mathbf{T} = \sum_{i=1}^{n}\sum_{j=i+1}^{n} \mathbf{x}_i\mathbf{x}_j = \mathbf{x}_1\mathbf{x}_2 + \mathbf{x}_1\mathbf{x}_3 + \cdots + \mathbf{x}_1\mathbf{x}_n$$
$$+ \mathbf{x}_2\mathbf{x}_3 + \cdots + \mathbf{x}_{n-1}\mathbf{x}_n.$$

No terms are repeated in the sum $\mathbf{T}$. Now we can write

$$n\hat{\mathbf{m}}^2 = n^{-1}(\mathbf{x}_1 + \mathbf{x}_2 + \cdots + \mathbf{x}_n)^2 = n^{-1}(\mathbf{S} + 2\mathbf{T}),$$

so that $\hat{\mathbf{v}}$ in (4.175) becomes

$$\hat{\mathbf{v}} = (n-1)^{-1}[\mathbf{S} - n^{-1}(\mathbf{S} + 2\mathbf{T})] = n^{-1}\mathbf{S} - 2[n(n-1)]^{-1}\mathbf{T}.$$

Remembering that we have taken the true means of the random variables $\mathbf{x}$ equal to zero, we form the characteristic function of the estimator $\hat{\mathbf{v}}$ as

$$\Phi_{\hat{\mathbf{v}}}(\omega) = E[\exp(j\omega\hat{\mathbf{v}})] = E \exp\left[\frac{j\omega\mathbf{S}}{n} - \frac{2j\omega\mathbf{T}}{n(n-1)}\right]$$
$$= (2\pi\sigma^2)^{-n/2}\int_{-\infty}^{\infty} dx_1 \cdots \int_{-\infty}^{\infty} dx_n \exp\left[-\frac{S}{2\sigma^2} + \frac{j\omega S}{n} - \frac{2j\omega T}{n(n-1)}\right]$$
$$= (2\pi\sigma^2)^{-n/2}\int_{-\infty}^{\infty} dx_1 \cdots \int_{-\infty}^{\infty} dx_n$$
$$\times \exp\left[-\tfrac{1}{2}\left(\frac{n - 2j\omega\sigma^2}{\sigma^2 n} S + \frac{4j\omega T}{n(n-1)}\right)\right].$$

Recalling the definitions of S and T, we see that we must evaluate the integral of an exponential function of a quadratic form whose $n \times n$ matrix $\mathbf{C}$ has diagonal elements

$$C_{ii} = a = \frac{n - 2j\omega\sigma^2}{\sigma^2 n} \qquad \forall\, i,$$

and off-diagonal elements

$$C_{ij} \equiv b = \frac{2j\omega}{n(n-1)} \qquad \forall\, (i \neq j).$$

Hence by (4.118) this characteristic function equals

$$\Phi_{\hat{v}}(\omega) = \sigma^{-n}(\det \mathbf{C})^{-1/2}.$$

The determinant $\det \mathbf{C}$ is

$$\begin{vmatrix} a & b & b & \cdots & b & b \\ b & a & b & \cdots & b & b \\ \cdot & \cdot & \cdot & \cdot & \cdot & \cdot \\ b & b & b & \cdots & a & b \\ b & b & b & \cdots & b & a \end{vmatrix} = \begin{vmatrix} a & b & b & \cdots & b & b \\ b-a & a-b & 0 & \cdots & 0 & 0 \\ \cdot & \cdot & \cdot & \cdot & \cdot & \cdot \\ b-a & 0 & 0 & \cdots & a-b & 0 \\ b-a & 0 & 0 & \cdots & 0 & a-b \end{vmatrix}$$

when we subtract the first row from each of the other rows. If we now add each other column to the first column, the determinant becomes

$$\det \mathbf{C} = \begin{vmatrix} a+(n-1)b & b & b & \cdots & b & b \\ 0 & a-b & 0 & \cdots & 0 & 0 \\ \cdot\,\cdot\,\cdot\,\cdot\,\cdot & \cdot & \cdot & \cdot & \cdot & \cdot \\ 0 & 0 & 0 & \cdots & a-b & 0 \\ 0 & 0 & 0 & \cdots & 0 & a-b \end{vmatrix}$$

$$= (a-b)^{n-1}[a+(n-1)b] = \sigma^{-2n}\left(1 - \frac{2j\omega\sigma^2}{n-1}\right)^{n-1}.$$

Hence the characteristic function of the estimator $\hat{\mathbf{v}}$ is

$$\Phi_{\hat{v}}(\omega) = \left(1 - \frac{2j\omega\sigma^2}{n-1}\right)^{-(n-1)/2} \tag{4.177}$$

This happens to be the characteristic function of the gamma distribution (2.26) with $c = \frac{1}{2}(n-1)$, if the random variable there is scaled by a factor $2\sigma^2/(n-1)$, as you can show by taking the Fourier transform of that density function. Thus we find for the probability density function of the sample variance $\hat{\mathbf{v}}$ of (4.169)

$$f_{\hat{v}}(v) = \frac{n-1}{2\sigma^2\Gamma\left(\dfrac{n-1}{2}\right)}\left[\frac{(n-1)v}{2\sigma^2}\right]^{(n-3)/2} \times \exp\left[-\frac{(n-1)v}{2\sigma^2}\right]U(v). \tag{4.178}$$

The mean and variance of our estimator $\hat{\mathbf{v}}$ are most easily determined by expanding the characteristic function $\Phi_{\hat{\mathbf{v}}}(\omega)$ in a power series about the origin $\omega = 0$ and using (2.88),

$$\Phi_{\hat{\mathbf{v}}}(\omega) = 1 + j\omega\sigma^2 + \tfrac{1}{2}\left(\frac{n+1}{n-1}\right)\sigma^4(j\omega)^2 + \cdots,$$

whence

$$E(\hat{\mathbf{v}}) = \sigma^2, \qquad E(\hat{\mathbf{v}}^2) = \left(\frac{n+1}{n-1}\right)\sigma^4 \tag{4.179}$$

and

$$\text{Var}\ \hat{\mathbf{v}} = \left(\frac{n+1}{n-1}\right)\sigma^4 - \sigma^4 = \frac{2\sigma^4}{n-1}.$$

As the number n of samples increases, the variance Var $\hat{\mathbf{v}}$ of the sample variance $\hat{\mathbf{v}}$ decreases, and it becomes a more and more accurate estimator of the true variance σ^2. The probability density function $f_{\hat{\mathbf{v}}}(v)$ in (4.178) becomes more and more narrowly peaked about its mean σ^2 as n grows larger.

This property, shared by both estimators $\hat{\mathbf{m}}$ and $\hat{\mathbf{v}}$, of becoming the more accurate, the more samples are utilized, is called *consistency*, and these estimators are said to be *consistent*.

D. Estimating the Covariance

If in the chance experiment E two random variables **x** and **y** are defined, their covariance can be estimated by conducting n independent trials of the experiment, whose outcomes provide $2n$ data,

$$\mathbf{x}_1, \mathbf{y}_1, \mathbf{x}_2, \mathbf{y}_2, \ldots, \mathbf{x}_n, \mathbf{y}_n.$$

The usual estimator of the covariance Cov (**x**, **y**) of **x** and **y** is

$$\begin{aligned}\hat{\mathbf{c}} &= (n-1)^{-1}\sum_{k=1}^{n}(\mathbf{x}_k - \hat{\mathbf{m}}_x)(\mathbf{y}_k - \hat{\mathbf{m}}_y) \\ &= (n-1)^{-1}\left(\sum_{k=1}^{n}\mathbf{x}_k\mathbf{y}_k - n\hat{\mathbf{m}}_x\hat{\mathbf{m}}_y\right),\end{aligned} \tag{4.180}$$

where

$$\hat{\mathbf{m}}_x = n^{-1}\sum_{k=1}^{n}\mathbf{x}_k, \qquad \hat{\mathbf{m}}_y = n^{-1}\sum_{k=1}^{n}\mathbf{y}_k$$

are the sample means of **x** and **y**, respectively. The proof that this estimator is unbiased,

$$E(\hat{\mathbf{c}}) = \text{Cov}\,(\mathbf{x}, \mathbf{y})$$

is similar to the proof that $\hat{\mathbf{v}}$ is an unbiased estimator of the variance Var $\mathbf{x}$. The other properties of this estimator $\hat{\mathbf{c}}$ are treated in texts on statistics, and we shall omit them.

From $\hat{\mathbf{c}}$ one can form the sample correlation coefficient

$$\hat{\mathbf{r}} = \frac{\hat{\mathbf{c}}}{(\hat{\mathbf{v}}_x\hat{\mathbf{v}}_y)^{1/2}},$$

where $\hat{\mathbf{v}}_x$ and $\hat{\mathbf{v}}_y$ are the estimators of Var $\mathbf{x}$ and Var $\mathbf{y}$. One can consider $\hat{\mathbf{r}}$ as an estimate of the true correlation coefficient r, defined in (3.92).

4-7 THE CENTRAL LIMIT THEOREM

The ubiquity of Gaussian random variables in science and engineering can be attributed in part to the fact that errors in measurement and noise in systems are often the sums of a great many small, random integrants, and by virtue of the central limit theorem, the distributions of such sums are to a good approximation Gaussian. Let $\mathbf{x}_1, \mathbf{x}_2, \ldots, \mathbf{x}_n$ be n independent and identically distributed random variables with means m and variances σ^2. The *central limit theorem* states that under conditions weak enough to cover random variables $\mathbf{x}$ with almost any distribution likely to be encountered in nature, the random variable

$$\mathbf{y} = n^{-1/2} \sum_{k=1}^{n} (\mathbf{x}_k - m), \tag{4.181}$$

whose mean is zero and whose variance is σ^2, has in the limit $n \to \infty$ a Gaussian probability density function.

Here is a crude proof of this important theorem. We form the characteristic function of $\mathbf{y}$ and show that as n goes to infinity, it approaches more and more closely to the characteristic function of a Gaussian random variable:

$$\begin{aligned}\Phi_{\mathbf{y}}(\omega) &= E(e^{j\omega\mathbf{y}}) = E \exp\left[j\omega n^{-1/2} \sum_{k=1}^{n} (\mathbf{x}_k - m)\right] \\ &= \prod_{k=1}^{n} E \exp\,[j\omega n^{-1/2}(\mathbf{x}_k - m)] \\ &= [\Phi'_{\mathbf{x}}\,(n^{-1/2}\omega)]^n,\end{aligned}$$

where

$$\Phi'_{\mathbf{x}}(\omega) = Ee^{j\omega(\mathbf{x}-m)}$$

is the characteristic function of the random variable $\mathbf{x} - m$. We assume that this function can be expanded in a power series about $\omega = 0$ with a remainder term of the order of $|\omega|^3$,

$$\Phi'_{\mathbf{x}}(\omega) = 1 - \tfrac{1}{2}\sigma^2\omega^2 + \theta(\omega)|\omega|^3,$$

where in the remainder term $|\theta(\omega)| < M$ for some constant M. Whether this expansion is possible can be checked in any particular instance. Then the characteristic function of $\mathbf{y}$ is

$$\Phi_{\mathbf{y}}(\omega) = \left[1 - \frac{\sigma^2\omega^2}{2n} + \theta(n^{-1/2}\omega)n^{-3/2}|\omega|^3\right]^n,$$

and by the rule

$$\lim_{n\to\infty}\left(1 + \frac{x}{n}\right)^n = e^x$$

with

$$x = -\tfrac{1}{2}\sigma^2\omega^2 + \theta(n^{-1/2}\omega)n^{-1/2}|\omega|^3 \to -\tfrac{1}{2}\sigma^2\omega^2$$

we obtain when n goes to infinity

$$\Phi_{\mathbf{y}}(\omega) = \exp\left(-\tfrac{1}{2}\sigma^2\omega^2\right), \tag{4.182}$$

which is the characteristic function of a Gaussian random variable with mean zero and variance σ^2.

More elaborate versions of the central limit theorem exist. It is unnecessary that the random variables $\mathbf{x}$ be identically distributed; the variable

$$\mathbf{y} = n^{-1/2}\sum_{k=1}^{n}[\mathbf{x}_k - E(\mathbf{x}_k)]$$

will in the limit $n \to \infty$ have a Gaussian density function provided that the variances $\text{Var}\ \mathbf{x}_k$ remain within fixed bounds,

$$0 < M' \le \text{Var}\ \mathbf{x}_k \le M'' < \infty$$

for all k, and provided that the third absolute central moments

$$E(|\mathbf{x}_k - E(\mathbf{x}_k)|^3)$$

remain bounded. Then $E(\mathbf{y}) = 0$ and

$$\sigma_{\mathbf{y}}^2 = \text{Var}\ \mathbf{y} = \lim_{n\to\infty} n^{-1}\sum_{k=1}^{n}\text{Var}\ \mathbf{x}_k \le M''$$

is finite, but not zero. In what follows we resume our assumption that the independent random variables $\mathbf{x}_k$ are identically distributed.

Because the random variable $\mathbf{y}$ defined in (4.181) has a Gaussian probability density function in the limit $n \to \infty$, we can assume that when the number n of terms in the sum in (4.181) is large, though finite, the probability density function of $\mathbf{y}$ will be closely approximated by the Gaussian form. The random variable

$$\mathbf{s} = \sum_{k=1}^{n}\mathbf{x}_k,$$

related to **y** by a linear transformation, has a probability density function that is also approximately Gaussian, with mean $\bar{s} = nm$ and variance Var $\mathbf{s} = n\sigma^2$,

$$f_{\mathbf{s}}(s) \simeq (2\pi n\sigma^2)^{-1/2} \exp\left[-\frac{(s - nm)^2}{2n\sigma^2} \right].$$

The sample mean $\hat{\mathbf{m}}$ of the independent and identically distributed random variables $\mathbf{x}_1, \mathbf{x}_2, \ldots, \mathbf{x}_n$, which was defined in (4.168) and is proportional to **s**, is therefore also approximately a Gaussian random variable.

The Gaussian approximation suggested by the central limit theorem can be considered as the first term of the representation of the probability density function by what is known as Edgeworth's series, the terms of which are Hermite functions. The accuracy of the approximation can be checked by calculating a few succeeding terms of this series. Details can be found in treatises on statistics. In general, the Gaussian approximation to the probability density function $f_{\mathbf{s}}(s)$ is the more accurate, the closer the value of s lies to the mean $E(\mathbf{s}) = nm$.

The integrated form of the central limit theorem,

$$\Pr\,(y_1 < \mathbf{y} < y_2) \rightarrow \operatorname{erfc}\left(\frac{y_1}{\sigma}\right) - \operatorname{erfc}\left(\frac{y_2}{\sigma}\right) \tag{4.183}$$

is useful even when the component variables $\mathbf{x}_k$ are discrete random variables. [The error-function integral erfc x is defined in (2.20).] Equivalently,

$$\Pr\left(s_1 < \sum_{k=1}^{n} \mathbf{x}_k < s_2\right) \simeq \operatorname{erfc}\left(\frac{s_1 - nm}{n^{1/2}\sigma}\right) - \operatorname{erfc}\left(\frac{s_2 - nm}{n^{1/2}\sigma}\right) \tag{4.184}$$

when n is large.

Suppose, for instance, that in a series of Bernoulli trials of a chance experiment E, the random variable **x** is defined as $\mathbf{x} = 1$ for a success and $\mathbf{x} = 0$ for a failure. Then

$$\mathbf{s} = \sum_{k=1}^{n} \mathbf{x}_k$$

equals the number of successes in n trials, and according to Sec. 1-8, it has a binomial distribution,

$$\Pr\,(\mathbf{s} = k) = \binom{n}{k} p^k q^{n-k}, \qquad 0 \le k \le n, \tag{4.185}$$

where $p = \Pr\,(\mathbf{x} = 1)$ is the probability of success and $q = 1 - p = \Pr\,(\mathbf{x} = 0)$ the probability of failure. When $n >> 1$ the probability

$$\Pr\,(k_1 \le \mathbf{s} \le k_2) = \sum_{k=k_1}^{k_2} \binom{n}{k} p^k q^{n-k} \tag{4.186}$$

that the number of successes lies between k_1 and k_2 requires a lengthy numerical calculation, but (4.184) provides a convenient approximation. Now $E(\mathbf{x}) = p$, and by (2.101),

$$\text{Var } \mathbf{x} = \sigma^2 = p(1 - p),$$

whence

$$\Pr(k_1 \le \mathbf{s} \le k_2) \simeq \text{erfc } x_1 - \text{erfc } x_2,$$

$$x_i = \frac{k_i - np}{\sqrt{npq}}, \qquad i = 1, 2. \tag{4.187}$$

The corresponding approximation for the binomial probabilities,

$$\binom{n}{k} p^k (1 - p)^{n-k} \simeq [2\pi np(1 - p)]^{-1/2} \exp\left[-\frac{(k - np)^2}{2np(1 - p)}\right] \tag{4.188}$$

is known as the de Moivre–Laplace approximation. It can also be derived by applying Stirling's formula (1.30) to the factorials in the binomial coefficient $\binom{n}{k}$.

EXAMPLE 4.18. The probability that a resistor manufactured by a certain process is outside tolerance limits equals 0.1. You have bought 1000 of these resistors. What is the probability that more than 109 of them are outside tolerance limits?

Here $p = 0.1$, $q = 0.9$, $n = 1000$, $np = 100$, $\sigma^2 = 90$,

$$k_1 = 110, \qquad x_1 = \frac{10}{\sqrt{90}} = 1.0541,$$

$$k_2 = 1000, \qquad x_2 = \frac{900}{\sqrt{90}} = 94.9.$$

Then by (4.187), since erfc x_2 is negligible,

$$\Pr(110 \le \mathbf{s} \le 1000) \simeq \text{erfc } x_1 = 0.146. \qquad \blacksquare$$

EXAMPLE 4.19. The probability that a certain diode will fail before 1000 hours' service equals 0.45. If 10,000 such diodes are tested, what is the probability that between 4480 and 4515 will have failed before 1000 hours?

Here $p = 0.45$, $q = 0.55$, $n = 10^4$, $np = 4500$, $\sigma^2 = 2475$,

$$k_1 = 4480, \qquad x_1 = \frac{4480 - 4500}{\sqrt{2475}} = -0.4020,$$

$$k_2 = 4515, \qquad x_2 = \frac{4515 - 4500}{\sqrt{2475}} = 0.3015.$$

Thus

$$\Pr(4480 \le \mathbf{s} \le 4515) \simeq \text{erfc}(-0.4020) - \text{erfc}(0.3015) = 0.275. \qquad \blacksquare$$

The central limit theorem can be used to approximate the Poisson distribution whenever its mean is much greater than 1; the variance of a Poisson distribution equals its mean.

EXAMPLE 4.20. When a certain light signal pulse falls on a photoelectric detector, the probability that k electrons are emitted and counted is given by a Poisson distribution with mean $\lambda = 150$. Because a strong background is present, the signal will be declared present only if more than 140 electrons are counted. What is the probability that the signal will be missed?

This probability

$$P = \sum_{k=0}^{\nu} \frac{\lambda^k e^{-\lambda}}{k!}, \qquad \lambda = 150, \quad \nu = 140,$$

can by (4.184) be approximated by

$$P \simeq 1 - \operatorname{erfc}\left(\frac{\nu - \lambda}{\sqrt{\lambda}}\right) = 1 - \operatorname{erfc}\left(-\frac{10}{\sqrt{150}}\right)$$

$$= \operatorname{erfc}\left(\frac{10}{\sqrt{150}}\right) = 0.207.$$

■

EXAMPLE 4.21. Assume that resistors that have been tested and found to have resistances within 5% of their rated value have random resistances **x** that may be anywhere within that range about $E(\mathbf{x}) = 1000\ \Omega$ with a uniform probability density function. One hundred of these are selected and connected in series. What is the probability that their total resistance deviates from $10^5\ \Omega$ by no more than 0.5%?

The uniform probability density function is easily shown to satisfy the conditions for applying the central limit theorem. The variance of a random variable uniformly distributed between a and b is

$$\sigma^2 = (b - a)^{-1} \int_a^b [x - \tfrac{1}{2}(a + b)]^2\, dx = \frac{(b - a)^2}{12},$$

so that here $\sigma^2 = 100^2/12$. Applying (4.184) with

$$s_1 = 0.995nm, \quad s_2 = 1.005nm, \quad n = 100, \quad m = 10^3,$$

we find that the probability sought is approximately

$$\operatorname{erfc}\left(-\frac{0.005nm}{n^{1/2}\sigma}\right) - \operatorname{erfc}\left(\frac{0.005nm}{n^{1/2}\sigma}\right) = 1 - 2 \operatorname{erfc} 1.732 = 0.917.$$

■

4-8 RANDOM SUMS

In a simple photomultiplier, incident light ejects photoelectrons from a thin cathode, and these, accelerated across a short distance, impinge on a second metallic surface. Each electron striking it knocks out a random number **x** of secondary electrons, which are collected at an anode and counted. If the kth primary electron ejects $\mathbf{x}_k$ secondaries, and if n primaries in all are produced by the incident light during an interval $(0, T)$, the total number of secondaries is $\mathbf{x}_1 + \mathbf{x}_2 + \cdots + \mathbf{x}_n$. The number of primary electrons, however, is a random variable **n**, for their ejection by the incident light is a random phenomenon. The total number **s** of secondaries is then a sum of a random number **n** of random variables,

$$\mathbf{s} = \sum_{k=1}^{\mathbf{n}} \mathbf{x}_k; \tag{4.189}$$

it is called a *random sum*. The random variables $\mathbf{x}_k$ are assumed independent and identically distributed, and the random number $\mathbf{n}$ of terms is assumed independent of them.

The mean and the variance of this random sum are most easily calculated by invoking the concepts of conditional probability. For instance,

$$E(\mathbf{s} \mid \mathbf{n} = n) = E \sum_{k=1}^{n} \mathbf{x}_k = nE(\mathbf{x}),$$

whence by (3.110) the mean value of the random sum is

$$E(\mathbf{s}) = E_{\mathbf{n}}[E(\mathbf{s} \mid \mathbf{n})] = E_{\mathbf{n}}[\mathbf{n}E(\mathbf{x})] = E(\mathbf{n})E(\mathbf{x}). \tag{4.190}$$

In the photomultiplier, $E(\mathbf{x}) = G$ is the mean gain, that is, the mean number of secondaries per primary electron, and $E(\mathbf{n}) = \lambda$ is the mean number of primary electrons ejected by the incident light. Thus the mean number of secondaries is equal to $G\lambda$.

Similarly,

$$\begin{aligned} E(\mathbf{s}^2 \mid \mathbf{n} = n) &= E\left(\sum_{k=1}^{n} \mathbf{x}_k\right)^2 \\ &= \operatorname{Var}\left(\sum_{k=1}^{n} \mathbf{x}_k\right) + \left[E\left(\sum_{k=1}^{n} \mathbf{x}_k\right)\right]^2 \\ &= n \operatorname{Var} \mathbf{x} + n^2[E(\mathbf{x})]^2, \end{aligned}$$

for the variance of a sum of independent random variables equals the sum of their variances. Hence

$$E(\mathbf{s}^2) = E_{\mathbf{n}}[E(\mathbf{s}^2 \mid \mathbf{n})] = E(\mathbf{n}) \operatorname{Var} \mathbf{x} + E(\mathbf{n}^2)[E(\mathbf{x})]^2,$$

and the variance of the random sum is

$$\operatorname{Var} \mathbf{s} = E(\mathbf{n}) \operatorname{Var} \mathbf{x} + \operatorname{Var} \mathbf{n}[E(\mathbf{x})]^2. \tag{4.191}$$

Besides a term equal to the variance $\operatorname{Var} \mathbf{x}$ of each term times the expected number $E(\mathbf{n})$ of terms, $\operatorname{Var} \mathbf{s}$ contains a term proportional to $\operatorname{Var} \mathbf{n}$, representing the uncertainty in the number of terms in the sum.

A similar analysis leads to the characteristic function $\Phi_{\mathbf{s}}(\omega)$ of the random sum $\mathbf{s}$. Let $\Phi_{\mathbf{x}}(\omega)$ be the common characteristic function of the random variables $\mathbf{x}_k$. Then

$$\Phi_{\mathbf{s}}(\omega) = E_{\mathbf{n}}[E_{\mathbf{x}}(e^{j\omega\mathbf{s}} \mid \mathbf{n})]$$

with

$$\begin{aligned} E_{\mathbf{x}}(e^{j\omega\mathbf{s}} \mid \mathbf{n} = n) &= E_{\mathbf{x}}\{\exp [j\omega(\mathbf{x}_1 + \cdots + \mathbf{x}_n) \mid \mathbf{n} = n]\} \\ &= [\Phi_{\mathbf{x}}(\omega)]^n \end{aligned}$$

by (4.99), the random variables $\mathbf{x}_k$ being independent and identically distributed.

Thus

$$\Phi_{\mathbf{s}}(\omega) = E_{\mathbf{n}}[\Phi_{\mathbf{x}}(\omega)]^{\mathbf{n}} = \sum_{k=0}^{\infty} \Pr(\mathbf{n} = k)[\Phi_{\mathbf{x}}(\omega)]^k$$
$$= h_{\mathbf{n}}[\Phi_{\mathbf{x}}(\omega)], \tag{4.192}$$

where

$$h_{\mathbf{n}}(z) = \sum_{k=0}^{\infty} \Pr(\mathbf{n} = k)z^k \tag{4.193}$$

is the probability generating function of the distribution of the number **n** of terms in the sum, as defined in Sec. 2-4G. If the **x**'s are discrete random variables taking on only nonnegative integral values, the random sum **s** also takes on only nonnegative integral values, and its probability generating function is, by a similar analysis,

$$h_{\mathbf{s}}(z) = h_{\mathbf{n}}(h_{\mathbf{x}}(z)) \tag{4.194}$$

in terms of the probability generating function $h_{\mathbf{x}}(z)$ of the random variables **x**.

EXAMPLE 4.22. The number of primary electrons in the photomultiplier has a Poisson distribution with mean λ,

$$\Pr(\mathbf{n} = k) = \frac{\lambda^k e^{-\lambda}}{k!}$$

and the number **x** of secondary electrons per primary electron also has a Poisson distribution, but with mean G. Then with **s** the total number of secondary electrons counted,

$$E(\mathbf{s}) = G\lambda, \qquad \operatorname{Var} \mathbf{s} = G(G + 1)\lambda.$$

By (2.102),

$$h_{\mathbf{x}}(z) = e^{G(z-1)}, \qquad h_{\mathbf{n}}(z) = e^{\lambda(z-1)},$$

so that the probability generating function of the number **s** is, by (4.194),

$$h_{\mathbf{s}}(z) = \exp[\lambda(e^{G(z-1)} - 1)]. \tag{4.195}$$

From (2.93) the probability of counting k secondary electrons is

$$p_k = \Pr(\mathbf{s} = k) = \frac{1}{k!}\frac{d^k}{dz^k} \exp[\lambda(e^{G(z-1)} - 1)]\bigg|_{z=0}$$

Thus

$$p_0 = h_{\mathbf{s}}(0) = \exp[-\lambda(1 - e^{-G})],$$

$$p_{k+1} = \frac{1}{(k+1)!}\frac{d^k}{dz^k}\frac{d}{dz} \exp[\lambda(e^{G(z-1)} - 1)]\bigg|_{z=0}$$

$$= \lambda G[(k+1)!]^{-1} \frac{d^k}{dz^k}[e^{G(z-1)}h_{\mathbf{s}}(z)]\bigg|_{z=0} \tag{4.196}$$

$$= \lambda G[(k+1)!]^{-1} \sum_{r=0}^{k} \binom{k}{r} \frac{d^r}{dz^r}(e^{G(z-1)}) \frac{d^{k-r}}{dz^{k-r}} h_{\mathbf{s}}(z)\bigg|_{z=0}$$

$$= \frac{\lambda G e^{-G}}{k+1} \sum_{r=0}^{k} \frac{G^r}{r!} p_{k-r}, \qquad k = 1, 2, 3, \ldots.$$

Here we used the rule for differentiating a product,

$$\frac{d^k}{dz^k}[f(z)g(z)] = \sum_{r=0}^{k}\binom{k}{r}\frac{d^r}{dz^r}f(z)\frac{d^{k-r}}{dz^{k-r}}g(z)$$

with

$$f(z) = e^{G(z-1)}, \qquad g(z) = h_s(z).$$

From this recurrent relation a computer can be programmed to calculate as many of the probabilities $p_k = \Pr(\mathbf{s} = k)$ as its memory will hold. Thus

$$p_1 = \lambda G p_0,$$

$$p_2 = \tfrac{1}{2}\lambda G(p_1 + Gp_0),$$

$$p_3 = \tfrac{1}{3}\lambda G(p_2 + Gp_1 + \tfrac{1}{2}G^2 p_0),$$

and so on. This probability distribution is known as the Neyman type A distribution. It first arose in describing observations in entomology and bacteriology and has found application in numerous other fields as well.* ∎

APPENDIX

The proof that the ratio of the volumes of the infinitesimal parallelepipeds V_x and V_z equals the Jacobian,

$$\frac{\text{Volume }(V_x)}{\text{Volume }(V_z)} = J\begin{pmatrix} x_1 & x_2 & \cdots & x_n \\ z_1 & z_2 & \cdots & z_n \end{pmatrix}, \tag{4.197}$$

rests on the fact that the volume of an n-dimensional parallelepiped spanned by the n linearly independent vectors $\mathsf{v}_1, \mathsf{v}_2, \ldots, \mathsf{v}_n$ equals the absolute value of the $n \times n$ determinant of their components v_{ij}, in whatever Cartesian coordinate system these are defined,

$$\text{Vol }(\mathsf{v}_1, \mathsf{v}_2, \ldots, \mathsf{v}_n) = |\det \| v_{ij} \| |. \tag{4.198}$$

This relation can be derived from (3.70) by induction, utilizing two facts: (a) the volume of a parallelepiped equals the product of its height times the volume of its base, which has one dimension fewer; and (b) the determinant in (4.198) is invariant to a rotation of the coordinate axes.

One assumes that (4.198) holds in n dimensions and proves from (a) and (b) that it holds in $(n + 1)$ dimensions by taking a Cartesian coordinate system having its $(n + 1)$th axis perpendicular to the base. The height of the parallelepiped in the $(n + 1)$ direction is then $v_{n+1,n+1}$, and the volume of the base, by hypothesis, is the $n \times n$ determinant $|\det_n \|v_{ij}\| |$, $1 \le (i, j) \le n$. Hence the

*For a review, see M. C. Teich, "Role of the doubly stochastic Neyman type-A and Thomas counting distributions in photon detection," *Applied Optics*, *20*, 2457–2467, July 15, 1981.

volume can be written

$$\text{Vol}\,(\mathbf{v}_1, \mathbf{v}_2, \ldots, \mathbf{v}_n, \mathbf{v}_{n+1})$$

$$= | v_{n+1,n+1} | \cdot | \det_n \| v_{ij} \| |$$

$$= \begin{vmatrix} v_{11} & v_{12} & \cdots & v_{1n} & 0 \\ \cdots & \cdots & \cdots & \cdots & \cdots \\ v_{n1} & v_{n2} & \cdots & v_{nn} & 0 \\ v_{n+1,1} & v_{n+1,2} & \cdots & v_{n+1,n} & v_{n+1,n+1} \end{vmatrix}, \tag{4.199}$$

as expanding this determinant with respect to the rightmost column shows. However, because in the coordinate system we adopted, the $(n + 1)$th axis is perpendicular to the n vectors $\mathbf{v}_k$, $1 \leq k \leq n$, that span the base, $v_{k,n+1} = 0$; and the last determinant in (4.199) is therefore $\det_{n+1}\|v_{ij}\|$ for the $n + 1$ vectors $\mathbf{v}_1, \mathbf{v}_2, \ldots, \mathbf{v}_n, \mathbf{v}_{n+1}$. Therefore,

$$\text{Vol}\,(\mathbf{v}_1, \ldots, \mathbf{v}_{n+1}) = | \det \| v_{ij} \| |,$$

the values of this determinant being the same in all Cartesian coordinate systems with the same scale. Since we know from Sec. 3-5 that (4.198) is true for $n = 2$, we have proved it true for all $n > 2$ as well.

The n vectors $\mathbf{v}_k$ spanning the infinitesimal parallelepiped V_x are

$$\mathbf{v}_k = \left(\frac{\partial h_k}{\partial z_1} dz_1, \ldots, \frac{\partial h_k}{\partial z_m} dz_m, \ldots, \frac{\partial h_k}{\partial z_n} dz_n \right), \qquad k = 1, 2, \ldots, n,$$

by the same argument as in Sec. 3-5, and by (4.198)

$$\text{Vol}\,(V_x) = | \det \| \frac{\partial h_i}{\partial z_j} \| | \cdot | dz_1\, dz_2 \cdots dz_n |,$$

whence (4.68).

PROBLEMS

If you have difficulty with a problem involving n random variables because it is hard to visualize relations in n dimensions, try working it first with $n = 2$ or $n = 3$. In a few of these problems the beta-function integral

$$B(m, n) = \int_0^1 x^{m-1}(1 - x)^{n-1} dx = \frac{\Gamma(m)\Gamma(n)}{\Gamma(m + n)}$$

may be useful.

4-1 The joint probability density function of the n nonnegative random variables $\mathbf{x}_1$, $\mathbf{x}_2, \ldots, \mathbf{x}_n$ has the form

$$f_{\mathbf{x}}(x_1, x_2, \ldots, x_n) = p(x_1 + x_2 + \cdots + x_n), \qquad x_k \geq 0, \quad \forall\, k,$$

where the function $p(\cdot)$ is nonnegative. Find a condition on this function such that this joint probability density function is properly normalized. Verify it for the joint density function

$$f_{\mathbf{x}}(x_1, x_2, \ldots, x_n) = \frac{N!}{(N-n)!}(1 - x_1 - x_2 - \cdots - x_n)^{N-n},$$

$$N > n, \quad 0 \le x_1 + x_2 + \cdots + x_n \le 1,$$

$$x_k > 0, \quad k = 1, 2, \ldots, n,$$

the density function being zero elsewhere in R_n.

Find the probability density function of

$$\mathbf{z} = \mathbf{x}_1 + \mathbf{x}_2 + \cdots + \mathbf{x}_n$$

in terms of the function $p(\cdot)$ and apply your result to the specific density function $f_{\mathbf{x}}(x_1, x_2, \ldots, x_n)$ given above.

4-2 The random variables $\mathbf{x}_1, \mathbf{x}_2, \ldots, \mathbf{x}_n$ have the joint probability density function

$$f_{\mathbf{x}}(x_1, x_2, \ldots, x_n) = (x_1 x_2 \cdots x_{n-1})^{-1}, \qquad 0 < x_n < x_{n-1} < \cdots < x_1 < 1,$$
$$\equiv 0 \qquad \text{elsewhere.}$$

This form holds for all $n > 1$;

$$f_{\mathbf{x}_1}(x) = \begin{cases} 1, & 0 < x < 1, \\ 0, & \text{elsewhere.} \end{cases}$$

Find the marginal probability density function $f_{\mathbf{x}_k}(x)$ for all $k > 1$. (*Hint:* Starting with $k = 2$, find these in the order $k = 2, 3, \ldots$. Use induction for general values of k.)

4-3 The five nonnegative random variables $\mathbf{x}_1$, $\mathbf{x}_2$, $\mathbf{x}_3$, $\mathbf{x}_4$, and $\mathbf{x}_5$ have the joint probability density function

$$f_{\mathbf{x}}(x_1, x_2, x_3, x_4, x_5)$$
$$= x_1 x_2 x_3 x_4 \exp[-(x_5 x_4 + x_4 x_3 + x_3 x_2 + x_2 x_1 + x_1)], \qquad x_k \ge 0, \quad \forall\, k.$$

The joint density function is zero if any x's are negative. Find:

(a) The marginal density function of $\mathbf{x}_2$
(b) The conditional density function $f_{\mathbf{x}_2}(x_2 \mid x_1)$
(c) The correlation $E(\mathbf{x}_2\mathbf{x}_3)$

4-4 The n random variables $\mathbf{x}_1, \mathbf{x}_2, \ldots, \mathbf{x}_n$ are independent and identically distributed, each having the density function $f_{\mathbf{x}}(x)$. Find the probability density function of the kth largest of these random variables. [*Hint:* First suppose a particular one of these is the kth largest, say $\mathbf{x}_1$. Then $k - 1$ of the rest must be larger than $\mathbf{x}_1$ and $n - k$ must be smaller. Write down the probability

Pr ($x_1 < \mathbf{x}_1 < x_1 + dx_1$ *and* any $k - 1$ $\mathbf{x}$'s are greater than x_1 *and* $n - k$ $\mathbf{x}$'s are smaller than x_1).

Then multiply by n since the kth largest might have been any one of the n $\mathbf{x}$'s.] Check your result by verifying that the integral of it over $(-\infty, \infty)$ equals 1.

4-5 Given an odd number $n = 2N + 1$ of random variables $\mathbf{x}_1, \mathbf{x}_2, \ldots, \mathbf{x}_n$, independent and identically distributed. The sample median $\mathbf{z}$ of these is the $(N + 1)$th largest; N of the $\mathbf{x}$'s are larger than $\mathbf{z}$, N are smaller. From the result of Problem 4-4, give the probability density function $f_{\mathbf{z}}(z)$ of the sample median $\mathbf{z}$ in terms of the probability density function $f_{\mathbf{x}}(x)$ of each random variable. Using Stirling's formula (1.30) for the factorial, show that when $n >> 1$ the sample median has approximately a Gaussian density function whose mean is the true median M of $\mathbf{x}$, that is, the value of x such that $\Pr(\mathbf{x} \le M) = \frac{1}{2}$. Assume that $f_{\mathbf{x}}(M) \ne 0$.

4-6 Let $\mathbf{x}_1, \mathbf{x}_2, \ldots, \mathbf{x}_n$ be independent and identically distributed random variables with the gamma density function,

$$f_{\mathbf{x}}(x) = [\Gamma(c)]^{-1} x^{c-1} e^{-x} U(x).$$

Determine the joint probability density function of the $n - 1$ random variables

$$\mathbf{z}_k = \frac{\mathbf{x}_k}{\mathbf{x}_1 + \mathbf{x}_2 + \cdots + \mathbf{x}_n}, \qquad 1 \le k \le n - 1.$$

(*Hint:* Introduce the dummy random variable

$$\mathbf{y} = \mathbf{x}_1 + \mathbf{x}_2 + \cdots + \mathbf{x}_n.$$

Then proceed as in Sec. 4-3.)

4-7 In Problem 4-6 find the probability density function of the random variable

$$\mathbf{s} = \mathbf{z}_1 + \mathbf{z}_2 + \cdots + \mathbf{z}_{n-1}.$$

(*Hint:* Express $\mathbf{s}$ in terms of $\mathbf{x}_n$ and $\mathbf{t} = \mathbf{x}_1 + \mathbf{x}_2 + \cdots + \mathbf{x}_{n-1}$.)

4-8 For the joint probability density function in (4.80), find the marginal density function of p_1 by integrating (4.80) over $0 < p_2 < \infty$. You can use the normalization integral for the density function in (3.61), after some manipulation, to evaluate the integral.

4-9 The discrete random variables $\mathbf{x}_1, \mathbf{x}_2, \ldots, \mathbf{x}_{n-1}$ have the multinomial distribution

$$\Pr(\mathbf{x}_1 = k_1, \mathbf{x}_2 = k_2, \ldots, \mathbf{x}_{n-1} = k_{n-1})$$
$$= \frac{N!}{k_1!k_2! \ldots k_n!} p_1^{k1} p_2^{k2} \cdots p_n^{kn},$$

where $k_1, k_2, \ldots, k_{n-1}$ are nonnegative integers, N is a nonnegative integer,

$$k_n = N - k_1 - k_2 - \cdots - k_{n-1},$$
$$p_1 + p_2 + \cdots + p_n = 1,$$
$$0 < p_j < 1, \quad 0 \le k_j \le N, \quad \forall j.$$

Find the covariance matrix of these $n - 1$ random variables. (See Problem 1-18.) [*Hint:* For a covariance a bivariate distribution suffices, and you can work out Cov $(\mathbf{x}_1, \mathbf{x}_2)$ from the distribution above for $n = 3$. Try using a multivariate characteristic function or probability generating function, applying the binomial theorem in order to carry out the summations.]

4-10 The random variables $\mathbf{x}_i$ have expected value $E(\mathbf{x}_i) \equiv 0$ and variance Var $\mathbf{x}_i \equiv 5$, $i = 1, 2, 3, 4$. Their covariances are all equal, Cov $(\mathbf{x}_i, \mathbf{x}_j) \equiv 2$, $i \ne j$. Find the variance of the random variable

$$\mathbf{y} = \mathbf{x}_1 + 2\mathbf{x}_2 + 3\mathbf{x}_3 + 4\mathbf{x}_4.$$

4-11 The three random variables $\mathbf{x}_1$, $\mathbf{x}_2$, and $\mathbf{x}_3$ have the covariance matrix

$$\mathsf{C} = \begin{bmatrix} 5 & 2 & 1 \\ 2 & 8 & -3 \\ 1 & -3 & 9 \end{bmatrix}.$$

Find the covariance matrix of the random variables

$$\mathbf{z}_1 = \mathbf{x}_1 - \mathbf{x}_2 + 2\mathbf{x}_3,$$
$$\mathbf{z}_2 = 3\mathbf{x}_1 + 2\mathbf{x}_2 - 4\mathbf{x}_3,$$
$$\mathbf{z}_3 = 2\mathbf{x}_1 - 5\mathbf{x}_2 + 3\mathbf{x}_3.$$

4-12 The covariance matrix of $\mathbf{x}_1$, $\mathbf{x}_2$, and $\mathbf{x}_3$ is

$$\mathsf{C} = \begin{bmatrix} 262 & -344 & 168 \\ -344 & 503 & -191 \\ 168 & -191 & 151 \end{bmatrix}.$$

Find three random variables $\mathbf{z}_1$, $\mathbf{z}_2$, and $\mathbf{z}_3$ that are distinct linear combinations of $\mathbf{x}_1$, $\mathbf{x}_2$, and $\mathbf{x}_3$ and are uncorrelated. Determine their variances.

4-13 For the probability density function of Example 4.1,

$$f_{\mathbf{x}}(x_1, x_2, \ldots, x_n) = \begin{cases} n!, & x_k \geq 0, \quad \forall\, k, \quad 0 \leq \sum_{k=1}^{n} x_k \leq 1, \\ 0, & \text{elsewhere,} \end{cases}$$

calculate the mean values $\bar{x}_k = E(\mathbf{x}_k)$, the variances Var $\mathbf{x}_k$, and the $n \times n$ covariance matrix of all the random variables.

4-14 Calculate the probability density function of the sum

$$\mathbf{z} = \mathbf{x}_1 + \mathbf{x}_2 + \cdots + \mathbf{x}_n$$

of the random variables in Problem 4-13. From it determine Var $\mathbf{z}$, and check the results of Problem 4-13 by also calculating Var $\mathbf{z}$ from the variances and the covariances of the $\mathbf{x}$'s as determined there.

4-15 The joint probability density function of the random variables $\mathbf{x}_1, \mathbf{x}_2, \ldots, \mathbf{x}_n$ is

$$f_{\mathbf{x}}(x_1, x_2, \ldots, x_n) = C(x_1^2 + x_2^2 + \cdots + x_n^2), \qquad 0 \leq x_1^2 + x_2^2 + \cdots + x_n^2 \leq R^2,$$
$$= 0 \qquad \text{elsewhere.}$$

Calculate the constant C and the variances Var $\mathbf{x}_k$ of these random variables. (*Hint:* Use the fact that by symmetry the variances are all equal.)

4-16 Assume that the independent random variables $\mathbf{x}_1, \mathbf{x}_2, \ldots, \mathbf{x}_n$ take on only positive values. Define their moment generating functions $\mu_k(p)$ as the Laplace transforms of their probability density functions,

$$\mu_k(p) = E[\exp(-p\mathbf{x}_k)] = \int_0^\infty e^{-px} f_{\mathbf{x}_k}(x)\, dx.$$

In terms of these, find the moment-generating function of the random variable

$$\mathbf{z} = c_1\mathbf{x}_1 + c_2\mathbf{x}_2 + \cdots + c_n\mathbf{x}_n.$$

4-17 The three independent random variables $\mathbf{x}_1$, $\mathbf{x}_2$, and $\mathbf{x}_3$ all have the density function

$$f_{\mathbf{x}}(x) = e^{-x}U(x).$$

Use the method of Problem 4-16 to find the probability density function of

$$\mathbf{z} = \tfrac{1}{2}\,\mathbf{x}_1 + \tfrac{1}{3}\,\mathbf{x}_2 + \mathbf{x}_3.$$

4-18 Calculate the moment generating function $E[\exp(-p\mathbf{z})]$ of the random variable

$$\mathbf{z} = \mathbf{x}_1^2 + \mathbf{x}_2^2 + \cdots + \mathbf{x}_n^2,$$

assuming that the random variables $\mathbf{x}_1, \mathbf{x}_2, \ldots, \mathbf{x}_n$ are independently Gaussian, with means m_k and equal variances σ^2,

$$f_{\mathbf{x}_k}(x) = (2\pi\sigma^2)^{-1/2}\exp\left[-\frac{(x - m_k)^2}{2\sigma^2}\right], \qquad k = 1, 2, \ldots, n.$$

For n an even number, determine the probability density function of $\mathbf{z}$ by looking the inverse Laplace transform of this moment generating function up in a table.

4-19 The n independent random variables $\mathbf{x}_1, \mathbf{x}_2, \ldots, \mathbf{x}_n$ have Cauchy distributions,

$$f_{\mathbf{x}_k}(x) = \frac{a_k/\pi}{x^2 + a_k^2}, \qquad k = 1, 2, \ldots, n.$$

Using the method of characteristic functions, find the probability density function of their sum

$$\mathbf{z} = \mathbf{x}_1 + \mathbf{x}_2 + \cdots + \mathbf{x}_n.$$

Will these random variables satisfy the conditions of the central limit theorem?

4-20 If we assume that we have already proved that any linear combination of Gaussian random variables has a Gaussian density function, we can derive the joint characteristic function of the n Gaussian random variables $\mathbf{x}_1, \mathbf{x}_2, \ldots, \mathbf{x}_n$ by evaluating

$$E\{\exp[j(\omega_1\mathbf{x}_1 + \omega_2\mathbf{x}_2 + \cdots + \omega_n\mathbf{x}_n)]\},$$

using the information that

$$\mathbf{z} = \omega_1\mathbf{x}_1 + \omega_2\mathbf{x}_2 + \cdots + \omega_n\mathbf{x}_n$$

is a Gaussian random variable, whose characteristic function is given in (2.89). Carry through the derivation in this way, assuming that the random variables $\mathbf{x}_j$ have mean values m_j and covariances Cov $(\mathbf{x}_i, \mathbf{x}_j) = c_{ij}$.

4-21 The three random variables $\mathbf{x}_1$, $\mathbf{x}_2$, and $\mathbf{x}_3$ are jointly Gaussian, with mean values zero and covariance matrix

$$\mathsf{C} = \begin{bmatrix} 3 & -1 & 2 \\ -1 & 6 & 4 \\ 2 & 4 & 10 \end{bmatrix}.$$

Determine the joint probability density function of the random variables

$$\mathbf{z}_1 = \mathbf{x}_1 + \mathbf{x}_2 - \mathbf{x}_3,$$
$$\mathbf{z}_2 = 2\mathbf{x}_1 - \mathbf{x}_2 + 3\mathbf{x}_3.$$

4-22 The covariance matrix of the Gaussian random variables $\mathbf{x}_1$, $\mathbf{x}_2$, $\mathbf{x}_3$, $\mathbf{x}_4$, and $\mathbf{x}_5$, whose means are zero, is

$$\mathsf{C} = \begin{bmatrix} 10 & -1 & 2 & 1 & 3 \\ -1 & 12 & -4 & -2 & -1 \\ 2 & -4 & 9 & 5 & 1 \\ 1 & -2 & 5 & 8 & 2 \\ 3 & -1 & 1 & 2 & 7 \end{bmatrix}.$$

Find the conditional means and the conditional covariance matrix of $\mathbf{x}_1$, $\mathbf{x}_2$, and $\mathbf{x}_3$, given that $\mathbf{x}_4$ has been observed to take on the value x_4 and $\mathbf{x}_5$ to take on the value x_5.

4-23 Show from the analysis of Sec. 4-5D that when a matrix C is divided into blocks as in (4.155), its determinant can be written

$$\det \mathsf{C} = \det(\mathsf{C}_{11} - \mathsf{C}_{12}\mathsf{C}_{22}^{-1}\mathsf{C}_{21}) \det \mathsf{C}_{22}.$$

This enables the determinant of a large matrix C to be evaluated by taking inverses of smaller block matrices. Apply the method to finding det C in Problem 4-22.

4-24 In an attempt to determine the probability that a biased coin shows heads, it is tossed 1000 times. If the true probability of heads is 0.4, what is the probability that the estimate obtained by dividing the number of heads by the total number (1000) of tosses is in error by more than 10%?

4-25 The variance of the Gaussian random variable $\mathbf{x}$ being unknown, the experiment that generates $\mathbf{x}$ is repeated 500 times to produce 500 data $\mathbf{x}_1, \mathbf{x}_2, \ldots, \mathbf{x}_{500}$, from which the variance is estimated by (4.169). Assuming that the probability density function in (4.178) can be approximated by a Gaussian density function, what is the probability that the estimated variance exceeds the true variance by more than 5%?

4-26 The random variables $\mathbf{x}_1$, $\mathbf{x}_2$, $\mathbf{x}_3$, and $\mathbf{x}_4$ are independent and uniformly distributed over $(-\frac{1}{2}, \frac{1}{2})$. Calculate the probability density function $f_{\mathbf{z}}(z)$ of their sum

$$\mathbf{z} = \mathbf{x}_1 + \mathbf{x}_2 + \mathbf{x}_3 + \mathbf{x}_4.$$

On the same graph plot both $f_{\mathbf{z}}(z)$ and the Gaussian density function having the same

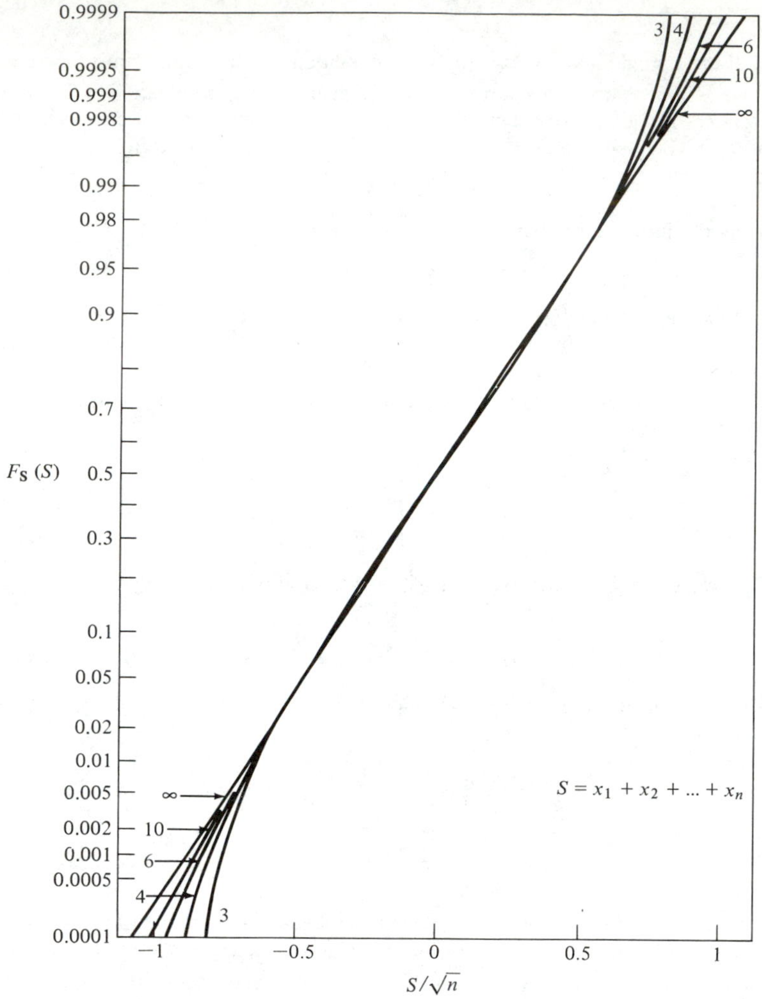

FIGURE 4P-26 Cumulative distribution function: sum of *n* uniformly distributed random variables. Curves are indexed with the number *n*.

mean and variance as **z** in order to see how close they are. Alternatively, calculate the cumulative distribution function $F_z(z)$ and plot it on normal probability graph paper and observe how it deviates from a straight line. See Fig. 4P.26 for the cumulative distribution of $\mathbf{x}_1 + \cdots + \mathbf{x}_n$.

4-27 When you enter a gambling den, the number of games of blackjack you will be allowed to play is determined by spinning a roulette wheel that stops at any integer from 1 to 100 with equal probability 0.01. The probability of winning $1 at blackjack equals 0.49; the probability of losing $1 equals 0.51. Calculate the mean and the variance of your total winnings, considered as a random variable.

4-28 Let the number **n** of terms in the random sum

$$\mathbf{s} = \sum_{k=1}^{\mathbf{n}} \mathbf{x}_k$$

depend on the values of $\mathbf{x}_1, \mathbf{x}_2, \ldots, \mathbf{x}_{\mathbf{n}-1}$, but not on $\mathbf{x}_k$ for $k \geq \mathbf{n}$. Once the sum is stopped, it is not resumed, no matter what the later values of $\mathbf{x}_k$, $k \geq \mathbf{n}$. Show that (4.190) still holds, even though $\mathbf{n}$ is not independent of the $\mathbf{x}$'s as assumed in Sec. 4-8. The random variables $\mathbf{x}_k$ are independent and identically distributed. [*Hint:* Introduce the random variable $\mathbf{e}_k$, which equals 1 if $\mathbf{x}_k$ is included in the sum and 0 if not, that is

$$\mathbf{e}_k = 1, \quad k \leq \mathbf{n}; \qquad \mathbf{e}_k = 0, \quad k > \mathbf{n}.$$

Then

$$\mathbf{s} = \sum_{k=1}^{\infty} \mathbf{e}_k \mathbf{x}_k.$$

Prove that $\mathbf{e}_k$ and $\mathbf{x}_k$ are independent random variables, and thence evaluate $E(\mathbf{s})$.]

4-29 Referring to part (b) of Problem 1-28, calculate the expected number $E(\mathbf{n})$ of boxes of cereal you need to buy in order to have a complete collection, that is, at least one picture of all M engineers. (*Hint:* Determine the expected value of the number of boxes bought in each state k of the collection, and sum over $1 \leq k < M$.)

4-30 In Example 4.22 assume that the number $\mathbf{x}$ of secondary electrons per primary electron still has a Poisson distribution with mean G, but that the number $\mathbf{n}$ of primary electrons has the geometric distribution

$$\Pr(\mathbf{n} = k) = (1 - v)v^k, \qquad 0 < v < 1.$$

Determine the probability generating function $h_{\mathbf{s}}(z)$ of the total number $\mathbf{s}$ of secondary electrons counted and evaluate the probability that $\mathbf{s} = 0$.

4-31 In Example 4.22, suppose that each primary electron gets through the surface on which it impinges, so that what is counted is the number of secondaries plus the number of primaries. Then (4.189) is replaced by

$$\mathbf{s} = \sum_{k=1}^{\mathbf{n}} (1 + \mathbf{x}_k),$$

where $\mathbf{x}_k$ is the random number of secondaries generated by the kth primary electron. The number $\mathbf{n}$ again has a Poisson distribution with mean λ, and the numbers $\mathbf{x}_k$ have a Poisson distribution with mean G; $\mathbf{n}$ and all the $\mathbf{x}_k$'s are statistically independent. Find the mean and variance of $\mathbf{s}$, its probability generating function $h_{\mathbf{s}}(z)$, and a recurrent relation similar to (4.196) for computing the probability distribution of $\mathbf{s}$, which is known as the "Thomas distribution."

CHAPTER 5

Stochastic Processes

5-1

PROBABILISTIC DESCRIPTION OF STOCHASTIC PROCESSES

A. Specifying the Basic Probability Measure

Random noise, as we said in the introduction to Chapter 1, is an erratically fluctuating voltage or current that appears at the input to a receiver and interferes with the reception of radio and radar signals. It exemplifies what is called a *random* or *stochastic process*. Other examples of stochastic processes are the acoustic pressure at some location on a noisy street, the height of ocean waves passing a given point, the velocity of a turbulent fluid, the density of charged particles in a plasma, the current through a photoelectric detector, and the load served by an electric power plant. All of these exhibit random fluctuations in the course of time, and our task now is to see how they can be analyzed in terms of the theory of probability.

The outcomes ζ of a chance experiment, as we have seen, may be labeled by a single number $\mathbf{x}$ or by a string of numbers $\mathbf{x}_1, \mathbf{x}_2, \ldots, \mathbf{x}_n$, and when a probability measure is assigned to events in the experiment, that is, to sets of outcomes, these numbers become random variables. Now we consider experiments in which the outcome ζ is labeled by a function $\mathbf{x}(t)$ of time t. This parameter t may be limited to an interval $a \leq t \leq b$ or to the half-line $0 \leq t < \infty$, or it may extend from $-\infty$ to ∞. When a probability measure is assigned to sets of such outcomes ζ, the functions $\mathbf{x}(t)$ labeling them constitute a *random* or *stochastic process*. The collection of labels $\mathbf{x}(t) = \mathbf{x}(t; \zeta)$ of all possible outcomes ζ of the experiment is called the *ensemble* of the stochastic process. The label $\mathbf{x}(t; \zeta) = \mathbf{x}(t)$ of a particular outcome ζ is called a *sample function* or a *realization* of the process.

How can the probability measure characterizing a stochastic process be specified? We imagine "sampling" the process $\mathbf{x}(t)$ at any finite number n of arbitrary times $t_1, t_2, \ldots, t_n$. The values $\mathbf{x}_1 = \mathbf{x}(t_1)$, $\mathbf{x}_2 = \mathbf{x}(t_2)$, $\ldots$, $\mathbf{x}_n = \mathbf{x}(t_n)$ of the samples are a partial labeling of the outcome ζ of the experiment and are therefore random variables. Their probability measure is specified, as described in Chapter 4, through a joint probability density function

$$f_{\mathbf{x}}(x_1, x_2, \ldots, x_n)$$

or—if they are discrete random variables—through the probabilities

$$\Pr(\mathbf{x}_1 = x_1, \mathbf{x}_2 = x_2, \ldots, \mathbf{x}_n = x_n),$$

that the samples assume particular sets $x_1, x_2, \ldots, x_n$ of values, these probabilities collectively making up the probability distribution of the samples. In order to indicate the times $t_1, t_2, \ldots, t_n$ at which the samples were taken, we write the joint probability density function as

$$f_{\mathbf{x}}(x_1, t_1; x_2, t_2; \ldots; x_n, t_n)$$

and the joint probability distribution as

$$\begin{aligned}\Pr(\mathbf{x}_1 &= x_1, t_1; \mathbf{x}_2 = x_2, t_2; \ldots; \mathbf{x}_n = x_n, t_n)\\ &= \Pr(\mathbf{x}(t_1) = x_1, \mathbf{x}(t_2) = x_2, \ldots, \mathbf{x}(t_n) = x_n)\end{aligned}$$

The number n of samples included is called the *order* of the density function or probability distribution. The collection of all such joint probability density functions or joint probability distributions for all values of n, $1 \leq n < \infty$, and for all possible sampling times $t_1, t_2, \ldots, t_n$, specifies the probability measure on the experiment within which the stochastic process $\mathbf{x}(t)$ is defined.

These joint probability density functions or joint probability distributions must have all properties associated with such functions in Chapter 4. In particular, they must be consistent: joint density functions of lower order must be obtainable from those of higher order by integrating out the excess variables, just as marginal probability density functions were defined in Sec. 4-1. For example,

$$\begin{aligned}f_{\mathbf{x}}(x_1, t_1; \ldots; x_{m-1}, t_{m-1}; x_{m+1}, t_{m+1}; \ldots; x_n, t_n)\\ = \int_{-\infty}^{\infty} dx_m\, f_{\mathbf{x}}(x_1, t_1; \ldots; x_{m-1}, t_{m-1}; x_m, t_m;\\ x_{m+1}, t_{m+1}; \ldots; x_n, t_n).\end{aligned} \tag{5.1}$$

Similarly, joint probability distributions of lower order are obtained when one sums over all possible values of the random variables to be eliminated,

$$\begin{aligned}\Pr(\mathbf{x}_1 = x_1, t_1; \ldots ; \mathbf{x}_{m-1} = x_{m-1}, t_{m-1};&\\ \mathbf{x}_{m+1} = x_{m+1}, t_{m+1}; \ldots ; \mathbf{x}_n = x_n, t_n)&\\ = \sum_{x_m} \Pr(\mathbf{x}_1 = x_1, t_1; \ldots ; \mathbf{x}_{m-1} = x_{m-1}, t_{m-1}; \mathbf{x}_m = x_m, t_m;&\\ \mathbf{x}_{m+1} = x_{m+1}, t_{m+1}; \ldots ; \mathbf{x}_n = x_n, t_n).&\end{aligned} \tag{5.2}$$

The times $t_1, t_2, \ldots, t_n$, it must be remembered, are not random variables; they appear only as parameters of the density functions and joint probability distributions in order to indicate the times at which the stochastic process was sampled. It is the sample values $\mathbf{x}(t_m) = \mathbf{x}_m$ that are the random variables. When a variable such as x_m is integrated out or summed over as in (5.1) and (5.2), the associated time parameter t_m also disappears from the expression.

Although in principle all the properties of a stochastic process are embodied in the array of probability density functions or probability distributions that, as just described, specify the fundamental probability measure on the experiment, there are many important kinds of stochastic process for which the general mathematical form of those joint density functions or joint probability distributions may be too complicated to write down for arbitrary n, or may even be unknown. Such processes are instead defined through some model of how they are generated and how they evolve in the course of time, and it is this model that determines the underlying probability measure. One may set out to deduce the joint density functions or probability distributions from the model, but sometimes their computation beyond a certain order may become too complicated, if not impossible. Useful conclusions can, however, be drawn from density functions or distributions of orders as low as one and two and from other characteristics of the random process that may even be determined experimentally. In this chapter we construct the theoretical framework within which stochastic processes can be characterized and analyzed, and we exemplify these concepts in terms of particular kinds of process.

Although here we consider the parameter t as the time, in important applications the parameter in the labeling function may represent a spatial coordinate or a number of spatial coordinates, together perhaps with a temporal coordinate. A two-dimensional random process $\mathbf{x}(\xi, \eta)$ associates a random variable $\mathbf{x}$ with each point (ξ, η) in a plane; $\mathbf{x}(\xi, \eta)$ might be the degree of darkening or the "density" of a photographic plate at point (ξ, η), and its statistical properties might characterize an ensemble of potential images to be subjected to some kind of image-processing technique. The heights of the surface of the ocean at all points (ξ, η) of some region and at all times t of some interval can be regarded as a spatiotemporal random process $\mathbf{x}(\xi, \eta, t)$ with three parameters ξ, η, and t. A particular component $\mathbf{B}_x$ of the magnetic intensity $\mathbf{B} = (\mathbf{B}_x, \mathbf{B}_y, \mathbf{B}_z)$ at all points (ξ, η, ζ) of some region of space and at all times t would constitute a spatiotemporal random process $\mathbf{B}_x(\xi, \eta, \zeta, t)$ with four parameters ξ, η, ζ, and t; such processes are considered in the statistical analysis of light. Associated with other spatiotemporal processes, such as the local density of charged particles and the components of the electric current density, the magnetic components $\mathbf{B}_x(\xi, \eta, \zeta, t)$, $\mathbf{B}_y(\xi, \eta, \zeta, t)$, and $\mathbf{B}_z(\xi, \eta, \zeta, t)$ figure in the study of ionospheric and interplanetary plasmas. The concepts we introduce in this chapter are readily extended to such spatial or spatiotemporal random processes.

B. Conditional Probabilities and Density Functions

From the joint probability density functions and probability distributions characterizing a stochastic process, conditional probability density functions and conditional probability distributions can be defined by the same rules as in Sec. 4-2. Such conditional density functions and distributions are often useful in analyzing models of stochastic processes and in deducing the joint density functions or distributions that specify their underlying probability measure. The conditional probability density function of $\mathbf{x}_2 = \mathbf{x}(t_2)$, given that the process was observed to take on the value x_1 at time t_1, for instance, is

$$\begin{aligned} f_{\mathbf{x}_2}(x_2, t_2 \mid x_1, t_1) &= f_{\mathbf{x}_2}(x_2, t_2 \mid \mathbf{x}(t_1) = x_1) \\ &= \frac{f_{\mathbf{x}}(x_1, t_1; x_2, t_2)}{f_{\mathbf{x}_1}(x_1, t_1)}, \end{aligned} \tag{5.3}$$

with

$$f_{\mathbf{x}_1}(x_1, t_1) = \int_{-\infty}^{\infty} f_{\mathbf{x}}(x_1, t_1; x_2, t_2)\, dx_2.$$

The joint probability density function of $\mathbf{x}_3 = \mathbf{x}(t_3)$, $\mathbf{x}_4 = \mathbf{x}(t_4)$, and $\mathbf{x}_5 = \mathbf{x}(t_5)$, given that the process took on values x_1 and x_2 at times t_1 and t_2, respectively, is

$$\begin{aligned} &f_{\mathbf{x}}(x_3, t_3; x_4, t_4; x_5, t_5 \mid x_1, t_1; x_2, t_2) \\ &\quad = f_{\mathbf{x}}(x_3, t_3; x_4, t_4; x_5, t_5 \mid \mathbf{x}(t_1) = x_1, \mathbf{x}(t_2) = x_2) \\ &\quad = \frac{f_{\mathbf{x}}(x_1, t_1; x_2, t_2; x_3, t_3; x_4, t_4; x_5, t_5)}{f_{\mathbf{x}}(x_1, t_1; x_2, t_2)}, \end{aligned} \tag{5.4}$$

with

$$\begin{aligned} &f_{\mathbf{x}}(x_1, t_1; x_2, t_2) \\ &\quad = \int_{-\infty}^{\infty}\int_{-\infty}^{\infty}\int_{-\infty}^{\infty} f_{\mathbf{x}}(x_1, t_1; x_2, t_2; x_3, t_3; x_4, t_4; x_5, t_5)\, dx_3\, dx_4\, dx_5, \end{aligned}$$

and so on. No particular order need be imposed on the times t_1, t_2, t_3, t_4, t_5 appearing in such definitions.

C. Some Simple Stochastic Processes

EXAMPLE 5.1. Constant Process

The sample functions $\mathbf{x}(t)$ are constant through time, $\mathbf{x}(t) \equiv \mathbf{x}$, but the value of the constant $\mathbf{x}$ is a random variable, the outcome of some chance experiment. Let the probability density function of $\mathbf{x}$ be $f_{\mathbf{x}}(x)$. Then for all t the first-order density function for this process is

$$f_{\mathbf{x}}(x, t) \equiv f_{\mathbf{x}}(x).$$

If we observe that $\mathbf{x}(t_1) = x_1$, we know that at any other time t this realization of the process takes on the same value x_1, whence the conditional probability density function

$$f_{\mathbf{x}_2}(x_2, t_2 \mid x_1, t_1) = f_{\mathbf{x}_2}(x_2, t_2 \mid \mathbf{x}(t_1) = x_1) = \delta(x_2 - x_1),$$

and by (5.3) the second-order joint density function of this process is

$$f_{\mathbf{x}}(x_1, t_1; x_2, t_2) = f_{\mathbf{x}}(x_1)\delta(x_2 - x_1). \tag{5.5}$$

By the same reasoning

$$f_{\mathbf{x}}(x_1, t_1; x_2, t_2; x_3, t_3) = f_{\mathbf{x}}(x_1)\delta(x_2 - x_1)\delta(x_3 - x_1), \tag{5.6}$$

and so on. Not a very exciting stochastic process! ■

EXAMPLE 5.2. Cosine Wave of Random Amplitude

The sample functions are cosines

$$\mathbf{x}(t) = \mathbf{A} \cos \omega t$$

whose amplitudes $\mathbf{A}$ are random variables with probability density function $f_{\mathbf{A}}(A)$. The amplitude $\mathbf{A}$ of a given realization is fixed for all time. Now, by Sec. 2-3,

$$f_{\mathbf{x}_1}(x_1, t_1) = |\cos \omega t_1|^{-1} f_{\mathbf{A}}\left(\frac{x_1}{\cos \omega t_1}\right) \tag{5.7}$$

provided that $\cos \omega t_1 \neq 0$. If $\cos \omega t_1 = 0$, that is, if ωt_1 equals $\frac{1}{2}\pi$ plus an integral multiple of π,

$$f_{\mathbf{x}_1}(x_1, t_1) = \delta(x_1).$$

The value $\mathbf{x}_1$ that the process is observed to take at any time t_1 determines its entire temporal behavior, provided that $\cos \omega t_1 \neq 0$. Thus

$$f_{\mathbf{x}_2}(x_2, t_2 \mid x_1, t_1) = \delta\left(x_2 - \frac{x_1 \cos \omega t_2}{\cos \omega t_1}\right) \tag{5.8}$$

because $\mathbf{A} = \mathbf{x}_1/\cos \omega t_1$, and the second-order joint probability density function of this process is

$$f_{\mathbf{x}}(x_1, t_1; x_2, t_2) = |\sec \omega t_1| \, f_{\mathbf{A}}\left(\frac{x_1}{\cos \omega t_1}\right) \delta\left(x_2 - \frac{x_1 \cos \omega t_2}{\cos \omega t_1}\right). \tag{5.9}$$

Density functions of higher order follow the same pattern. ■

EXAMPLE 5.3. Cosine Wave of Random Amplitude and Phase

The sample functions are cosines of random, but constant, amplitudes and phases,

$$\mathbf{x}(t) = \mathbf{A} \cos (\omega t + \boldsymbol{\psi}), \tag{5.10}$$

the random variables $\mathbf{A}$ and $\boldsymbol{\psi}$ having the joint density function $f_{\mathbf{A}\boldsymbol{\psi}}(A, \psi)$. This process might be the output of one of a large collection of oscillators of the same kind, turned on at random times and, because of random differences in the parameters of their electronic components, settling down to amplitudes $\mathbf{A}$ with some degree of variation from one oscillator to another.

Given the values $\mathbf{x}_1 = \mathbf{x}(t_1)$ and $\mathbf{x}_2 = \mathbf{x}(t_2)$ at two different times, we can calculate the values of $\mathbf{A}$ and $\boldsymbol{\psi}$ by solving the equations

$$\mathbf{x}_1 = \mathbf{A} \cos (\omega t_1 + \boldsymbol{\psi}),$$
$$\mathbf{x}_2 = \mathbf{A} \cos (\omega t_2 + \boldsymbol{\psi}), \tag{5.11}$$

an exercise we leave to the reader. These two equations specify a transformation from the pair $(\mathbf{A}, \boldsymbol{\psi})$ of random variables to the pair $(\mathbf{x}_1, \mathbf{x}_2)$ in the sense of Sec. 3-5, and the method given there enables us to determine the joint probability density function

$$f_{\mathbf{x}}(x_1, t_1; x_2, t_2) = \frac{f_{\mathbf{A}\boldsymbol{\psi}}(A, \psi)}{\left| J\begin{pmatrix} x_1 & x_2 \\ A & \psi \end{pmatrix} \right|}, \tag{5.12}$$

where the Jacobian is

$$J\begin{pmatrix} x_1 & x_2 \\ A & \psi \end{pmatrix} = \begin{vmatrix} \cos (\omega t_1 + \psi) & -A \sin (\omega t_1 + \psi) \\ \cos (\omega t_2 + \psi) & -A \sin (\omega t_2 + \psi) \end{vmatrix} = A \sin \omega(t_1 - t_2). \tag{5.13}$$

Into (5.12) we must substitute for A and ψ the solutions of (5.11) in terms of x_1 and x_2. The first-order density function $f_{\mathbf{x}_1}(x_1, t_1)$ can then be determined by integrating (5.12) over $-\infty < x_2 < \infty$.

Now the conditional probability density function of $\mathbf{x}_3 = \mathbf{x}(t_3)$, given that $\mathbf{x}(t_1) = x_1$, $\mathbf{x}(t_2) = x_2$, is a delta function,

$$f_{\mathbf{x}_3}(x_3, t_3 \mid x_1, t_1; x_2, t_2) = \delta(x_3 - A(x_1, x_2) \cos [\omega t_3 + \psi(x_1, x_2)]),$$

where $A(x_1, x_2)$ and $\psi(x_1, x_2)$ are the solutions of (5.11). From this we can write down the third-order joint density function $f_{\mathbf{x}}(x_1, t_1; x_2, t_2; x_3, t_3)$, and all density functions of higher order can be determined in like manner. ■

EXAMPLE 5.4. Pulse Code Modulation

A transmitter sends messages coded into binary digits 0 and 1, which emerge from a source every T seconds. Let the probability of 0 be p and the probability of 1 be $1 - p$. The successive digits are statistically independent. For each 1 a pulse of amplitude A and duration T seconds is sent along a cable; for each 0, nothing is sent. A sample function of the output of the transmitter, corresponding to a segment

$$\ldots 1\ 1\ 0\ 1\ 0\ 0\ 1\ 0 \ldots$$

of the message, would then look like that depicted in Fig. 5.1. The sample values of this stochastic process are discrete random variables. Their first-order probability distribution is

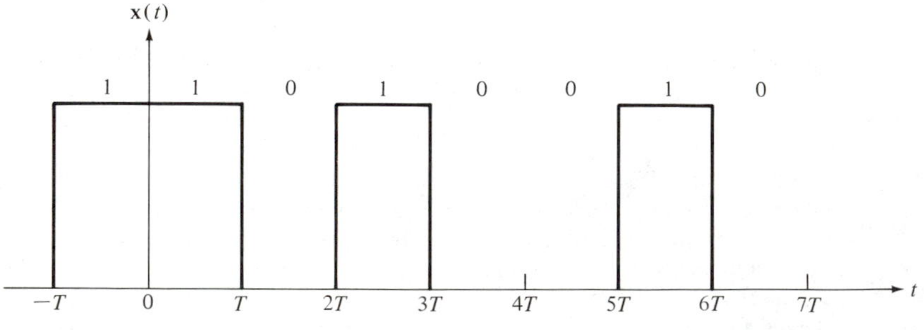

FIGURE 5.1 Pulse code modulation: on-off system.

$$\Pr(\mathbf{x}_1 = 0, t_1) = p,$$
$$\Pr(\mathbf{x}_1 = A, t_1) = 1 - p.$$

The second-order joint distribution for times t_1 and t_2 depends on whether t_1 and t_2 lie in the same interval of duration T or not:

$$\left.\begin{aligned}
&\Pr(\mathbf{x}_1 = 0, t_1; \mathbf{x}_2 = 0, t_2) = p,\\
&\Pr(\mathbf{x}_1 = 0, t_1; \mathbf{x}_2 = A, t_2) = 0,\\
&\Pr(\mathbf{x}_1 = A, t_1; \mathbf{x}_2 = 0, t_2) = 0,\\
&\Pr(\mathbf{x}_1 = A, t_1; \mathbf{x}_2 = A, t_2) = 1 - p.
\end{aligned}\right\} mT < (t_1, t_2) < (m + 1)T,$$

$$\left.\begin{aligned}
&\Pr(\mathbf{x}_1 = 0, t_1; \mathbf{x}_2 = 0, t_2) = p^2,\\
&\Pr(\mathbf{x}_1 = 0, t_1; \mathbf{x}_2 = A, t_2) = p(1 - p),\\
&\Pr(\mathbf{x}_1 = A, t_1; \mathbf{x}_2 = 0, t_2) = p(1 - p),\\
&\Pr(\mathbf{x}_1 = A, t_1; \mathbf{x}_2 = A, t_2) = (1 - p)^2,
\end{aligned}\right\} \begin{aligned} &mT < t_1 < (m + 1)T,\\ &nT < t_2 < (n + 1)T,\\ &m \neq n, \end{aligned} \tag{5.14}$$

for all integers m and n. The third-order joint distribution similarly depends on whether some or all of the times t_1, t_2, and t_3 lie in the same or different intervals.

More generally, the transmitter might send a signal $f_1(t)$ for each 1 and a different signal $f_0(t)$ for each 0. A sample function might look as shown in Fig. 5.2. The values taken on by this stochastic process are different from before, but the array of probabilities is the same. ■

The sample functions in these examples do not have the erratic appearance that we often imagine when thinking about stochastic processes in general, and they illustrate the breadth of the concept. In order to model random noise and other chaotic processes, we must bring in mechanisms that at least potentially can induce alterations in $\mathbf{x}(t)$ at any and all instants of time. We take up examples of such more erratic stochastic processes later in this chapter.

D. Stationary Stochastic Processes

Many important stochastic processes manifest the same statistical properties at one time as another, and the ensemble of processes distinguishes no particular time over any other time. The first-order probability density function $f_{\mathbf{x}}(x, t)$ of the process, that is, the density function of a sample $\mathbf{x} = \mathbf{x}(t)$ taken at an arbitrary time t, then does not depend on t,

$$f_{\mathbf{x}}(x, t) = f_{\mathbf{x}}(x), \qquad \forall\, t.$$

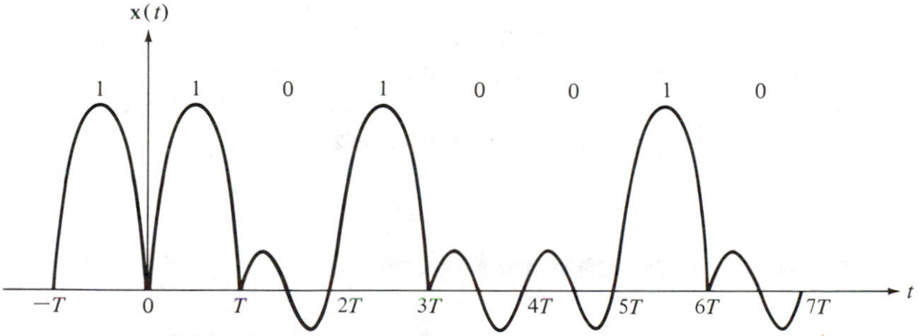

FIGURE 5.2 Pulse code modulation: arbitrary pulse shapes.

The second-order joint density function depends only on the interval between sampling times,

$$f_{\mathbf{x}}(x_1, t_1; x_2, t_2) = f_{\mathbf{x}_1\mathbf{x}_2}(x_1, x_2; t_2 - t_1), \qquad \forall\,(t_1, t_2),$$

and not on t_1 and t_2 individually. Thus the joint density function of samples $\mathbf{x}_1$ and $\mathbf{x}_2$ taken at times 1:15 P.M. and 1:17 P.M. today is the same as the joint density function of samples $\mathbf{x}_1$ and $\mathbf{x}_2$ taken at times 5:01 A.M. and 5:03 A.M. yesterday. What matters is the interval between the sampling times. In general, the joint probability density function of n samples $\mathbf{x}_1, \mathbf{x}_2, \ldots, \mathbf{x}_n$ depends only on the differences between the sampling times:

$$f_{\mathbf{x}}(x_1, t_1; x_2, t_2; \ldots; x_n, t_n)$$
$$= f_{\mathbf{x}}(x_1, x_2, \ldots, x_n; t_2 - t_1, t_3 - t_1, \ldots, t_n - t_1).$$

Such a stochastic process is said to be *stationary*.

The random noise produced by an electronic device, for instance, can be considered as a stationary stochastic process once the transients associated with turning on the device have died away, provided that it is kept at constant temperature, none of the applied voltages is changed, and none of the components of the device is altered. During an interval of a few hours the height of the waves passing a given point in the ocean might be described as a segment of a stationary process, but diurnal and annual variations of tides and weather exclude stationarity from a long-term model. In analyzing the performance of a communication system, the signal emitted by the transmitter and conveying messages from a stable though random source of message digits is generally taken as a stationary stochastic process.

The stochastic process in Example 5.1 is trivially stationary. Those in Examples 5.2 and 5.3 are in general nonstationary. If in Example 5.3, however, the joint density of amplitude and phase has the form

$$f_{\mathbf{A}\boldsymbol{\psi}}(A, \psi) = (2\pi)^{-1} f_{\mathbf{A}}(A), \qquad 0 \le \psi < 2\pi,$$

the phase ψ now being uniformly distributed over $(0, 2\pi)$ and independent of the amplitude $\mathbf{A}$, the stochastic process $\mathbf{x}(t) = \mathbf{A}\cos(\omega t + \boldsymbol{\psi})$ is stationary. The solution of (5.11) for $\mathbf{A}$ in terms of $\mathbf{x}_1$ and $\mathbf{x}_2$ will depend on the times t_1 and t_2 only through $t_2 - t_1$, and the Jacobian in (5.13) also depends on the times only through $t_2 - t_1$. Shifting all the sampling times by the same amount δ is equivalent to changing the phase by $\omega\delta$ (modulo 2π), but since $f_{\mathbf{A}\boldsymbol{\psi}}(A, \psi)$ now does not depend on the phase, none of the joint probability density functions of sample values of $\mathbf{x}(t)$ will be altered by such a shift. The process of Example 5.4 is nonstationary because of its synchronism with the times $0, T, 2T, \ldots$ when the message digits emerge from the source.

E. Multiple Stochastic Processes

In some chance experiments two or more stochastic processes may arise. One stochastic process $\mathbf{x}(t)$ may be the input to a linear filter, for instance, and a second process $\mathbf{y}(t)$ may be its output, and we may be concerned with the statistical relations between them, or we may need to determine the statistical properties of the output $\mathbf{y}(t)$ from those of the input $\mathbf{x}(t)$. In a turbulent fluid the three components $\mathbf{v}_x(t)$, $\mathbf{v}_y(t)$, and $\mathbf{v}_z(t)$ of the velocity vector constitute three processes that are defined in the same experiment and whose relations may be under study. Limiting ourselves now to two processes $\mathbf{x}(t)$ and $\mathbf{y}(t)$ for simplicity, we can consider both of these as labeling the outcome ζ of the basic experiment. The specification of the probability measure in the experiment would then involve not only the probability density functions of samples of $\mathbf{x}(t)$ and $\mathbf{y}(t)$ individually, such as

$$f_{\mathbf{x}}(x_1, t_1; x_2, t_2; \ldots; x_n, t_n)$$

and

$$f_{\mathbf{y}}(y_1, t_1; y_2, t_2; \ldots; y_n, t_n),$$

but also the joint density functions of samples of both processes, possibly at the same times, possibly at different times,

$$f_{\mathbf{xy}}(x_1, t_1; x_2, t_2; \ldots; x_n, t_n; y_1, t'_1; y_2, t'_2; \ldots; y_m, t'_m)$$

for all positive integers m and n and all possible sampling times $t_1, t_2, \ldots, t_n$, $t'_1, t'_2, \ldots, t'_m$. If all the joint probability density functions of this kind factor into joint probability density functions of $\mathbf{x}(t)$ and $\mathbf{y}(t)$ separately,

$$f_{\mathbf{xy}}(x_1, t_1; x_2, t_2; \ldots; x_n, t_n; y_1, t'_1; y_2, t'_2; \ldots; y_m, t'_m)$$
$$= f_{\mathbf{x}}(x_1, t_1; x_2, t_2; \ldots; x_n, t_n) f_{\mathbf{y}}(y_1, t'_1; y_2, t'_2; \ldots; y_m, t'_m), \quad (5.15)$$

the processes $\mathbf{x}(t)$ and $\mathbf{y}(t)$ are said to be *statistically independent*.

If the probability density functions $f_{\mathbf{x}}$ and $f_{\mathbf{y}}$ depend only on the intervals between sampling times, $\mathbf{x}(t)$ and $\mathbf{y}(t)$ are stationary processes. If, in addition, $f_{\mathbf{xy}}$ depends only on the intervals between sampling times, and not on those times individually, the processes $\mathbf{x}(t)$ and $\mathbf{y}(t)$ are said to be *stationarily related*. If the input $\mathbf{x}(t)$ to a time-invariant filter is stationary, for instance, and has been turned on for all time, the output $\mathbf{y}(t)$ will be stationary and stationarily related to it.

5-2

EXPECTED VALUES AND AUTOCOVARIANCE FUNCTIONS

In the course of time the first-order probability density function $f_{\mathbf{x}}(x, t)$ of a stochastic process might change its form and move about along the x-axis. Where it is located can be specified by the mean or expected value

$$\overline{x(t)} = E[\mathbf{x}(t)] = m(t) = \int_{-\infty}^{\infty} x f_{\mathbf{x}}(x, t)\, dx \quad (5.16)$$

of the sample $\mathbf{x} = \mathbf{x}(t)$ taken at time t. This expected value may depend on t. Often $m(t)$ represents a signal, and the difference $\mathbf{x}(t) - m(t)$ represents random noise, which is ordinarily taken to have expected value zero.

The range of values about the mean $m(t)$ over which the random variable $\mathbf{x}(t)$ is dispersed is measured by the standard deviation $\sigma(t)$, which is as usual the square root of the variance,

$$\begin{aligned}\sigma^2(t) &= \operatorname{Var} \mathbf{x}(t) = E\{[\mathbf{x}(t) - m(t)]^2\} \\ &= \int_{-\infty}^{\infty} [x - m(t)]^2 f_{\mathbf{x}}(x, t)\, dx \\ &= E\{[\mathbf{x}(t)]^2\} - \{E[\mathbf{x}(t)]\}^2 \\ &= \int_{-\infty}^{\infty} x^2 f_{\mathbf{x}}(x, t)\, dx - [m(t)]^2.\end{aligned} \tag{5.17}$$

This parameter of the first-order probability density function may also vary with the time. When the stochastic process $\mathbf{x}(t)$ is thought of as the sum of signal and noise, Var $\mathbf{x}(t)$ represents, in suitable units, the average power of the noise.

How the value of the process $\mathbf{x}(t)$ at one time is statistically related to that at another time is indicated by the *correlation function* $R_{\mathbf{x}}(t_1, t_2)$, defined by

$$\begin{aligned}R_{\mathbf{x}}(t_1, t_2) &= E[\mathbf{x}(t_1)\, \mathbf{x}(t_2)] \\ &= \int_{-\infty}^{\infty}\int_{-\infty}^{\infty} x_1 x_2\, f_{\mathbf{x}}(x_1, t_1; x_2, t_2)\, dx_1\, dx_2.\end{aligned} \tag{5.18}$$

For most processes this function contains a term readily identified as the product of the mean values $E[\mathbf{x}(t_1)]E[\mathbf{x}(t_2)]$, which is uninformative and may as well be dropped. The result is the *autocovariance function* of this process, defined by

$$\begin{aligned}C_{\mathbf{x}}(t_1, t_2) &= \operatorname{Cov}\{\mathbf{x}(t_1), \mathbf{x}(t_2)\} \\ &= E\{[\mathbf{x}(t_1) - m(t_1)][\mathbf{x}(t_2) - m(t_2)]\} \\ &= R_{\mathbf{x}}(t_1, t_2) - m(t_1)m(t_2) \\ &= \int_{-\infty}^{\infty}\int_{-\infty}^{\infty} [x_1 - m(t_1)][x_2 - m(t_2)]\, f_{\mathbf{x}}(x_1, t_1; x_2, t_2)\, dx_1\, dx_2.\end{aligned} \tag{5.19}$$

In particular,

$$\operatorname{Var} \mathbf{x}(t) = C_{\mathbf{x}}(t, t)$$

is the average noise power in the process $\mathbf{x}(t)$, and

$$C_{\mathbf{x}}(t_1, t_2) = C_{\mathbf{x}}(t_2, t_1)$$

always.

We prefer to characterize random processes by their autocovariance function $C_{\mathbf{x}}(t_1, t_2)$ rather than by their correlation function, for when $\mathbf{x}(t)$ is the sum of signal and noise, the autocovariance $C_{\mathbf{x}}(t_1, t_2)$ depends only on the noise component and informs us about the fluctuations superposed on the usually deterministic signal $m(t)$. Altering the mean value $m(t)$ does not change the autocovariance function $C_{\mathbf{x}}(t_1, t_2)$. The correlation function is the more directly measurable, the autocovariance function the more informative. When the mean is zero, $m(t) \equiv 0$, they are identical.

Moments of higher order of the sample values of a stochastic process can be defined as in Sec. 4-4B; there is nothing new to be said about them in this context, except that the times at which the samples are taken must be specified. The nth-order joint characteristic function

$$\begin{aligned}\Phi_{\mathbf{x}}(\omega_1, t_1; \omega_2, t_2; \ldots; \omega_n, t_n)\\ = E(\exp\{j[\omega_1\mathbf{x}(t_1) + \omega_2\mathbf{x}(t_2) + \cdots + \omega_n\mathbf{x}(t_n)]\})\end{aligned} \tag{5.20}$$

is defined as in (4.94) as the n-dimensional Fourier transform of the nth order joint probability density function $f_{\mathbf{x}}(x_1, t_1; x_2, t_2; \ldots; x_n, t_n)$. The integration is carried out over the space R_n of $x_1, x_2, \ldots, x_n$; the sampling times $t_1, t_2, \ldots, t_n$ remain as parameters of the joint characteristic function. As before, the joint characteristic function of a subset of the random variables is obtained by setting to zero the ω's associated with the variables or sample values to be eliminated; the corresponding sampling times will drop out as well. For instance,

$$\begin{aligned}\Phi_{\mathbf{x}}(\omega_1, t_1; \omega_4, t_4)\\ = \Phi_{\mathbf{x}}(\omega_1, t_1; 0, t_2; 0, t_3; \omega_4, t_4)\end{aligned} \tag{5.21}$$

is the Fourier transform of the second-order probability density function $f_{\mathbf{x}}(x_1, t_1; x_4, t_4)$ in terms of the fourth-order characteristic function

$$\begin{aligned}\Phi_{\mathbf{x}}(\omega_1, t_1; \omega_2, t_2; \omega_3, t_3; \omega_4, t_4)\\ = E(\exp\{j[\omega_1\mathbf{x}(t_1) + \omega_2\mathbf{x}(t_2) + \omega_3\mathbf{x}(t_3) + \omega_4\mathbf{x}(t_4)]\}).\end{aligned}$$

Conditional expected values are defined as in Sec. 4-4D in terms of conditional probability density functions. For example,

$$E[\mathbf{x}(t_2) \mid \mathbf{x}(t_1) = x_1] = \int_{-\infty}^{\infty} x_2 f_{\mathbf{x}_2}(x_2, t_2 \mid \mathbf{x}(t_1) = x_1)\, dx_2 \tag{5.22}$$

in terms of the conditional probability density function defined in (5.3). Furthermore,

$$\begin{aligned}E[\mathbf{x}(t_3)\mathbf{x}(t_4) \mid \mathbf{x}(t_1) = x_1, \mathbf{x}(t_2) = x_2]\\ = \int_{-\infty}^{\infty}\int_{-\infty}^{\infty} x_3x_4\, f_{\mathbf{x}}(x_3, t_3; x_4, t_4 \mid \mathbf{x}(t_1) = x_1, \mathbf{x}(t_2) = x_2)\, dx_3\, dx_4,\end{aligned} \tag{5.23}$$

with

$$f_{\mathbf{x}}(x_3, t_3; x_4, t_4 \mid \mathbf{x}(t_1) = x_1, \mathbf{x}(t_2) = x_2) = \frac{f_{\mathbf{x}}(x_1, t_1; x_2, t_2; x_3, t_3; x_4, t_4)}{f_{\mathbf{x}}(x_1, t_1; x_2, t_2)}$$

Counterparts of (3.110) and (4.130) are often useful. In particular,

$$\begin{aligned} &E[g(\mathbf{x}(t_1), \mathbf{x}(t_2))] \\ &= \int_{-\infty}^{\infty}\int_{-\infty}^{\infty} g(x_1, x_2) f_{\mathbf{x}}(x_1, t_1; x_2, t_2)\, dx_1\, dx_2 \\ &= E_{\mathbf{x}_1}\{E[g(\mathbf{x}(t_1), \mathbf{x}(t_2)) \mid \mathbf{x}(t_1) = \mathbf{x}_1]\} \\ &= E_{\mathbf{x}_1}\{E_{\mathbf{x}_2}[g(\mathbf{x}_1, \mathbf{x}_2) \mid \mathbf{x}(t_1) = \mathbf{x}_1]\}, \qquad \mathbf{x}_2 = \mathbf{x}(t_2), \end{aligned} \tag{5.24}$$

can conveniently be employed in evaluating the correlation function of a random process:

$$\begin{aligned} R_{\mathbf{x}}(t_1, t_2) &= E[\mathbf{x}(t_1)\mathbf{x}(t_2)] \\ &= E_{\mathbf{x}_1}\{E[\mathbf{x}(t_1)\mathbf{x}(t_2) \mid \mathbf{x}(t_1) = \mathbf{x}_1]\} \\ &= E_{\mathbf{x}_1}\{\mathbf{x}_1 E[\mathbf{x}(t_2) \mid \mathbf{x}(t_1) = \mathbf{x}_1]\}, \end{aligned} \tag{5.25}$$

as in (3.111). One first evaluates the conditional expected value

$$E[\mathbf{x}(t_2) \mid \mathbf{x}(t_1) = x_1] = \int_{-\infty}^{\infty} x_2 f_{\mathbf{x}_2}(x_2, t_2 \mid \mathbf{x}(t_1) = x_1)\, dx_2$$

given that at time t_1 the process takes on the value x_1. In the resulting function of x_1, one replaces x_1 by the random variable $\mathbf{x}_1$, multiplies by $\mathbf{x}_1$, and takes the expected value of the product, utilizing the first-order probability density function $f_{\mathbf{x}}(x_1, t_1)$. We shall meet several examples of this technique.

When the random process $\mathbf{x}(t)$ is stationary, its expected value $E[\mathbf{x}(t)]$ and its variance Var $\mathbf{x}(t)$ must be independent of the time t. The autocovariance function $C_{\mathbf{x}}(t_1, t_2)$ will depend only on the interval $t_2 - t_1$ because the joint probability density function $f_{\mathbf{x}}(x_1, t_1; x_2, t_2)$ in (5.19) depends on the sampling times t_1 and t_2 only through $t_2 - t_1$, and the mean values $m(t_1)$, $m(t_2)$ appearing there are independent of t_1 and t_2. We then write

$$C_{\mathbf{x}}(t_1, t_2) = C_{\mathbf{x}}(t_2 - t_1), \tag{5.26}$$

the autocovariance function now being a function $C_{\mathbf{x}}(\tau)$ of a single variable $\tau = t_2 - t_1$. Furthermore,

$$C_{\mathbf{x}}(0) = \text{Var}\, \mathbf{x}(t), \qquad C_{\mathbf{x}}(-\tau) = C_{\mathbf{x}}(\tau). \tag{5.27}$$

Sometimes it is known that the expected value $E[\mathbf{x}(t)]$ is constant and that the autocovariance function depends only on the interval $\tau = t_2 - t_1$, as in (5.26). The process $\mathbf{x}(t)$ is then said to be *wide-sense* stationary. Stationary processes are wide-sense stationary, but the converse is not necessarily true; one may know nothing about expected values or density functions of order higher than two. For most wide-sense stationary random processes the autocovariance function $C_{\mathbf{x}}(\tau)$ drops off to zero as $|\tau| \to \infty$; it may look like that depicted in Fig. 5.3. Samples of most processes taken at times separated by a very long interval tend to be uncorrelated.

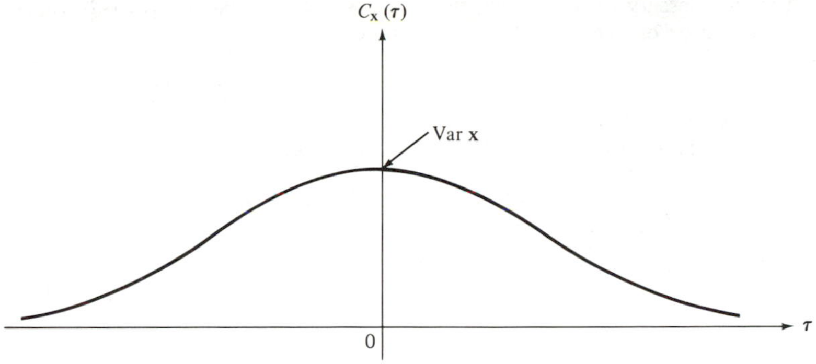

FIGURE 5.3 Typical autocovariance function, stationary process.

EXAMPLE 5.5. Constant Process (continued)

For the process in Example 5.1 the expected value is constant,

$$E[\mathbf{x}(t)] = \int_{-\infty}^{\infty} x f_{\mathbf{x}}(x)\, dx = E(\mathbf{x}).$$

Now by (5.5),

$$E[\mathbf{x}(t_2) \mid \mathbf{x}(t_1) = x_1] = x_1,$$

whence the correlation function is, by (5.25),

$$R_{\mathbf{x}}(t_1, t_2) = E(\mathbf{x}_1^2),$$

and the autocovariance function of this process is simply

$$C_{\mathbf{x}}(t_1, t_2) = \text{Var}\,\mathbf{x} = \int_{-\infty}^{\infty} [x - E(\mathbf{x})]^2 f_{\mathbf{x}}(x)\, dx.$$

■

EXAMPLE 5.6. Cosine Wave of Random Amplitude (continued)

For the process in Example 5.2 the expected value is sinusoidal,

$$E[\mathbf{x}(t)] = E(\mathbf{A}) \cos \omega t,$$

$$E(\mathbf{A}) = \int_{-\infty}^{\infty} A f_A(A)\, dA,$$

unless $E(\mathbf{A}) = 0$. The correlation function is

$$R_{\mathbf{x}}(t_1, t_2) = E[\mathbf{x}(t_1)\mathbf{x}(t_2)] = E(\mathbf{A}^2) \cos \omega t_1 \cos \omega t_2,$$

and the autocovariance function is

$$C_{\mathbf{x}}(t_1, t_2) = (\text{Var}\,\mathbf{A}) \cos \omega t_1 \cos \omega t_2. \tag{5.28}$$

■

EXAMPLE 5.7. Cosine Wave of Random Amplitude and Phase (continued)

For the process in Example 5.3 the expected value is, by (3.82),

$$E[\mathbf{x}(t)] = \int_{-\infty}^{\infty} dA \int_0^{2\pi} d\psi\, f_{\mathbf{A}\psi}(A, \psi)\, A \cos(\omega t + \psi). \tag{5.29}$$

It is unnecessary to have calculated the first-order probability density function of this process in order to evaluate its expected value. If

$$f_{\mathbf{A}\boldsymbol{\psi}}(A, \psi) = (2\pi)^{-1} f_{\mathbf{A}}(A), \tag{5.30}$$

the expected value is zero.

The correlation function of this process is similarly

$$R_{\mathbf{x}}(t_1, t_2) = \int_{-\infty}^{\infty} dA \int_0^{2\pi} d\psi \, f_{\mathbf{A}\boldsymbol{\psi}}(A, \psi)\, A^2 \cos(\omega t_1 + \psi) \cos(\omega t_2 + \psi).$$

If the phase is uniformly distributed over $(0, 2\pi)$ and independent of the amplitude, as in (5.30),

$$\begin{aligned} R_{\mathbf{x}}(t_1, t_2) &\\ &= \tfrac{1}{2}\int_{-\infty}^{\infty} A^2 f_{\mathbf{A}}(A)\, dA \int_0^{2\pi} \{\cos \omega(t_2 - t_1) + \cos[\omega(t_2 + t_1) + 2\psi]\} \frac{d\psi}{2\pi} \\ &= \tfrac{1}{2}E(\mathbf{A}^2) \cos \omega(t_2 - t_1). \end{aligned}$$

The process $\mathbf{x}(t)$, with (5.30), is stationary, and its autocovariance function can be written

$$C_{\mathbf{x}}(\tau) = \tfrac{1}{2}E(\mathbf{A}^2) \cos \omega\tau. \tag{5.31}$$

Unlike that in Fig. 5.3, this autocovariance function does not vanish as $|\tau| \to \infty$. ■

EXAMPLE 5.8. Pulse Code Modulation (continued)

The expected value of the stochastic process in Example 5.4 is

$$E[\mathbf{x}(t)] = 0 \cdot \Pr\{\mathbf{x}(t) = 0\} + A \cdot \Pr\{\mathbf{x}(t) = A\} = A(1 - p).$$

It is constant and resembles none of the sample functions of this process.

When the times t_1 and t_2 lie in the same signal interval,

$$E[\mathbf{x}(t_2) \mid \mathbf{x}(t_1) = x_1] = x_1$$

since the process changes its value only at the end of such an interval. Hence by (5.25)

$$R_{\mathbf{x}}(t_1, t_2) = E(\mathbf{x}^2) = (1 - p)A^2 + p \cdot 0^2, \tag{5.32}$$

and the autocovariance function is

$$C_{\mathbf{x}}(t_1, t_2) = p(1 - p)A^2, \qquad mT < (t_1, t_2) < (m + 1)T.$$

If, on the other hand, the times t_1 and t_2 lie in different signal intervals, $\mathbf{x}(t_1)$ and $\mathbf{x}(t_2)$ are statistically independent, and the autocovariance function vanishes,

$$C_{\mathbf{x}}(t_1, t_2) = 0, \qquad mT < t_1 < (m + 1)T, \qquad nT < t_2 < (n + 1)T, \qquad m \neq n.$$ ■

EXAMPLE 5.9. Randomly Delayed Pulse Code Modulation

From the ensemble of processes in Example 5.4 we can construct a larger ensemble by considering all possible delayed versions of the process $\mathbf{x}(t)$. That is, we define a new process $\mathbf{y}(t)$ as

$$\mathbf{y}(t) = \mathbf{x}(t - \mathbf{d}), \qquad 0 \le \mathbf{d} < T,$$

in which the random delay $\mathbf{d}$ is uniformly distributed over $(0, T)$. The synchronism with the particular sequence of times $0, T, 2T, \ldots$ is now abolished, and the process $\mathbf{y}(t)$, distinguishing no origin of time, is stationary. Its expected value is the same as for the process $\mathbf{x}(t)$ in Example 5.4. How about its autocovariance function?

If the sampling times t_1 and t_2 are separated by more than T, $|t_2 - t_1| > T$, the values of $\mathbf{y}(t_1)$ and $\mathbf{y}(t_2)$ will always be statistically independent, and $C_{\mathbf{y}}(t_1, t_2) = 0$. Now suppose that $t_2 > t_1$ and $t_2 - t_1 < T$. Then $\mathbf{y}(t_1)$ and $\mathbf{y}(t_2)$ will be independent if and only if a switch time $nT + \mathbf{d}$ lies between t_1 and t_2 for some integer n; denote this event by M. On the other hand, $\mathbf{y}(t_1) = \mathbf{y}(t_2)$ if no such switch time lies between t_1 and t_2; this is event M'. The probability of the former event M is $|t_2 - t_1|/T$; that of the latter M' is

$$1 - \frac{|t_2 - t_1|}{T}.$$

In the former event

$$E[\mathbf{y}(t_1)\mathbf{y}(t_2) \mid M] = E[\mathbf{y}(t_1)]\, E[\mathbf{y}(t_2)] = [(1 - p)A]^2$$

by (5.32); in the latter event,

$$E[\mathbf{y}(t_1)\mathbf{y}(t_2) \mid M'] = E[\mathbf{y}(t_1)]^2 = (1 - p)A^2.$$

Hence the correlation function is

$$R_{\mathbf{y}}(t_1, t_2) = \frac{|t_2 - t_1|}{T}[(1 - p)A]^2 + \left[1 - \frac{|t_2 - t_1|}{T}\right](1 - p)A^2$$

and after subtracting the square of the mean, we find for the autocovariance function of this process, with $\tau = t_2 - t_1$,

$$C_{\mathbf{y}}(\tau) = p(1 - p)A^2\left(1 - \frac{|\tau|}{T}\right), \qquad |\tau| < T.$$

It is sketched in Fig. 5.4. ■

If two random processes $\mathbf{x}(t)$ and $\mathbf{y}(t)$ arise in the same experiment, as described in Sec. 5-1D, their cross-correlation function is defined as

$$R_{\mathbf{xy}}(t_1, t_2) = E[\mathbf{x}(t_1)\mathbf{y}(t_2)] \tag{5.33}$$

$$= \int_{-\infty}^{\infty}\int_{-\infty}^{\infty} xy f_{\mathbf{xy}}(x, t_1; y, t_2)\, dx\, dy,$$

where $f_{\mathbf{xy}}(x, t_1; y, t_2)$ is the joint probability density function of the sample $\mathbf{x}(t_1)$ of the first process at time t_1 and the sample $\mathbf{y}(t_2)$ of the second process at time t_2. The cross-covariance of the two processes is then defined by

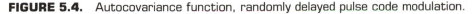

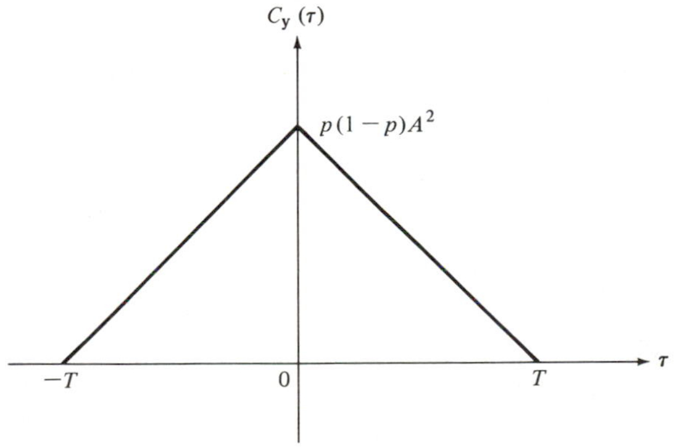

FIGURE 5.4. Autocovariance function, randomly delayed pulse code modulation.

$$C_{\mathbf{xy}}(t_1, t_2) = \text{Cov}\{\mathbf{x}(t_1), \mathbf{y}(t_2)\} \\ = R_{\mathbf{xy}}(t_1, t_2) - E[\mathbf{x}(t_1)]E[\mathbf{y}(t_2)]. \tag{5.34}$$

This function gives some information about how the fluctuations of one process are related to those of the other. If the processes are statistically independent, $C_{\mathbf{xy}}(t_1, t_2) = 0$. If they are stationary and stationarily related, the cross-covariance function $C_{\mathbf{xy}}(t_1, t_2)$ is a function only of the interval $t_2 - t_1$ and is written $C_{\mathbf{xy}}(t_2 - t_1) = C_{\mathbf{xy}}(\tau)$ in terms of a single variable $\tau = t_2 - t_1$ for the interval between sampling times.

5-3

POISSON AND AFFILIATED STOCHASTIC PROCESSES

A. Poisson Counting Process and Its Distributions

Light of constant intensity falls on the cathode of a photoelectric detector. The number $\mathbf{n}(t)$ of electrons emitted between time 0 and time t is a discrete stochastic process, the value of which increases by 1 at each instant $\boldsymbol{\tau}_1, \boldsymbol{\tau}_2, \ldots, \boldsymbol{\tau}_m, \ldots$ when an electron is ejected from the cathode. A typical sample function is depicted in Fig. 5.5. In order to determine the probability distributions of this stochastic process $\mathbf{n}(t)$, we must make some assumptions about the random times $\boldsymbol{\tau}_1, \boldsymbol{\tau}_2, \ldots, \boldsymbol{\tau}_m, \ldots$ at which electrons are emitted.

The total number $\boldsymbol{\nu} = \mathbf{n}(t + T) - \mathbf{n}(t)$ of electrons emitted in an interval $(t, t + T)$ of duration T was shown in Sec. 1-9 to have a Poisson distribution with a mean value proportional to the total radiant energy W absorbed by the surface of the cathode during that interval, and that total energy is in turn proportional to the duration T of the interval. The mean value of the number $\boldsymbol{\nu}$ of electrons emitted between t and $t + T$ is therefore

$$E(\boldsymbol{\nu}) = \lambda T,$$

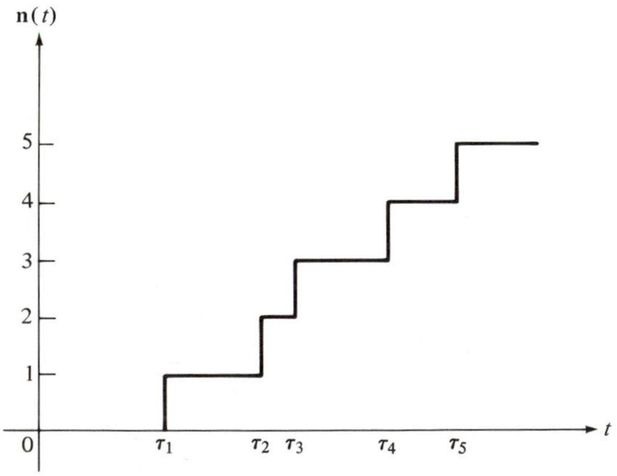

FIGURE 5.5. Poisson counting process.

where λ is a constant rate of emission proportional to the intensity of the incident light. The probability that k electrons are emitted between times t and $t + T$ is

$$\Pr(\boldsymbol{\nu} = k) = \frac{(\lambda T)^k e^{-\lambda T}}{k!}, \tag{5.35}$$

and this is the probability that there are k emission times, or "points," $\boldsymbol{\tau}_m$ during the interval $(t, t + T)$. The numbers $\boldsymbol{\nu}_1$ and $\boldsymbol{\nu}_2$ of such points $\boldsymbol{\tau}_m$ in any two disjoint intervals, furthermore, are statistically independent, for it is assumed that the light is not so intense as to deplete the supply of available electrons or to alter the physical properties of the cathode, and the emission of one photoelectron does not influence the emission of any other. A sequence of random points $\boldsymbol{\tau}_1, \boldsymbol{\tau}_2, \ldots, \boldsymbol{\tau}_m, \ldots$ subject to these two conditions—that their number in any interval have a Poisson distribution and that their numbers in disjoint intervals be independent—is said to constitute a *Poisson point process*. The parameter λ is called the *rate* of the process. When the instants $\boldsymbol{\tau}_1, \boldsymbol{\tau}_2, \ldots, \boldsymbol{\tau}_m, \ldots$ at which the stochastic process $\mathbf{n}(t)$ jumps up by one unit form a Poisson point process, the process $\mathbf{n}(t)$ is called a *Poisson counting process*.

The incidence of calls at a telephone exchange, the arrival of customers at a bank or other service facility, and the submission of jobs to a computer terminal are frequently modeled as Poisson point processes for the purposes of analyzing the performance of various modes of handling calls, customers, or computer jobs. In "queuing theory," which treats such problems, the Poisson counting process plays a central role.

The samples $\mathbf{n}_1 = \mathbf{n}(t_1)$, $\mathbf{n}_2 = \mathbf{n}(t_2), \ldots$ of the Poisson counting process are discrete random variables. The first-order probability distribution governing it, that is, the probability that k electrons are emitted in $(0, t)$, is

$$\Pr[\mathbf{n}(t) = k] = \frac{(\lambda t)^k e^{-\lambda t}}{k!}, \qquad k = 0, 1, 2, \ldots, \tag{5.36}$$

by (5.35). The conditional probability that for $t_2 > t_1$, $\mathbf{n}(t_2) = k_2$, given $\mathbf{n}(t_1) = k_1$, is the probability that $k_2 - k_1$ electrons are emitted between t_1 and t_2,

$$\begin{aligned} \Pr[\mathbf{n}(t_2) = k_2 \mid \mathbf{n}(t_1) = k_1] &= \frac{[\lambda(t_2 - t_1)]^{k_2 - k_1} \exp[-\lambda(t_2 - t_1)]}{(k_2 - k_1)!}, \qquad k_2 \geq k_1, \\ &= 0, \qquad k_2 < k_1. \end{aligned} \tag{5.37}$$

Hence the second-order joint probability distribution of the Poisson counting process is, for $t_2 > t_1$,

$$\begin{aligned} \Pr[\mathbf{n}(t_1) = k_1, \mathbf{n}(t_2) = k_2] &= \Pr[\mathbf{n}(t_2) = k_2 \mid \mathbf{n}(t_1) = k_1] \Pr[\mathbf{n}(t_1) = k_1] \\ &= \frac{(\lambda t_1)^{k_1} [\lambda(t_2 - t_1)]^{k_2 - k_1} e^{-\lambda t_2}}{k_1!(k_2 - k_1)!}, \qquad k_2 \geq k_1, \\ &= 0, \qquad k_2 < k_1. \end{aligned} \tag{5.38}$$

Let $t_3 > t_2 > t_1$. Then

$$\begin{aligned}
&\Pr[\mathbf{n}(t_3) = k_3 \mid \mathbf{n}(t_1) = k_1, \mathbf{n}(t_2) = k_2] \\
&\quad = \Pr[\mathbf{n}(t_3) = k_3 \mid \mathbf{n}(t_2) = k_2] \\
&\quad = \frac{[\lambda(t_3 - t_2)]^{k_3-k_2} \exp[-\lambda(t_3 - t_2)]}{(k_3 - k_2)!}, \qquad k_3 \geq k_2, \\
&\quad = 0, \qquad k_3 < k_2.
\end{aligned} \tag{5.39}$$

for this is just the probability that $k_3 - k_2$ electrons were emitted in the interval (t_2, t_3). The value $\mathbf{n}(t_3)$ of the process at time t_3 is thus independent of its value $\mathbf{n}(t_1)$ at time t_1, once its value at an intermediate time t_2 is known. By what steps the process $\mathbf{n}(t)$ reached the value $\mathbf{n}(t_2)$ is irrelevant to its behavior after time t_2. Stochastic processes having this feature are very common and are called *Markov processes*. The third-order joint probability for this Poisson counting process is, therefore, for $t_1 < t_2 < t_3$,

$$\begin{aligned}
&\Pr[\mathbf{n}(t_1) = k_1, \mathbf{n}(t_2) = k_2, \mathbf{n}(t_3) = k_3] \\
&\quad = \frac{(\lambda t_1)^{k_1}[\lambda(t_2 - t_1)]^{k_2-k_1}[\lambda(t_3 - t_2)]^{k_3-k_2}e^{-\lambda t_3}}{k_1!(k_2 - k_1)!(k_3 - k_2)!}
\end{aligned} \tag{5.40}$$

for $k_1 \leq k_2 \leq k_3$; it equals zero otherwise. Joint probabilities for more than three samples can be written down by the same procedure, but there is no point in doing so.

B. Expected Value and Autocovariance Function of the Poisson Counting Process

The expected value of the Poisson counting process $\mathbf{n}(t)$ at time t is

$$E[\mathbf{n}(t)] = \lambda t \tag{5.41}$$

by (5.36) and Example 2.24. This is just the rate λ times the duration of the interval $(0, t)$ since counting began.

The correlation function $R_{\mathbf{n}}(t_1, t_2)$ of this process can easily be obtained by means of (5.25). For $t_2 > t_1$,

$$\begin{aligned}
&E[\mathbf{n}(t_2) \mid \mathbf{n}(t_1) = k] \\
&\quad = k + \text{expected number of emissions in } (t_1, t_2) \\
&\quad = k + \lambda(t_2 - t_1).
\end{aligned}$$

Hence for $t_2 > t_1$, with $\mathbf{n}(t_1) = \mathbf{n}_1$,

$$\begin{aligned}
R_{\mathbf{n}}(t_1, t_2) &= E[\mathbf{n}(t_1)\mathbf{n}(t_2)] \\
&= E\{\mathbf{n}_1[\mathbf{n}_1 + \lambda(t_2 - t_1)]\} \\
&= E(\mathbf{n}_1^2) + \lambda(t_2 - t_1)E(\mathbf{n}_1).
\end{aligned} \tag{5.42}$$

Since we know from Example 2.24 that the variance of a Poisson-distributed random variable equals its mean

$$E(\mathbf{n}_1^2) = \text{Var } \mathbf{n}_1 + [E(\mathbf{n}_1)]^2 = \lambda t_1 + \lambda^2 t_1^2.$$

Putting this into (5.42) we find that

$$R_{\mathbf{n}}(t_1, t_2) = \lambda t_1 + \lambda^2 t_1^2 + \lambda^2 t_1 t_2 - \lambda^2 t_1^2.$$

Subtracting the product of the means

$$E[\mathbf{n}(t_1)]E[\mathbf{n}(t_2)] = (\lambda t_1)(\lambda t_2),$$

we obtain the autocovariance function

$$C_{\mathbf{n}}(t_1, t_2) = \lambda t_1, \; t_1 < t_2.$$

Had we taken $t_2 < t_1$ instead, the same argument would have yielded λt_2. Hence the autocovariance function of the Poisson counting process is

$$C_{\mathbf{n}}(t_1, t_2) = \lambda \min (t_1, t_2). \tag{5.43}$$

C. Random Telegraph Signal

The process $\mathbf{x}(t)$ starts at the value $\mathbf{x}(0) = 1$ at time $t = 0$. At random times $\boldsymbol{\tau}_1, \boldsymbol{\tau}_2, \ldots, \boldsymbol{\tau}_m, \ldots$ thereafter it changes sign. The "switching times" $\boldsymbol{\tau}_1, \boldsymbol{\tau}_2, \ldots, \boldsymbol{\tau}_m, \ldots$ constitute a Poisson point process with rate λ; that is, the number $\boldsymbol{\nu}$ of changes of sign between any two times t_1 and t_2 has a Poisson distribution with rate λ as in (5.35), and the numbers of changes of sign in disjoint intervals are statistically independent. Figure 5.6 illustrates a typical sample function of this process, which is called the *random telegraph signal*.

At time $t > 0$, $\mathbf{x}(t) = +1$ if there have been an even number of changes of sign in $(0, t)$, and the probability of this event is

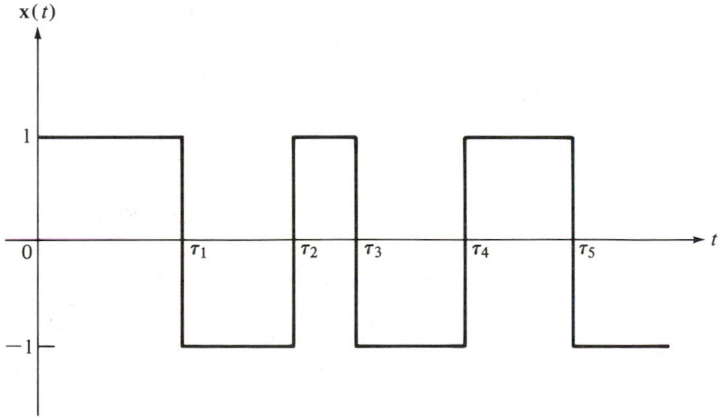

FIGURE 5.6. Random telegraph signal.

$$\Pr[\mathbf{x}(t) = +1] = \left(1 + \frac{(\lambda t)^2}{2!} + \frac{(\lambda t)^4}{4!} + \cdots\right)e^{-\lambda t}. \qquad (5.44)$$
$$= \cosh(\lambda t)e^{-\lambda t}.$$

If there have been an odd number of changes of sign, $\mathbf{x}(t) = -1$, and

$$\Pr[\mathbf{x}(t) = -1] = \sinh(\lambda t)e^{-\lambda t}. \qquad (5.45)$$

These equations specify the first-order distribution of the random telegraph signal $\mathbf{x}(t)$.

To derive the second-order distribution we again utilize conditional probability. For $t_2 > t_1 > 0$,

$$\Pr[\mathbf{x}(t_2) = 1 \mid \mathbf{x}(t_1) = 1] = \Pr[\mathbf{x}(t_2) = -1 \mid \mathbf{x}(t_1) = -1]$$
$$= \Pr[\text{even number of sign changes in } (t_1, t_2)] \qquad (5.46)$$
$$= \exp[-\lambda(t_2 - t_1)] \cosh \lambda(t_2 - t_1).$$

Similarly,

$$\Pr[\mathbf{x}(t_2) = -1 \mid \mathbf{x}(t_1) = 1] = \Pr[\mathbf{x}(t_2) = 1 \mid \mathbf{x}(t_1) = -1]$$
$$= \Pr[\text{odd number of sign changes in } (t_1, t_2)] \qquad (5.47)$$
$$= \exp[-\lambda(t_2 - t_1)] \sinh \lambda(t_2 - t_1).$$

Hence the joint probability distribution of a pair of samples is, for $t_2 > t_1$,

$$\begin{aligned} \Pr[\mathbf{x}(t_1) = 1, \mathbf{x}(t_2) = 1] &= \exp(-\lambda t_2) \cosh \lambda t_1 \cosh \lambda(t_2 - t_1), \\ \Pr[\mathbf{x}(t_1) = 1, \mathbf{x}(t_2) = -1] &= \exp(-\lambda t_2) \cosh \lambda t_1 \sinh \lambda(t_2 - t_1), \\ \Pr[\mathbf{x}(t_1) = -1, \mathbf{x}(t_2) = 1] &= \exp(-\lambda t_2) \sinh \lambda t_1 \sinh \lambda(t_2 - t_1), \\ \Pr[\mathbf{x}(t_1) = -1, \mathbf{x}(t_2) = -1] &= \exp(-\lambda t_2) \sinh \lambda t_1 \cosh \lambda(t_2 - t_1). \end{aligned} \qquad (5.48)$$

It is easy to show that these probabilities sum to 1.

The random telegraph signal is also a Markov process: with $t_1 < t_2 < t_3$,

$$\Pr[\mathbf{x}(t_3) = 1 \mid \mathbf{x}(t_1) = 1, \mathbf{x}(t_2) = 1]$$
$$= \Pr[\mathbf{x}(t_3) = 1 \mid \mathbf{x}(t_2) = 1] \qquad (5.49)$$
$$= \exp[-\lambda(t_3 - t_2)] \cosh \lambda(t_3 - t_2),$$

and so on. From this and similar equations the third-order joint distribution can be worked out, but we forbear to do so.

The mean value of this process is, by (5.44) and (5.45),

$$E[\mathbf{x}(t)] = 1 \cdot \Pr[\mathbf{x}(t) = 1] + (-1) \cdot \Pr[\mathbf{x}(t) = -1] \qquad (5.50)$$
$$= e^{-\lambda t}(\cosh \lambda t - \sinh \lambda t) = e^{-2\lambda t}.$$

This function resembles none of the sample functions of the stochastic process.

The correlation function is again most easily obtained from the conditional expected value by (5.25). For $t_2 > t_1$,

$$E[\mathbf{x}(t_2) \mid \mathbf{x}(t_1) = 1] = \exp[-2\lambda(t_2 - t_1)],$$

$$E[\mathbf{x}(t_2) \mid \mathbf{x}(t_1) = -1] = -\exp[-2\lambda(t_2 - t_1)],$$

by (5.46) and (5.47); that is,

$$E[\mathbf{x}(t_2) \mid \mathbf{x}(t_1) = x_1] = x_1 \exp[-2\lambda(t_2 - t_1)]. \tag{5.51}$$

Hence by (5.25),

$$\begin{aligned} R_{\mathbf{x}}(t_1, t_2) &= E\{[\mathbf{x}(t_1)]^2 \exp[-2\lambda(t_2 - t_1)]\} \\ &= \exp[-2\lambda(t_2 - t_1)], \qquad t_2 > t_1. \end{aligned} \tag{5.52}$$

The autocovariance function of this process is, therefore,

$$C_{\mathbf{x}}(t_1, t_2) = \exp[-2\lambda \mid t_2 - t_1 \mid] - \exp(-2\lambda t_1 - 2\lambda t_2), \tag{5.53}$$

the condition $t_2 > t_1$ having been dropped.

All these results bear the tacit condition $\mathbf{x}(0) = 1$. If the process had instead started at $t = -\infty$, its value at $t = 0$, or indeed at any other finite time, would just as likely be negative as positive. Let us denote such a process by $\mathbf{y}(t)$. Then at any time t,

$$\Pr[\mathbf{y}(t) = 1] = \Pr[\mathbf{y}(t) = -1] = \tfrac{1}{2}. \tag{5.54}$$

From (5.46) and (5.47), which hold for the new process as well, the second-order joint distribution of $\mathbf{y}(t)$ is, for $t_2 > t_1$,

$$\begin{aligned} &\Pr[\mathbf{y}(t_1) = 1, \mathbf{y}(t_2) = 1] = \Pr[\mathbf{y}(t_1) = -1, \mathbf{y}(t_2) = -1] \\ &\quad = \tfrac{1}{2}\exp[-\lambda(t_2 - t_1)] \cosh \lambda(t_2 - t_1) \\ &\Pr[\mathbf{y}(t_1) = 1, \mathbf{y}(t_2) = -1] = \Pr[\mathbf{y}(t_1) = -1, \mathbf{y}(t_2) = 1] \\ &\quad = \tfrac{1}{2}\exp[-\lambda(t_2 - t_1)] \sinh \lambda(t_2 - t_1). \end{aligned} \tag{5.55}$$

The process $\mathbf{y}(t)$ is stationary, as one can see by considering this distribution and those of higher order that would result from conditional probabilities such as (5.49), which hold for $\mathbf{y}(t)$ as well and depend only on intervals such as $t_3 - t_2$, never on the instants $t_1, t_2, \ldots$ by themselves. The process having started at $t = -\infty$ has lost all memory of its origin and distinguishes no particular point on the time axis.

The argument leading to (5.52) holds also for the process $\mathbf{y}(t)$, whose correlation function is therefore

$$R_{\mathbf{y}}(t_1, t_2) = \exp[-2\lambda(t_2 - t_1)].$$

Since $E[\mathbf{y}(t)] = 0$, this is also the autocovariance function of the stationary process $\mathbf{y}(t)$; that is,

$$C_{\mathbf{y}}(\tau) = e^{-2\lambda|\tau|}. \tag{5.56}$$

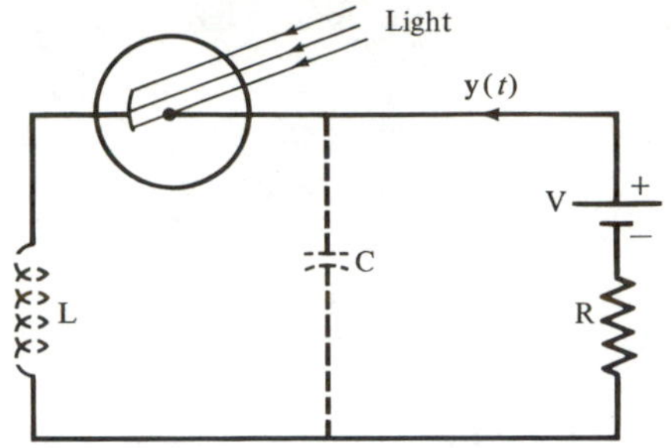

FIGURE 5.7 Photodetection circuit.

D. Shot Noise

In the circuit of Fig. 5.7 a current pulse passes every time a photoelectron is ejected by the light falling on the cathode. The shape $h(t)$ of such a pulse might look like that in Fig. 5.8; it depends on the resistance R in the circuit and on the amount of distributed interelectrode and interwiring inductance and capacitance. If photoelectrons are emitted at random times $\boldsymbol{\tau}_k$, $-\infty < k < \infty$, the total current in the circuit is

$$\mathbf{y}(t) = \sum_{k=-\infty}^{\infty} h(t - \boldsymbol{\tau}_k). \tag{5.57}$$

As before, the emission times τ_k of the photoelectrons form a Poisson point process, which we now suppose, however, to have been going on for all time with fixed rate λ. Because there is no distinguishing origin of time, the process $\mathbf{y}(t)$ is stationary. It is called *shot noise*.

Calculating even the first-order probability density function of $\mathbf{y}(t)$ is a most difficult mathematical problem.* The mean and the autocovariance function of $\mathbf{y}(t)$ are, however, within our present capabilities. We simplify the calculation by an artifice. We divide all time into equal intervals Δt that are so short that $\lambda\Delta t << 1$, and we define the random variable ε_m to be such that for all integers m

$$\varepsilon_m = \begin{cases} 0 & \text{if no electron is emitted in } m\,\Delta t < t < (m+1)\,\Delta t; \\ 1 & \text{if one electron is emitted in } m\,\Delta t < t < (m+1)\,\Delta t. \end{cases}$$

We neglect the probability

$$1 - (1 + \lambda\,\Delta t)e^{-\lambda\Delta t} = \tfrac{1}{2}(\lambda\,\Delta t)^2 + O(\Delta t)^3$$

that more than one electron would be emitted during so brief an interval. By the definition of the Poisson point process, the random variables ε_m and ε_k are statistically independent for $m \neq k$. The probability distribution of ε_m is

*O. C. Yue, R. Lugannani, and S. O. Rice, "Series approximations for the amplitude distribution and density of shot processes," *IEEE Transactions on Communications, COM-26,* 45–54, Jan. 1978.

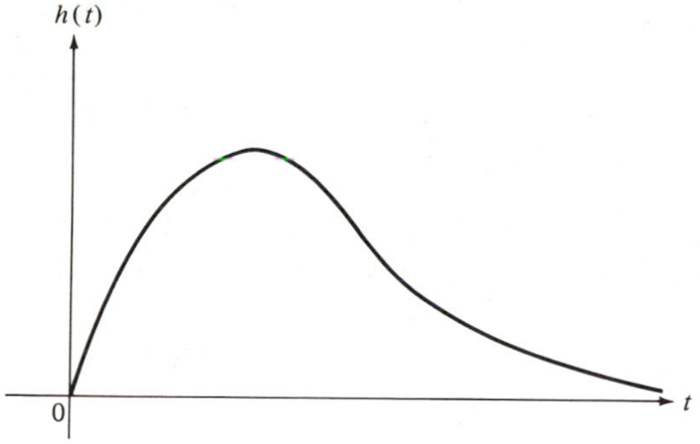

FIGURE 5.8 Current pulse.

$$\Pr(\varepsilon_m = 0) = e^{-\lambda \Delta t} \doteq 1 - \lambda\, \Delta t,$$

$$\Pr(\varepsilon_m = 1) = \lambda\, \Delta t\, e^{-\lambda \Delta t} \doteq \lambda\, \Delta t,$$

and its expected value is

$$E(\boldsymbol{\varepsilon}_m) = \Pr(\boldsymbol{\varepsilon}_m = 1) = \lambda\, \Delta t.$$

Now we approximate the process $\mathbf{y}(t)$ by the random process

$$\tilde{\mathbf{y}}(t) = \sum_{m=-\infty}^{\infty} \boldsymbol{\varepsilon}_m h(t - m\, \Delta t). \tag{5.58}$$

This new process, which allows at most one pulse to start in each interval Δt, becomes the more nearly identical to the shot-noise process $\mathbf{y}(t)$, the shorter the intervals Δt.

The expected value of $\tilde{\mathbf{y}}(t)$ is

$$\begin{aligned} E[\tilde{\mathbf{y}}(t)] &= \sum_{m=-\infty}^{\infty} E(\boldsymbol{\varepsilon}_m) h(t - m\Delta t) \\ &= \sum_{m=-\infty}^{\infty} \lambda\, \Delta t\, h(t - m\Delta t). \end{aligned} \tag{5.59}$$

Upon passing to the limit $\Delta t \to 0$, the summation becomes an integration, and we obtain

$$E[\mathbf{y}(t)] = \lambda \int_{-\infty}^{\infty} h(t - \tau)\, d\tau = \lambda \int_{-\infty}^{\infty} h(u)\, du \tag{5.60}$$

as the expected value of the shot-noise process $\mathbf{y}(t)$.

The autocovariance function of the shot-noise process can be approximated by that of the process $\tilde{\mathbf{y}}(t)$. Its correlation function, first, is

$$\begin{aligned} R_{\tilde{y}}(t_1, t_2) &= E[\tilde{\mathbf{y}}(t_1)\tilde{\mathbf{y}}(t_2)] \\ &= E \sum_{m=-\infty}^{\infty} \sum_{n=-\infty}^{\infty} \boldsymbol{\varepsilon}_m \boldsymbol{\varepsilon}_n\, h(t_1 - m\Delta t)\, h(t_2 - n\Delta t). \end{aligned}$$

Now

$$E(\boldsymbol{\varepsilon}_m \boldsymbol{\varepsilon}_n) = \begin{cases} E(\boldsymbol{\varepsilon}_m^2) = E(\boldsymbol{\varepsilon}_m) = \lambda\, \Delta t, & m = n, \\ E(\boldsymbol{\varepsilon}_m)E(\boldsymbol{\varepsilon}_n) = (\lambda\, \Delta t)^2, & m \neq n, \end{cases}$$

because of the independence of the random variables $\boldsymbol{\varepsilon}_m$ and $\boldsymbol{\varepsilon}_n$. Subtracting the product of the means, as given by (5.59), we find for the autocovariance function of the approximating random process $\tilde{\mathbf{y}}(t)$,

$$C_{\tilde{\mathbf{y}}}(t_1, t_2) = \sum_{m=-\infty}^{\infty} [\lambda\, \Delta t - (\lambda\, \Delta t)^2] h(t_1 - m\Delta t) h(t_2 - m\Delta t),$$

and when $\Delta t \to 0$ this becomes the autocovariance function of the shot-noise process,

$$C_{\mathbf{y}}(t_1, t_2) = \lambda \int_{-\infty}^{\infty} h(t_1 - \tau) h(t_2 - \tau)\, d\tau$$

$$= \lambda \int_{-\infty}^{\infty} h(u) h(t_2 - t_1 + u)\, du.$$

Since the process $\mathbf{y}(t)$ is stationary, we can write its autovariance function as

$$C_{\mathbf{y}}(\tau) = \lambda \int_{-\infty}^{\infty} h(u) h(\tau + u)\, du, \tag{5.61}$$

a result known as *Campbell's theorem.*

If the shot pulses are rectangles of height A and duration T, for instance, the autocovariance function is

$$C_{\mathbf{y}}(\tau) = \begin{cases} \lambda A^2 (T - |\tau|), & |\tau| < T, \\ 0, & |\tau| > T, \end{cases} \tag{5.62}$$

which resembles that in Fig. 5.4. If they are exponential functions,

$$h(t) = Ae^{-\mu t} U(t),$$

the autocovariance function is

$$C_{\mathbf{y}}(\tau) = \lambda \left(\frac{A^2}{2\mu} \right) e^{-\mu|\tau|}, \tag{5.63}$$

which is depicted in Fig. 5.9.

The first-order characteristic function of the shot-noise process can also be calculated in terms of that of the approximating process $\tilde{\mathbf{y}}(t)$:

$$\begin{aligned} \Phi_{\tilde{\mathbf{y}}}(\omega) &= E\{\exp [j\omega \tilde{\mathbf{y}}(t)]\} \\ &= E\{\exp [j\omega \sum_{m=-\infty}^{\infty} \boldsymbol{\varepsilon}_m h(t - m\Delta t)]\} \\ &= \prod_{m=-\infty}^{\infty} E\{\exp [j\omega \boldsymbol{\varepsilon}_m h(t - m\Delta t)]\} \end{aligned} \tag{5.64}$$

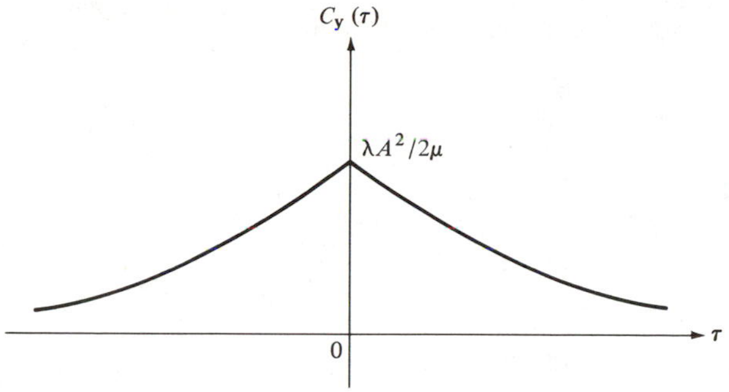

FIGURE 5.9 Autocovariance function, exponential shot-noise pulses.

because the random variables $\boldsymbol{\varepsilon}_m$ are independent. Continuing,

$$\begin{aligned}\Phi_{\bar{\mathbf{y}}}(\omega) &= \prod_{m=-\infty}^{\infty} \{\Pr(\boldsymbol{\varepsilon}_m = 0)\cdot 1 \\ &\qquad + \Pr(\boldsymbol{\varepsilon}_m = 1)\exp[j\omega h(t - m\Delta t)]\} \\ &= \prod_{m=-\infty}^{\infty} \{(1 - \lambda\Delta t) + \lambda\Delta t \exp[j\omega h(t - m\Delta t)]\} \\ &= \prod_{m=-\infty}^{\infty} (1 + \lambda\Delta t\{\exp[j\omega h(t - m\Delta t)] - 1\}) \\ &\doteq \prod_{m=-\infty}^{\infty} \exp(\lambda\Delta t\{\exp[j\omega h(t - m\Delta t)] - 1\}) \\ &= \exp\left(\sum_{m=-\infty}^{\infty} \lambda\Delta t\{\exp[j\omega h(t - m\Delta t)] - 1\}\right).\end{aligned} \tag{5.65}$$

Now letting Δt go to zero, we convert the sum to an integral and obtain the characteristic function of the shot-noise process $\mathbf{y}(t)$,

$$\begin{aligned}\Phi_{\mathbf{y}}(\omega) &= \exp\left\{\lambda\int_{-\infty}^{\infty} [e^{j\omega h(t-\tau)} - 1]\,d\tau\right\} \\ &= \exp\left\{\lambda\int_{-\infty}^{\infty} [e^{j\omega h(u)} - 1]\,du\right\}.\end{aligned} \tag{5.66}$$

The first-order probability density function of $\mathbf{y} = \mathbf{y}(t)$ is the inverse Fourier transform

$$f_{\mathbf{y}}(y, t) = \int_{-\infty}^{\infty} \Phi_{\mathbf{y}}(\omega)e^{-j\omega y}\,\frac{d\omega}{2\pi}, \tag{5.67}$$

but evaluating it is in general a formidable mathematical task, and one must resort to various approximation techniques.

An exception is when the pulses are rectangular,

$$h(t) = \begin{cases} 1, & 0 < t < T, \\ 0, & t < 0, \quad t > T, \end{cases}$$

whereupon

$$\Phi_{\mathbf{y}}(\omega) = \exp\,[\lambda T(e^{j\omega} - 1)],$$

and comparison with (2.102) with $z = e^{j\omega}$ shows that $\mathbf{y} = \mathbf{y}(t)$ in this case has a Poisson distribution with mean λT,

$$\Pr\,[\mathbf{y}(t) = k] = \frac{(\lambda T)^k e^{-\lambda T}}{k!}, \tag{5.68}$$

as is to be expected, for this is the probability that k shot pulses started during the T seconds before time t, and only those contribute to the value of $\mathbf{y}(t)$ at time t.

E. Random Pulses of Random Amplitude

When an airplane flying over the ocean sends out a radar pulse, it receives a multitude of echoes, which are Doppler-shifted copies of its transmitted pulse reflected from numerous points on the surface of the water. Those points reflect the pulse with random amplitudes and delays that, at least with a turbulent, windblown surface, can be considered statistically independent. The echoes constitute a random signal known as *clutter* and taking the form

$$\mathbf{x}(t) = \sum_{k=-\infty}^{\infty} \mathbf{A}_k h(t - \boldsymbol{\tau}_k), \tag{5.69}$$

in which $h(t)$ represents the shape of the reflected radar pulse, $\boldsymbol{\tau}_k$ the time when the kth pulse reaches the receiver, and $\mathbf{A}_k$ its amplitude. We study the statistical properties of this stochastic process, assuming that it lasts from $-\infty$ to ∞, that the times $\boldsymbol{\tau}_k$ constitute a Poisson point process, and that the times $\boldsymbol{\tau}_k$ and the amplitudes $\mathbf{A}_k$ are all statistically independent. We denote the probability density function of the independent pulse amplitudes by $f_{\mathbf{A}}(A)$. This *random-pulse process* is fundamental to our definition of random noise in Sec. 5-6.

As in part D we divide time into intervals Δt so short that $\lambda\,\Delta t << 1$, where λ is the rate of arrival of the pulses. Then we can approximate $\mathbf{x}(t)$ by the random process

$$\tilde{\mathbf{x}} = \sum_{m=-\infty}^{\infty} \boldsymbol{\varepsilon}_m \mathbf{A}_m h(t - m\Delta t), \tag{5.70}$$

where the random variables $\boldsymbol{\varepsilon}_m$ have the same meaning and the same probabilities as in part D. Thus the expected value of the approximating process $\tilde{\mathbf{x}}(t)$ is

$$E[\tilde{\mathbf{x}}(t)] = \sum_{m=-\infty}^{\infty} \lambda\Delta t\, E(\mathbf{A}) h(t - m\Delta t),$$

and passing to the limit $\Delta t \to 0$, we find

$$E[\mathbf{x}(t)] = \lambda E(\mathbf{A})\int_{-\infty}^{\infty} h(t - \tau)\, d\tau = \lambda E(\mathbf{A})\int_{-\infty}^{\infty} h(u)\, du \qquad (5.71)$$

as the expected value of the clutter process $\mathbf{x}(t)$. For high-frequency radar signals, the integral equals zero.

The calculation of the autocovariance function of this stochastic process is much like that leading to (5.61). The correlation function of the approximating process is

$$\begin{aligned} R_{\mathbf{x}}(t_1, t_2) &= \sum_{m=-\infty}^{\infty} \sum_{n=-\infty}^{\infty} E(\boldsymbol{\varepsilon}_m \boldsymbol{\varepsilon}_n) E(\mathbf{A}_m \mathbf{A}_n) h(t_1 - m\Delta t) h(t_2 - n\Delta t) \\ &= E[\tilde{\mathbf{x}}(t_1)]E[\tilde{\mathbf{x}}(t_2)] \\ &\quad + \sum_{m=-\infty}^{\infty} \sum_{n=-\infty}^{\infty} [E(\boldsymbol{\varepsilon}_m \boldsymbol{\varepsilon}_n)E(\mathbf{A}_m \mathbf{A}_n) - E(\boldsymbol{\varepsilon}_m)E(\boldsymbol{\varepsilon}_n)E(\mathbf{A}_m)E(\mathbf{A}_n)] \\ &\qquad \times h(t_1 - m\Delta t)h(t_2 - n\Delta t). \end{aligned}$$

Since for $m \neq n$ all the variables are independent, we find that all the terms with $m \neq n$ drop out. The remainder are evaluated as in part D, and

$$R_{\mathbf{x}}(t_1, t_2) = E[\tilde{\mathbf{x}}(t_1)]E[\tilde{\mathbf{x}}(t_2)]$$
$$+ \sum_{m=-\infty}^{\infty} [\lambda \Delta t\, E(\mathbf{A}^2) - (\lambda \Delta t)^2 [E(\mathbf{A})]^2] h(t_1 - m\Delta t) h(t_2 - m\Delta t).$$

Subtracting the product of the expected values and passing to the limit $\Delta t \to 0$, we obtain for the autocovariance of the clutter process $\mathbf{x}(t)$,

$$C_{\mathbf{x}}(t_1, t_2) = \lambda E(\mathbf{A}^2)\int_{-\infty}^{\infty} h(t_1 - \tau)h(t_2 - \tau)\, d\tau.$$

Since $\mathbf{x}(t)$ is a stationary process, we can write its autocovariance function as

$$C_{\mathbf{x}}(\tau) = \lambda E(\mathbf{A}^2)\int_{-\infty}^{\infty} h(u)h(\tau + u)\, du. \qquad (5.72)$$

It differs from that in (5.61) only by the factor $E(\mathbf{A}^2)$, the expected value of the square of the pulse amplitudes.

We shall return to this random-pulse process later, taking $E(\mathbf{A}) = 0$ and considering the limit in which the pulses arrive at an enormous rate, $\lambda >> 1$, but with such infinitesimal amplitudes, $E(\mathbf{A}^2) << 1$, that the product $\lambda E(\mathbf{A}^2)$ remains at some finite value proportional to the average power in the process. The resulting process serves as a model of random noise.

5-4 TRANSFORMATIONS OF STOCHASTIC PROCESSES

The signals and noise picked up by a radio or radar receiver are filtered and subjected to various nonlinear operations such as mixing and rectification. The stochastic process $\mathbf{x}(t)$ comprising the input signals and noise is transformed by the receiver into an output stochastic process $\mathbf{y}(t)$. The designer and analyst would like to determine the statistical properties of the output $\mathbf{y}(t)$ from those of the input $\mathbf{x}(t)$. In general, such problems are very difficult, and only limited results can be obtained. We consider two special types of transformation for which the calculations are relatively simple: (a) transformation by a memoryless nonlinearity and (b) transformation by a linear filter.

A. Memoryless Nonlinear Transformations

If the output $\mathbf{y}(t)$ of a device depends at each time t on the input $\mathbf{x}(t)$ only at the same time t, the device is said to effect a *memoryless* transformation of $\mathbf{x}(t)$. The relation between the output and the input can be written

$$\mathbf{y}(t) = g(\mathbf{x}(t)) \tag{5.73}$$

for some function $y = g(x)$ assigning a unique value of y to each value of x. If input and output are sampled at times $t_1, t_2, \ldots, t_n$ to yield samples $\mathbf{x}_1, \mathbf{x}_2, \ldots, \mathbf{x}_n$ and $\mathbf{y}_1, \mathbf{y}_2, \ldots, \mathbf{y}_n$, then

$$\mathbf{y}_k = g(\mathbf{x}_k), \qquad k = 1, 2, \ldots, n, \tag{5.74}$$

specifies a transformation of the random variables $\mathbf{x}_k$ into the random variables $\mathbf{y}_k$, $1 \le k \le n$, and it falls into the category of transformations treated in Sec. 4-3B. The Jacobian of this transformation is especially simple,

$$J\begin{pmatrix} y_1 & y_2 & \cdots & y_n \\ x_1 & x_2 & \cdots & x_n \end{pmatrix} = g'(x_1)g'(x_2) \cdots g'(x_n), \tag{5.75}$$

where the primes indicate differentiation. When the transformation $y = g(x)$ is monotone, the joint probability density function of the samples of the output is

$$f_{\mathbf{y}}(y_1, t_1; y_2, t_2; \cdots; y_n, t_n) = \frac{f_{\mathbf{x}}(x_1, t_1; x_2, t_2; \ldots; x_n, t_n)}{|g'(x_1)g'(x_2) \cdots g'(x_n)|}, \tag{5.76}$$

into which one must substitute the solutions

$$x_k = g^{-1}(y_k), \qquad 1 \le k \le n, \tag{5.77}$$

of (5.74) for each x_k in terms of y_k. If $y = g(x)$ has several roots x when solved, the right-hand side of (5.76) must be summed over those roots, as in (2.69) and (3.71).

EXAMPLE 5.10. Quadratic Rectifier

The stationary process $\mathbf{x}(t)$ is passed through a quadratic rectifier, whose output is

$$\mathbf{y}(t) = [\mathbf{x}(t)]^2.$$

We assume that the joint probability density functions of samples of the input process have the Gaussian form, as in (4.134). Then, for instance, the first-order density function for the input process is

$$f_{\mathbf{x}}(x, t) = (2\pi\sigma_x^2)^{-1/2} \exp\left[-\frac{(x-m)^2}{2\sigma_x^2}\right], \tag{5.78}$$

where $\sigma_x^2 = \text{Var}\,\mathbf{x}(t) = C_{\mathbf{x}}(0)$ and $m = E[\mathbf{x}(t)]$, and the first-order probability density function of the output is, by the method of Sec. 2-3,

$$\begin{aligned} f_{\mathbf{y}}(y, t) = \tfrac{1}{2}(2\pi\sigma_x^2 y)^{-1/2} \bigg\{ &\exp\left[-\frac{(y^{1/2}-m)^2}{2\sigma_x^2}\right] \\ &+ \exp\left[-\frac{(y^{1/2}+m)^2}{2\sigma_x^2}\right]\bigg\} U(y). \end{aligned} \tag{5.79}$$

Referring to (3.40), we write the second-order joint probability density function of samples of the input at times t_1 and t_2 as

$$\begin{aligned} f_{\mathbf{x}}(x_1, t_1; x_2, t_2) &= (2\pi\sigma_x^2)^{-1}(1-r^2)^{-1/2} \\ &\times \exp\left[-\frac{(x_1-m)^2 - 2r(x_1-m)(x_2-m) + (x_2-m)^2}{2\sigma_x^2(1-r^2)}\right], \end{aligned} \tag{5.80}$$

where

$$r = \frac{C_{\mathbf{x}}(t_2 - t_1)}{C_{\mathbf{x}}(0)}$$

is the correlation coefficient of the samples $\mathbf{x}_1 = \mathbf{x}(t_1)$ and $\mathbf{x}_2 = \mathbf{x}(t_2)$. Then the second-order density function of the output is

$$\begin{aligned} f_{\mathbf{y}}(y_1, t_1; y_2, t_2) &= \tfrac{1}{4}(2\pi\sigma_x^2)^{-1}[(1-r^2)y_1y_2]^{-1/2} \\ &\times \bigg\{\exp\left[-\frac{(y_1^{1/2}-m)^2 - 2r(y_1^{1/2}-m)(y_2^{1/2}-m) + (y_2^{1/2}-m)^2}{2\sigma_x^2(1-r^2)}\right] \\ &+ \exp\left[-\frac{(y_1^{1/2}-m)^2 + 2r(y_1^{1/2}-m)(y_2^{1/2}+m) + (y_2^{1/2}+m)^2}{2\sigma_x^2(1-r^2)}\right] \\ &+ \exp\left[-\frac{(y_1^{1/2}+m)^2 + 2r(y_1^{1/2}+m)(y_2^{1/2}-m) + (y_2^{1/2}-m)^2}{2\sigma_x^2(1-r^2)}\right] \\ &+ \exp\left[-\frac{(y_1^{1/2}+m)^2 - 2r(y_1^{1/2}+m)(y_2^{1/2}+m) + (y_2^{1/2}+m)^2}{2\sigma_x^2(1-r^2)}\right]\bigg\} \\ &\times U(y_1)U(y_2). \end{aligned} \tag{5.81}$$

Joint density functions of higher order for samples of the output can be determined from those of the input in the same way. ■

The mean, the autocovariance function, and whatever other moments of the output $\mathbf{y}(t)$ are needed can be calculated from the joint density functions of the input process $\mathbf{x}(t)$ as in (4.82) without first determining the joint density functions for the output process $\mathbf{y}(t)$. Thus the mean value of the output is

$$\begin{aligned} \overline{y(t)} &= E[\mathbf{y}(t)] = E[g(\mathbf{x}(t))] \\ &= \int_{-\infty}^{\infty} g(x) f_{\mathbf{x}}(x, t)\, dx, \end{aligned} \tag{5.82}$$

its correlation function is

$$R_{\mathbf{y}}(t_1, t_2) = E[g(\mathbf{x}(t_1))g(\mathbf{x}(t_2))]$$
$$= \int_{-\infty}^{\infty}\int_{-\infty}^{\infty} g(x_1)g(x_2)f_{\mathbf{x}}(x_1, t_1; x_2, t_2)\,dx_1\,dx_2, \tag{5.83}$$

and its autocovariance function is

$$C_{\mathbf{y}}(t_1, t_2) = R_{\mathbf{y}}(t_1, t_2) - E[\mathbf{y}(t_1)]E[\mathbf{y}(t_2)]. \tag{5.84}$$

EXAMPLE 5.10. Quadratic Rectifier (continued)

For simplicity we write

$$\mathbf{x}(t) = \mathbf{z}(t) + m,$$

where $\mathbf{z}(t)$ is a stationary Gaussian random process of mean zero. Then the output of the quadratic rectifier is

$$\mathbf{y}(t) = [\mathbf{z}(t)]^2 + 2m\mathbf{z}(t) + m^2. \tag{5.85}$$

Its expected value is

$$E[\mathbf{y}(t)] = E(\mathbf{z}^2) + m^2 = \sigma_x^2 + m^2 \tag{5.86}$$

since $E(\mathbf{z}^2) = \text{Var}\,\mathbf{z} = \text{Var}\,\mathbf{x} = \sigma_x^2$.

The autocovariance function of the output $\mathbf{y}(t) = [\mathbf{x}(t)]^2$ is the same as the correlation function of

$$\mathbf{y}^1(t) = \mathbf{y}(t) - E[\mathbf{y}(t)]$$
$$= [\mathbf{z}(t)]^2 + 2m\mathbf{z}(t) - \sigma_x^2,$$

whose expected value is zero, that is,

$$C_{\mathbf{y}}(t_1, t_2) = E\left\{\left[[\mathbf{z}(t_1)]^2 + 2m\mathbf{z}(t_1) - \sigma_x^2\right]\left[[\mathbf{z}(t_2)]^2 + 2m\mathbf{z}(t_2) - \sigma_x^2\right]\right\}$$
$$= E\left\{[\mathbf{z}(t_1)]^2[\mathbf{z}(t_2)]^2 - \sigma_x^2[\mathbf{z}(t_1)]^2 - \sigma_x^2[\mathbf{z}(t_2)]^2 + 4m^2\mathbf{z}(t_1)\mathbf{z}(t_2) + \sigma_x^4\right\},$$

from which we have dropped terms with an odd number of factors of $\mathbf{z}$, whose expected values are zero. Using (4.147) we obtain

$$C_{\mathbf{y}}(t_1, t_2) = E\{[\mathbf{z}(t_1)]^2\}E\{[\mathbf{z}(t_2)]^2\} + 2\{E[\mathbf{z}(t_1)\mathbf{z}(t_2)]\}^2$$
$$- 2\sigma_x^2\,\text{Var}\,\mathbf{z} + 4m^2E[\mathbf{z}(t_1)\mathbf{z}(t_2)] + \sigma_x^4$$
$$= 2[C_{\mathbf{x}}(t_1, t_2)]^2 + 4m^2C_{\mathbf{x}}(t_1, t_2).$$

Since we assumed the input process $\mathbf{x}(t)$ stationary, the output $\mathbf{y}(t)$ is also stationary, and its autocovariance function is

$$C_{\mathbf{y}}(\tau) = 2[C_{\mathbf{x}}(\tau)]^2 + 4m^2C_{\mathbf{x}}(\tau). \tag{5.87}$$

■

Memoryless nonlinear transformations of stochastic processes can be handled by applying to the samples $\mathbf{y}_k = \mathbf{y}(t_k) = g(\mathbf{x}_k)$ of the output process the methods of Secs. 2-3 and 2-4, Secs. 3-4 and 3-6, and Secs. 4-3 and 4-4. Nothing more needs to be said about those methods in this context.

B. Linear Filtering

Arbitrary time-variable linear filtering of the stochastic process $\mathbf{x}(t)$ can be expressed in the form

$$\mathbf{y}(t) = \int_{-\infty}^{\infty} k(t, s)\mathbf{x}(s)\, ds \tag{5.88}$$

with a suitable kernel $k(t, s)$. If the input were a delta function $\delta(t - s)$ entering at time s, the output of the filter would be $k(t, s)$; the kernel $k(t, s)$ is therefore the impulse response of the filter. The output of a time-invariant linear filter whose impulse response is $h(\tau)$ is

$$\mathbf{y}(t) = \int_{-\infty}^{\infty} h(\tau)\mathbf{x}(t - \tau)\, d\tau, \tag{5.89}$$

and this can be put into the form (5.88) if we take

$$k(t, s) = h(t - s). \tag{5.90}$$

For a causal filter, that is, one whose output at time t depends only on the input $\mathbf{x}(s)$ for times s earlier than t,

$$k(t, s) \equiv 0, \qquad s > t. \tag{5.91}$$

We treat the general form (5.88), confident that readers can work out special cases for themselves by either following the same pattern or simply substituting the appropriate kernel $k(t, s)$ into our results.

An integral like those in (5.88) and (5.89) is a weighted sum of infinitely dense samples of $\mathbf{x}(s)$, and the expected value of a sum equals the sum of the expected values of the components, as in (4.85), whether those components are statistically independent or not. Hence the expected value of the output $\mathbf{y}(t)$ of the linear filter specified by (5.88) is

$$\begin{aligned} E[\mathbf{y}(t)] &= E\int_{-\infty}^{\infty} k(t, s)\mathbf{x}(s)\, ds = \int_{-\infty}^{\infty} k(t, s)E[\mathbf{x}(s)]\, ds \\ &= \int_{-\infty}^{\infty} k(t, s)m(s)\, ds \end{aligned} \tag{5.92}$$

in terms of the expected value $m(t) = E[\mathbf{x}(t)]$ of the input. If we think of the input $\mathbf{x}(t)$ as the sum of a "deterministic" signal $m(t)$ and random noise of mean value zero, the signal and the noise are transformed independently because of the linearity of the filter, and (5.92) represents the signal component of the output.

The cross-covariance function of input and output, defined by (5.34), is, in terms of the autocovariance function $C_{\mathbf{x}}(t_1, t_2)$ of the input,

$$\begin{aligned} C_{\mathbf{xy}}(t_1, t_2) &= E\{[\mathbf{x}(t_1) - m(t_1)] \int_{-\infty}^{\infty} k(t_2, s)[\mathbf{x}(s) - m(s)]\, ds\} \\ &= \int_{-\infty}^{\infty} k(t_2, s)E\{[\mathbf{x}(t_1) - m(t_1)][\mathbf{x}(s) - m(s)]\}\, ds \\ &= \int_{-\infty}^{\infty} k(t_2, s)C_{\mathbf{x}}(t_1, s)\, ds, \end{aligned} \tag{5.93}$$

again by the principle that the expected value of a weighted sum (or integral) of random variables equals the same weighted sum (or integral) of their expected values.

Finally, the autocovariance function of the output process $\mathbf{y}(t)$ is

$$\begin{aligned} C_{\mathbf{y}}(t_1, t_2) &= E\left\{\left[\mathbf{y}(t_1) - E[\mathbf{y}(t_1)]\right]\left[\mathbf{y}(t_2) - E[\mathbf{y}(t_2)]\right]\right\} \\ &= E\left\{\int_{-\infty}^{\infty} k(t_1, s)[\mathbf{x}(s) - m(s)]\, ds \,\{\mathbf{y}(t_2) - E[\mathbf{y}(t_2)]\}\right\} \\ &= \int_{-\infty}^{\infty} k(t_1, s)C_{\mathbf{xy}}(s, t_2)\, ds \\ &= \int_{-\infty}^{\infty}\int_{-\infty}^{\infty} k(t_1, s_1)C_{\mathbf{x}}(s_1, s_2)k(t_2, s_2)\, ds_1\, ds_2. \end{aligned} \tag{5.94}$$

In calculating the autocovariance of the output of a linear filter, we can take the signal component $E[\mathbf{x}(t)]$ of the input equal to zero without altering the result, but often sparing ourselves much writing.

Joint moments of higher order of the output process $\mathbf{y}(t)$ can be calculated by an obvious continuation of this procedure, but they are seldom needed.

To calculate the joint probability density functions of samples of the output $\mathbf{y}(t)$ of a linear filter from the density functions characterizing the input is generally most difficult, for (5.88) shows that the value of $\mathbf{y}(t)$ at even a single time t depends on the values of the input $\mathbf{x}(s)$ at an infinitude of densely sampled times s over $(-\infty, \infty)$. The only kind of stochastic process for which this calculation is simple is the Gaussian process, which will be treated in detail in Sec. 5-6. The joint probability density function of any number of samples of a Gaussian process has the multivariate Gaussian form in (4.134). Under this circumstance the output $\mathbf{y}(t)$ in (5.88) is a linear combination of Gaussian random variables, and from what was said in Sec. 4-5, it must itself be a Gaussian random variable. As we shall see in Sec. 5-6, the output of a linear filter whose input is a Gaussian stochastic process is also a Gaussian stochastic process.

EXAMPLE 5.11. Perfect Integrator

The output of a "perfect integrator" whose input is a random process $\mathbf{x}(t)$ is

$$\mathbf{y}(t) = \frac{1}{T}\int_{t-T}^{t} \mathbf{x}(s)\, ds; \tag{5.95}$$

the integrator sums over the past T seconds of its input, forgetting the values of the input that are older than T seconds. Let us suppose that the input $\mathbf{x}(t)$ is a stationary process with mean m and covariance function $C_{\mathbf{x}}(\tau)$. The equivalent impulse response of this filter is

$$k(t, s) = \begin{cases} T^{-1}, & t - T < s < t, \\ 0, & s < t - T, \quad s > t. \end{cases}$$

The mean value of the output, either by direct inspection of (5.95) or by (5.92), is

$$E[\mathbf{y}(t)] = \frac{1}{T}\int_{t-T}^{t} m\, ds = m. \tag{5.96}$$

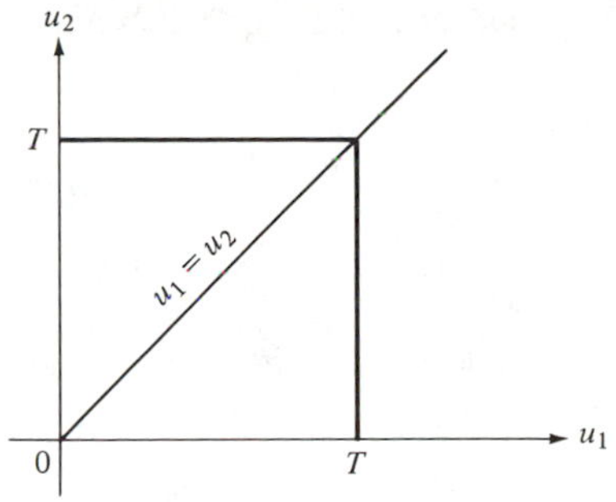

FIGURE 5.10 First region of integration.

The filter can be thought of as producing an unbiased estimate of the mean m of its input $\mathbf{x}(t)$ when its output is observed at an arbitrary time t.

The autocovariance function of the output of the perfect integrator is, by (5.94),

$$\begin{aligned} C_{\mathbf{y}}(t_1, t_2) &= T^{-2}\int_{t_1-T}^{t_1}\int_{t_2-T}^{t_2} C_{\mathbf{x}}(s_2 - s_1)\, ds_1\, ds_2 \\ &= T^{-2}\int_0^T\int_0^T C_{\mathbf{x}}(t_2 - t_1 - u_2 + u_1)\, du_1\, du_2. \end{aligned} \tag{5.97}$$

Because the input $\mathbf{x}(t)$ is stationary, and because this filter is time invariant and has been processing the input forever, the output $\mathbf{y}(t)$ is a stationary random process, and its autocovariance function can be written

$$C_{\mathbf{y}}(\tau) = T^{-2}\int_0^T\int_0^T C_{\mathbf{x}}(\tau - u_2 + u_1)\, du_1\, du_2. \tag{5.98}$$

In particular, the variance of the output is

$$\operatorname{Var} \mathbf{y}(t) = C_{\mathbf{y}}(0) = T^{-2}\int_0^T\int_0^T C_{\mathbf{x}}(u_1 - u_2)\, du_1\, du_2. \tag{5.99}$$

Such double integrals of an even function over a square often occur, and we shall therefore show how to reduce this integral to a single integral. Figure 5.10 depicts the square in the (u_1, u_2)-plane over which the integration is carried.

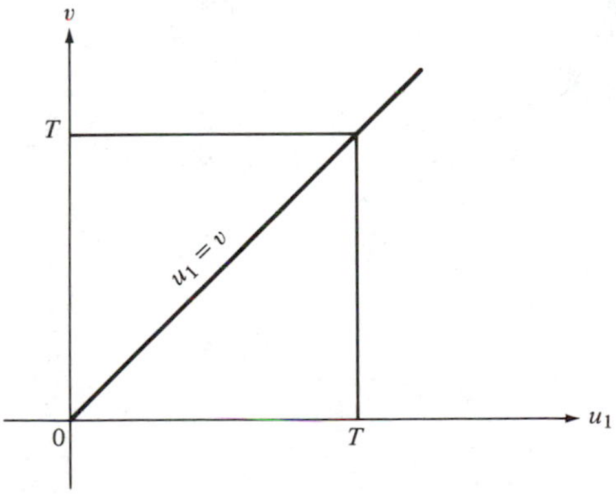

FIGURE 5.11 Second region of integration.

Because $C_{\mathbf{x}}(u_1 - u_2) = C_{\mathbf{x}}(u_2 - u_1)$, the halves of the square above and below the diagonal $(u_1 = u_2)$ contribute equally. Hence

$$\text{Var } \mathbf{y}(t) = 2T^{-2}\int_0^T du_1 \int_0^{u_1} du_2 \, C_{\mathbf{x}}(u_1 - u_2)$$

$$= 2T^{-2}\int_0^T du_1 \int_0^{u_1} dv \, C_{\mathbf{x}}(v)$$

when we put $v = u_1 - u_2$. Looking at the square in the (u_1, v)-plane shown in Fig. 5.11, we see that this integral, which is carried over the lower-right triangle, equals

$$\begin{aligned}
\text{Var } \mathbf{y}(t) &= 2T^{-2}\int_0^T dv \int_v^T du_1 \, C_{\mathbf{x}}(v) \\
&= 2T^{-2}\int_0^T (T - v)C_{\mathbf{x}}(v) \, dv \\
&= T^{-2} \int_0^T (T - |v|)[C_{\mathbf{x}}(v) + C_{\mathbf{x}}(-v)] \, dv \\
&= T^{-1} \int_{-T}^T \left(1 - \frac{|v|}{T}\right) C_{\mathbf{x}}(v) \, dv
\end{aligned} \tag{5.100}$$

when we again use the symmetry $C_{\mathbf{x}}(v) = C_{\mathbf{x}}(-v)$.

A typical autocovariance function $C_{\mathbf{x}}(v)$ of a stationary process looks like that in Fig. 5.12, upon which we have superimposed the triangular function $(1 - |v|/T)$ that multiplies it in the integrand of (5.100). Because the absolute value of an integral is at most equal to the integral of the absolute value of the integrand, we find for the ratio of the mean-square fluctuation at the output of the integrator to that of the input

$$\frac{\text{Var } \mathbf{y}(t)}{\text{Var } \mathbf{x}(t)} \le \int_{-T}^{T} \left| \frac{C_{\mathbf{x}}(v)}{C_{\mathbf{x}}(0)} \right| \frac{dv}{T} \le \int_{-\infty}^{\infty} \left| \frac{C_{\mathbf{x}}(v)}{C_{\mathbf{x}}(0)} \right| \frac{dv}{T}. \tag{5.101}$$

This bound on the ratio will be of the order of magnitude of τ_c/T, where τ_c, the *correlation time* of the input, is roughly the width of its autocovariance function $C_{\mathbf{x}}(v)$, as indicated in the figure. Samples of the process $\mathbf{x}(t)$ separated by much more than τ_c are nearly uncorrelated. If, for instance, the autocovariance function of the input is

$$C_{\mathbf{x}}(v) = C_{\mathbf{x}}(0)e^{-\mu|v|}, \tag{5.102}$$

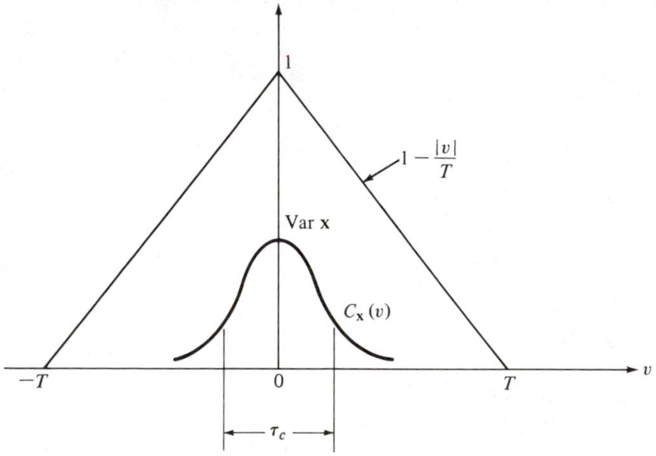

FIGURE 5.12 Factors in integrand of (5.100).

as in Fig. 5.9, $\tau_c \sim 1/\mu$, and

$$\frac{\text{Var } \mathbf{y}(t)}{\text{Var } \mathbf{x}(t)} \le \frac{2}{\mu T}. \tag{5.103}$$

The longer the time T over which the input is integrated, compared with its correlation time τ_c, the smaller are the fluctuations of the output $\mathbf{y}(t)$ about its mean value, relative to the fluctuations of the input. This principle is often useful. ■

C. Linear Differential Equations with Random Forcing Terms

Suppose that the input to a linear, time-invariant filter is a random process $\mathbf{x}(t)$, and that the output $\mathbf{y}(t)$ is related to the input through the linear differential equation with constant coefficients,

$$\begin{aligned} a_0 \frac{d^n\mathbf{y}}{dt^n} + a_1 \frac{d^{n-1}\mathbf{y}}{dt^{n-1}} + \cdots + a_{n-1}\frac{d\mathbf{y}}{dt} + a_n\mathbf{y} \\ = b_0 \frac{d^m\mathbf{x}}{dt^m} + b_1 \frac{d^{m-1}\mathbf{x}}{dt^{m-1}} + \cdots + b_{m-1}\frac{d\mathbf{x}}{dt} + b_m\mathbf{x}, \qquad m \le n, \end{aligned} \tag{5.104}$$

which characterizes the linear system. The input is applied from time $t = 0$, and at time $t = 0$ the output and its first $n - 1$ derivatives take on the n initial values

$$\begin{aligned} \mathbf{y}(0) = c_0, \quad \left.\frac{d\mathbf{y}}{dt}\right|_{t=0} = c_1, \quad \ldots, \\ \left.\frac{d^k\mathbf{y}}{dt^k}\right|_{t=0} = c_k, \quad \ldots, \quad \left.\frac{d^{n-1}\mathbf{y}}{dt^{n-1}}\right|_{t=0} = c_{n-1}. \end{aligned} \tag{5.105}$$

Equations of this type are called *Langevin equations*.

We seek the mean value and the autocovariance function of the output for $t > 0$. If the input $\mathbf{x}(t)$ is a Gaussian random process, then so is the output $\mathbf{y}(t)$, and these suffice, as we shall see, to determine its probability measure. Otherwise finding the probability density functions of samples of the output is generally an insurmountable problem, but as many moment functions can be calculated as necessary, provided that those of $\mathbf{x}(t)$ are known.

Because (5.104) is linear, we can apply the principle of superposition. We put

$$\mathbf{x}(t) = m_x(t) + \mathbf{x}^1(t), \tag{5.106}$$

with $m_x(t) = E[\mathbf{x}(t)]$, so that the process $\mathbf{x}^1(t)$ has mean zero. Thus $m_x(t)$ represents the signal component of the input and $\mathbf{x}^1(t)$ the noise. Let $m_y(t)$ be the solution of the linear differential equation

$$\begin{aligned} a_0 \frac{d^n m_y}{dt^n} + a_1 \frac{d^{n-1}m_y}{dt^{n-1}} + \cdots + a_n m_y \\ = b_0 \frac{d^m m_x}{dt^m} + b_1 \frac{d^{m-1}m_x}{dt^{m-1}} + \cdots + b_m m_x, \end{aligned} \tag{5.107}$$

with the same initial conditions,

$$m_y(0) = c_0, \quad \left.\frac{dm_y}{dt}\right|_{t=0} = c_1, \quad \ldots, \quad \left.\frac{d^{n-1}m_y}{dt^{n-1}}\right|_{t=0} = c_{n-1}. \tag{5.108}$$

Then because of the linearity of (5.104), the total output is

$$\mathbf{y}(t) = m_y(t) + \mathbf{y}^1(t), \tag{5.109}$$

where $\mathbf{y}^1(t)$ obeys the same linear differential equation as $\mathbf{y}(t)$, that is, it obeys (5.104), but with zero initial conditions,

$$\left.\frac{d^k\mathbf{y}^1}{dt^k}\right|_{t=0} \equiv 0, \qquad k = 0, 1, \ldots, n-1. \tag{5.110}$$

Thus we have broken our solution into two parts: a deterministic part $m_y(t)$ that depends only on the signal component $m_x(t)$ of the input and on the initial conditions (5.108), and a noise component $\mathbf{y}^1(t)$ that depends only on the noise component $\mathbf{x}^1(t)$ of the input. The noise component of the output is given by

$$\mathbf{y}^1(t) = \int_0^t h(s)\mathbf{x}^1(t-s)\,ds = \int_0^t h(t-s)\,\mathbf{x}^1(s)\,ds, \qquad t > 0, \tag{5.111}$$

where $h(s)$, the impulse response of the system, is the inverse Laplace transform of its transfer function

$$\begin{aligned} H(p) = \mathcal{L}h(s) &= \int_0^\infty h(s)e^{-ps}\,ds \\ &= \frac{b_0p^m + b_1p^{m-1} + \cdots + b_{m-1}p + b_m}{a_0p^n + a_1p^{n-1} + \cdots + a_{n-1}p + a_n}. \end{aligned} \tag{5.112}$$

The mean value of the component $\mathbf{y}^1(t)$ is zero, and its autocovariance function, which is also the autocovariance function of the entire output $\mathbf{y}(t)$, is given by (5.94), which when applied to (5.111) with

$$k(t, s) = \begin{cases} h(t-s), & 0 < s < t, \\ 0, & s < 0, \quad s > t, \end{cases}$$

yields

$$\begin{aligned} C_\mathbf{y}(t_1, t_2) &= \int_0^{t_1} ds_1 \int_0^{t_2} ds_2\, h(t_1 - s_1)C_\mathbf{x}(s_1, s_2)h(t_2 - s_2) \\ &= \int_0^{t_1} d\tau_1 \int_0^{t_2} d\tau_2\, h(\tau_1)C_\mathbf{x}(t_1 - \tau_1, t_2 - \tau_2)h(\tau_2). \end{aligned} \tag{5.113}$$

Even when the input $\mathbf{x}(t)$ is a stationary process, the output is in general nonstationary because of the transient behavior of the mean value $m_y(t)$ from (5.107) and (5.108) and because of the finite limits of integration in (5.113), which are related to transients in the noise component $\mathbf{y}^1(t)$.

EXAMPLE 5.12. Let the input $\mathbf{x}(t)$ be a stochastic process with mean

$$E[\mathbf{x}(t)] = Ae^{-at}U(t)$$

and autocovariance function

$$C_{\mathbf{x}}(t_1, t_2) = \sigma^2 \exp[-\mu|t_2 - t_1|]. \tag{5.114}$$

The output $\mathbf{y}(t)$ obeys the linear differential equation

$$\frac{d\mathbf{y}}{dt} + \beta\mathbf{y} = \mathbf{x}(t) \tag{5.115}$$

and at time $t = 0$ the output is $\mathbf{y}(0) = C$.

The expected value $m_y(t)$ of the output then obeys the differential equation

$$\frac{dm_y}{dt} + \beta m_y = Ae^{-at}U(t)$$

with initial condition $m_y(0) = C$, and by Laplace transforms or otherwise it is simple to show that the solution of this differential equation is

$$\begin{aligned} m_y(t) = E[\mathbf{y}(t)] &= \mathcal{L}^{-1}\left[\frac{C}{p+\beta} + \frac{A}{(p+\beta)(p+a)}\right] \\ &= Ce^{-\beta t} + \frac{A}{\beta - a}(e^{-at} - e^{-\beta t}), \qquad t > 0. \end{aligned} \tag{5.116}$$

The noise component $\mathbf{y}^1(t)$ of the output obeys the same differential equation, but with initial value $\mathbf{y}^1(0) = 0$. The impulse response is

$$h(s) = \mathcal{L}^{-1}[(p+\beta)^{-1}] = e^{-\beta s}U(s),$$

and hence the noise component of the output is

$$\mathbf{y}^1(t) = \int_0^t e^{-\beta s}\mathbf{x}^1(t-s)\, ds, \tag{5.117}$$

and by (5.113) the autocovariance of the output is

$$C_{\mathbf{y}}(t_1, t_2) = \sigma^2 \int_0^{t_1} ds_1 \int_0^{t_2} ds_2 e^{-\beta(t_1 - s_1)} e^{-\mu|s_1 - s_2|} e^{-\beta(t_2 - s_2)}.$$

Let us evaluate this for $t_2 > t_1$, whereupon the region of integration looks like that in Fig. 5.13. Taking account of the absolute value of the exponent of the autocovariance function, we find

$$\begin{aligned} C_{\mathbf{y}}(t_1, t_2) &= \sigma^2 e^{-\beta t_1 - \beta t_2} \int_0^{t_1} e^{\beta s_1} ds_1 \left[\int_0^{s_1} e^{-\mu(s_1 - s_2) + \beta s_2}\, ds_2 + \int_{s_1}^{t_2} e^{-\mu(s_2 - s_1) + \beta s_2}\, ds_2\right] \\ &= \sigma^2 e^{-\beta t_1 - \beta t_2} \int_0^{t_1} e^{\beta s_1} ds_1 \left[\frac{e^{\beta s_1} - e^{-\mu s_1}}{\beta + \mu} + \frac{e^{(\beta - \mu)t_2 + \mu s_1} - e^{\beta s_1}}{\beta - \mu}\right] \\ &= \frac{\sigma^2}{\beta^2 - \mu^2}\left\{-\frac{\mu}{\beta}\left[e^{\beta(t_1 - t_2)} - e^{-\beta t_1 - \beta t_2}\right] + e^{\mu(t_1 - t_2)}\right. \\ &\qquad \left. - e^{-\beta t_1 - \mu t_2} - e^{-\mu t_1 - \beta t_2} + e^{-\beta t_1 - \beta t_2}\right\}, \qquad t_2 > t_1. \end{aligned}$$

Because the autocovariance function of the output must be symmetrical in t_1 and t_2, we can write this in general as

$$\begin{aligned} C_{\mathbf{y}}(t_1, t_2) = \frac{\sigma^2}{\beta^2 - \mu^2}\Bigg[& e^{-\mu|t_2 - t_1|} - \frac{\mu}{\beta} e^{-\beta|t_2 - t_1|} \\ & - e^{-\beta t_1 - \mu t_2} - e^{-\mu t_1 - \beta t_2} + \frac{\beta + \mu}{\beta} e^{-\beta t_1 - \beta t_2}\Bigg]. \end{aligned} \tag{5.118}$$

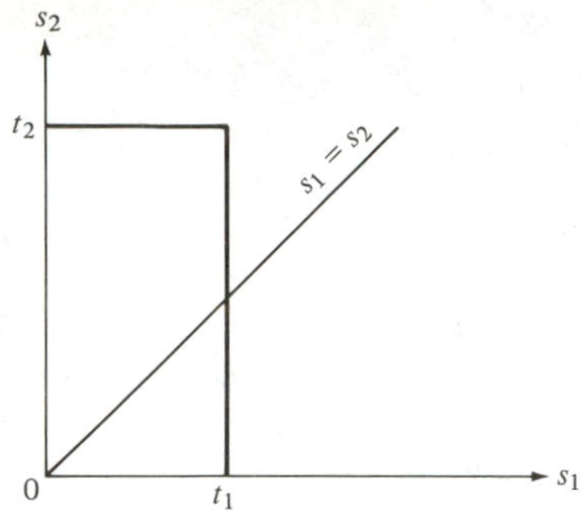

FIGURE 5.13. Region of integration.

After a long time the transients die away, the output $\mathbf{y}(t)$ becomes a stationary process, and the last three terms in (5.118) vanish. ∎

If the coefficients in the linear differential equation (5.104) describing the system vary with the time t, the noise component of the output will depend on that of the input by an integral like that in (5.88) whose kernel is the Green's function of the differential equation, and the autocovariance of the output can be calculated by (5.94). The signal component of the output, that is, the expected value $E[\mathbf{y}(t)]$, can again be determined as though the noise component were zero, by solving the differential equation for the system with the signal component of the input as driving force and with the given initial conditions on the output.

5-5

SPECTRAL DENSITY OF A STATIONARY PROCESS

Let us imagine that for all time a stationary stochastic process $\mathbf{x}(t)$ has been passing into a time-invariant linear system whose impulse response is $h(\tau)$. The output is

$$\mathbf{y}(t) = \int_0^\infty h(\tau)\mathbf{x}(t - \tau)\,d\tau = \int_{-\infty}^t h(t - s)\mathbf{x}(s)\,ds, \tag{5.119}$$

and because any transients due to turning on the input have long ago died away, this output is also a stationary process. Applying (5.94) with the kernel

$$k(t, s) = \begin{cases} h(t - s), & -\infty < s < t, \\ 0, & s > t, \end{cases} \tag{5.120}$$

we find that the autocovariance function of the output process is

$$\begin{aligned} C_{\mathbf{y}}(t_1, t_2) &= \int_{-\infty}^{t_1}\int_{-\infty}^{t_2} h(t_1 - s_1)C_{\mathbf{x}}(s_2 - s_1)h(t_2 - s_2)\,ds_1\,ds_2 \\ &= \int_0^\infty\int_0^\infty h(u_1)C_{\mathbf{x}}(t_2 - u_2 - t_1 + u_1)h(u_2)\,du_1\,du_2 \end{aligned} \tag{5.121}$$

in terms of the autocovariance function $C_{\mathbf{x}}(\tau)$ of the input. The output being stationary, we can write its autocovariance function as

$$C_{\mathbf{y}}(\tau) = \int_0^\infty \int_0^\infty h(u_1) C_{\mathbf{x}}(\tau - u_2 + u_1) h(u_2)\, du_1\, du_2. \tag{5.122}$$

This formula looks much like a convolution, and we are tempted to see what it becomes in the Fourier domain. We therefore write the autocovariance function $C_{\mathbf{x}}(\tau)$ of the input as the Fourier transform of some function $S_{\mathbf{x}}(\omega)$ of angular frequency ω,

$$C_{\mathbf{x}}(\tau) = \int_{-\infty}^\infty S_{\mathbf{x}}(\omega) e^{j\omega\tau} \frac{d\omega}{2\pi}, \tag{5.123}$$

whereupon we find that

$$\begin{aligned} C_{\mathbf{y}}(\tau) &= \int_{-\infty}^\infty S_{\mathbf{x}}(\omega) \int_0^\infty \int_0^\infty h(u_1) h(u_2) \exp\,[j\omega(\tau - u_2 + u_1)]\, du_1\, du_2 \frac{d\omega}{2\pi} \\ &= \int_{-\infty}^\infty S_{\mathbf{x}}(\omega) H(j\omega) H^*(j\omega) e^{j\omega\tau} \frac{d\omega}{2\pi} \end{aligned} \tag{5.124}$$

in terms of the transfer function

$$H(j\omega) = \int_0^\infty h(u) e^{-j\omega u}\, du \tag{5.125}$$

of our filter. Hence, defining the Fourier transform $S_{\mathbf{y}}(\omega)$ of the autocovariance function of the output as

$$S_{\mathbf{y}}(\omega) = \int_{-\infty}^\infty C_{\mathbf{y}}(\tau) e^{-j\omega\tau}\, d\tau, \tag{5.126}$$

we find that

$$S_{\mathbf{y}}(\omega) = |H(j\omega)|^2 S_{\mathbf{x}}(\omega). \tag{5.127}$$

This important result much simplifies calculating the autocovariance function of the output of a linear system whose stationary input has been turned on for such a long time that all the initial transients have died away. The useful Fourier transform

$$S_{\mathbf{x}}(\omega) = \int_{-\infty}^\infty C_{\mathbf{x}}(\tau) e^{-j\omega\tau}\, d\tau \tag{5.128}$$

of the autocovariance function is called the *spectral density* of the stochastic process $\mathbf{x}(t)$.

Because $C_{\mathbf{x}}(\tau) = C_{\mathbf{x}}(-\tau)$, we can write (5.128) as

$$S_{\mathbf{x}}(\omega) = \int_{-\infty}^\infty C_{\mathbf{x}}(\tau) \cos \omega\tau\, d\tau, \tag{5.129}$$

which shows that the spectral density is an even function of the angular frequency,

$$S_{\mathbf{x}}(\omega) = S_{\mathbf{x}}(-\omega). \tag{5.130}$$

EXAMPLE 5.13. The stationary process $\mathbf{x}(t)$ has the autocovariance function

$$C_{\mathbf{x}}(\tau) = \sigma^2 e^{-\mu|\tau|}. \tag{5.131}$$

It has been forever passing into an RC filter whose impulse response is

$$h(\tau) = \beta e^{-\beta\tau} U(\tau), \qquad \beta = \frac{1}{RC}. \tag{5.132}$$

We seek the autocovariance function of the output $\mathbf{y}(t)$ of the filter.

The spectral density of the input is

$$\begin{aligned} S_{\mathbf{x}}(\omega) &= \sigma^2 \int_{-\infty}^{\infty} e^{-\mu|\tau| - j\omega\tau}\, d\tau \\ &= \sigma^2 \int_{-\infty}^{0} e^{(\mu - j\omega)\tau}\, d\tau + \sigma^2 \int_{0}^{\infty} e^{-(\mu + j\omega)\tau}\, d\tau \\ &= \sigma^2 \left(\frac{1}{\mu - j\omega} + \frac{1}{\mu + j\omega} \right) = \frac{2\mu\sigma^2}{\mu^2 + \omega^2}. \end{aligned} \tag{5.133}$$

The transfer function of the filter is

$$H(j\omega) = \beta \int_{0}^{\infty} e^{-\beta\tau - j\omega\tau}\, d\tau = \frac{\beta}{\beta + j\omega}. \tag{5.134}$$

Hence by (5.127) the spectral density of the output $\mathbf{y}(t)$ is

$$\begin{aligned} S_{\mathbf{y}}(\omega) &= \frac{2\mu\sigma^2\beta^2}{(\mu^2 + \omega^2)|\beta + j\omega|^2} = \frac{2\mu\sigma^2\beta^2}{(\mu^2 + \omega^2)(\beta^2 + \omega^2)} \\ &= \frac{2\mu\sigma^2\beta^2}{\beta^2 - \mu^2} \left(\frac{1}{\mu^2 + \omega^2} - \frac{1}{\beta^2 + \omega^2} \right). \end{aligned} \tag{5.135}$$

Using (5.133) to guide us toward the inverse transform, we obtain for the autocovariance function of the output

$$C_{\mathbf{y}}(\tau) = \frac{\sigma^2\beta^2}{\beta^2 - \mu^2} \left(e^{-\mu|\tau|} - \frac{\mu}{\beta} e^{-\beta|\tau|} \right) \tag{5.136}$$

provided that $\mu \neq \beta$, in agreement with the stable terms of (5.118). In particular, the variance of the output is

$$\operatorname{Var} \mathbf{y}(t) = C_{\mathbf{y}}(0) = \frac{\sigma^2\beta}{\beta + \mu}. \tag{5.137}$$

Here the time T_{int} over which the filter effectively integrates its past input is the time constant $T_{\text{int}} \sim RC = 1/\beta$, and when $T_{\text{int}} >> \tau_c \sim 1/\mu$, $\beta << \mu$, we find as in Example 5.11 that

$$\frac{\operatorname{Var} \mathbf{y}(t)}{\operatorname{Var} \mathbf{x}(t)} \sim \frac{\beta}{\mu} \sim \frac{\tau_c}{T_{\text{int}}}.$$

■

In general, the variance of the output of a linear filter with stationary input $\mathbf{x}(t)$ and transfer function $H(j\omega)$ is, by (5.124),

$$\operatorname{Var} \mathbf{y}(t) = C_{\mathbf{y}}(0) = \int_{-\infty}^{\infty} |\, H(j\omega)\, |^2 S_{\mathbf{x}}(\omega)\, \frac{d\omega}{2\pi}. \tag{5.138}$$

In order to perceive the physical significance of the spectral density $S_{\mathbf{x}}(\omega)$, imagine that the stationary process $\mathbf{x}(t)$ is the input to a narrowband pass filter passing only frequencies in the neighborhood of angular frequency Ω,

$$|H(j\omega)| = \begin{cases} 1, & \Omega - \pi W < |\omega| < \Omega + \pi W, \\ 0, & \text{elsewhere,} \end{cases} \tag{5.139}$$

as sketched in Fig. 5.14. Here W is the bandwidth of the filter in hertz, and we suppose that $W << \Omega/2\pi$. Then by (5.138) the variance of the output of this filter is

$$\begin{aligned} \operatorname{Var} \mathbf{y}(t) &= 2\int_{\Omega-\pi W}^{\Omega+\pi W} S_{\mathbf{x}}(\omega)\,\frac{d\omega}{2\pi} \\ &\doteq 2WS_{\mathbf{x}}(\Omega) = W[S_{\mathbf{x}}(\Omega) + S_{\mathbf{x}}(-\Omega)], \end{aligned} \tag{5.140}$$

and the narrower the passband of the filter, the more nearly proportional $\operatorname{Var} \mathbf{y}(t)$ is to the spectral density $S_{\mathbf{x}}(\Omega)$ of the input at frequency Ω. $\operatorname{Var} \mathbf{y}(t)$ is the average power in those components of $\mathbf{x}(t)$ that get through the filter (5.139).

We can therefore state that at any angular frequency ω the spectral density $S_{\mathbf{x}}(\omega)$ is the average density of "power" in the frequency components of $\mathbf{x}(t)$ in the neighborhood of ω. The factor 2 appears in (5.140) because we always conceive the frequency components as equally divided between positive and negative frequencies. Our argument furthermore shows that because $\operatorname{Var} \mathbf{y}(t) \geq 0$, the spectral density is a nonnegative function of the frequency,

$$S_{\mathbf{x}}(\omega) \geq 0, \qquad \forall\ \omega. \tag{5.141}$$

Because the sample function $\mathbf{x}(t)$ stretches from $-\infty$ to $+\infty$, it does not have a Fourier transform in the usual sense. We can, however, imagine it as cut off outside the interval $(-\frac{1}{2}T, \frac{1}{2}T)$, take the Fourier transform of the truncated sample function, and ask for its mean-square value in the limit $T \to \infty$. For simplicity we assume that $E[\mathbf{x}(t)] = 0$. (The expected value of a stationary process corresponds to its "dc component.") Then we consider the expected value of the absolute square,

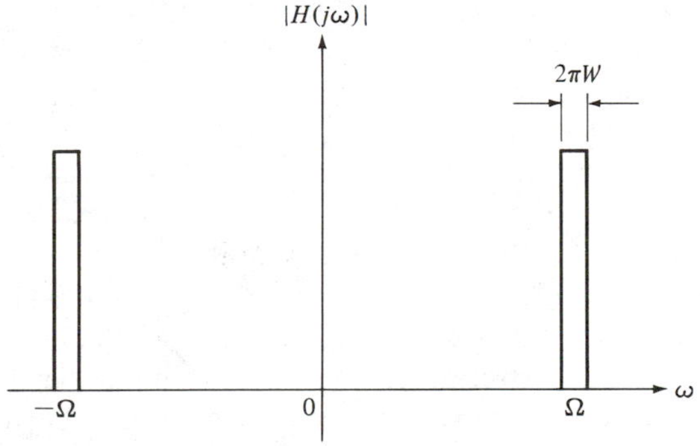

FIGURE 5.14 Narrowband pass-filter.

$$E\left|\int_{-T/2}^{T/2} \mathbf{x}(t)e^{-j\omega t}\,dt\right|^2$$

$$= E\int_{-T/2}^{T/2}\int_{-T/2}^{T/2} \mathbf{x}(t_1)\mathbf{x}(t_2)e^{-j\omega(t_1-t_2)}\,dt_1\,dt_2$$

$$= \int_{-T/2}^{T/2}\int_{-T/2}^{T/2} C_{\mathbf{x}}(t_1 - t_2)e^{-j\omega(t_1-t_2)}\,dt_1\,dt_2$$

$$= \int_{-T/2}^{T/2}\int_{-T/2}^{T/2} C_{\mathbf{x}}(t_1 - t_2)\cos\omega(t_1 - t_2)\,dt_1\,dt_2;$$

in the final step we used the symmetry of the covariance function, $C_{\mathbf{x}}(t_1 - t_2) = C_{\mathbf{x}}(t_2 - t_1)$, to eliminate the term in $\sin\omega(t_1 - t_2)$. The resulting integral of an even function is carried over a square, and we can use the technique introduced in Example 5.11 to write the double integral as

$$\int_{-T}^{T} (T - |v|)C_{\mathbf{x}}(v)\cos\omega v\,dv.$$

When the duration T of the segment $\mathbf{x}(t)$ of the process is much longer than the correlation time τ_c of the process, this integral will be proportional to T, and we therefore modify the quantity under study by dividing by T before passing to the limit $T \to \infty$:

$$E\,T^{-1}\left|\int_{-T/2}^{T/2} \mathbf{x}(t)e^{-j\omega t}\,dt\right|^2 = \int_{-T}^{T}\left(1 - \frac{|v|}{T}\right) C_{\mathbf{x}}(v)\cos\omega v\,dv.$$

When the duration T of the segment goes to infinity, the term in $|v|/T$ vanishes, and we find

$$\lim_{T\to\infty}\frac{1}{T}E\left|\int_{-T/2}^{T/2} \mathbf{x}(t)e^{-j\omega t}\,dt\right|^2 = \int_{-\infty}^{\infty} C_{\mathbf{x}}(v)\cos\omega v\,dv = S_{\mathbf{x}}(\omega), \qquad (5.142)$$

a result known as the *Wiener–Khinchin theorem*. It provides both an interpretation of the spectral density and a clue to estimating it from an empirical sample function, a topic we defer to Sec. 5-8. It confirms, furthermore, that the spectral density $S_{\mathbf{x}}(\omega)$ must be nonnegative, for it is proportional to the expected value of a nonnegative quantity.

When the spectral density resembles that depicted in Fig. 5.15(a), low frequencies predominate in the stochastic process $\mathbf{x}(t)$. The autocovariance function $C_{\mathbf{x}}(\tau)$ of the process, as the Fourier transform of its spectral density, then resembles the function in Fig. 5.15(b). The width τ_c of the autocovariance function, which we called the correlation time of $\mathbf{x}(t)$, is inversely proportional to the width W of the spectral density. We can call W the bandwidth of the process.

If, on the other hand, the spectral density $S_{\mathbf{x}}(\omega)$ peaks in the neighborhood of some high frequency Ω, as in Fig. 5.16(a), the autocovariance function $C_{\mathbf{x}}(\tau)$ will look somewhat like that in Fig. 5.16(b) when the bandwidth W is much smaller than $\Omega/2\pi$. Frequencies in the neighborhood of Ω predominate in $\mathbf{x}(t)$, which looks like a carrier of frequency Ω with random amplitude and phase

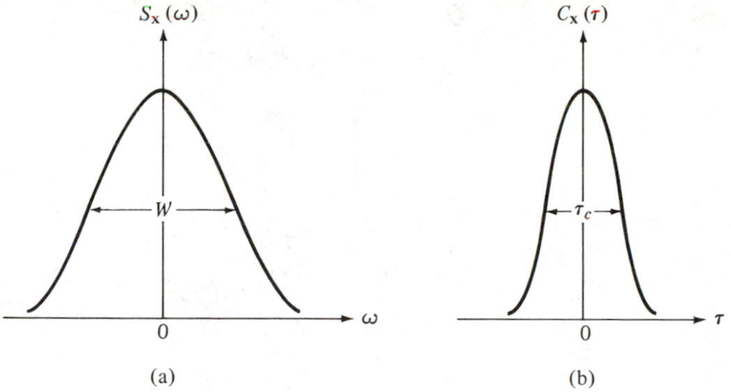

FIGURE 5.15 Spectral density and autocovariance function of a "low-pass" process.

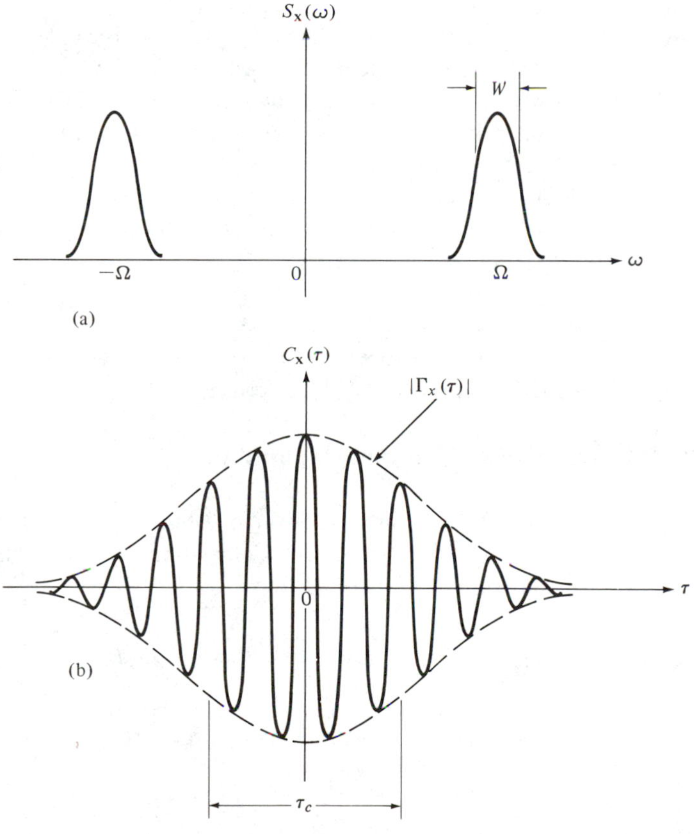

FIGURE 5.16 Spectral density and autocovariance function of a narrowband, high-frequency process.

modulations,

$$\mathbf{x}(t) \sim \mathbf{A}(t) \cos [\Omega t + \boldsymbol{\phi}(t)]. \tag{5.143}$$

The autovariance function can then be written approximately as

$$C_{\mathbf{x}}(\tau) = \text{Re}\,(\Gamma_{\mathbf{x}}(\tau)e^{j\Omega\tau}), \tag{5.144}$$

where $\Gamma_{\mathbf{x}}(\tau)$ may be complex, but

$$\Gamma_{\mathbf{x}}(-\tau) = \Gamma_{\mathbf{x}}^{*}(\tau); \tag{5.145}$$

$|\Gamma_{\mathbf{x}}(\tau)|$ represents the envelope of the autocovariance function.

If, for example, the spectral density is

$$\begin{aligned} S_{\mathbf{x}}(\omega) &= \tfrac{1}{2}B(2\pi W^2)^{-1/2} \\ &\quad \times \left(\exp\left[-\frac{(\omega - \Omega)^2}{2W^2}\right] + \exp\left[-\frac{(\omega + \Omega)^2}{2W^2}\right]\right), \end{aligned} \tag{5.146}$$

we find for the autocovariance function

$$\begin{aligned} C_{\mathbf{x}}(\tau) &= \tfrac{1}{2}B[\exp\,(j\Omega\tau - \tfrac{1}{2}W^2\tau^2) + \exp\,(-j\Omega\tau - \tfrac{1}{2}W^2\tau^2)] \\ &= Be^{-W^2\tau^2/2}\cos\Omega\tau. \end{aligned} \tag{5.147}$$

The overall width τ_c of the autocovariance function is also $\tau_c \sim 1/W$ for a narrowband high-frequency stochastic process, where W is the bandwidth of the spectral density, now defined for the peak in the neighborhood of the carrier frequency Ω, as in Fig. 5.16; but the autocovariance function exhibits sinusoidal oscillations of "frequency" Ω in the delay variable τ. The random amplitude $\mathbf{A}(t)$ and the random phase $\boldsymbol{\phi}(t)$ in (5.143) vary significantly only in times of the order of $\tau_c \sim 1/W$, which is much longer than the period $2\pi/\Omega$ of the carrier when $W << \Omega$.

5-6 GAUSSIAN STOCHASTIC PROCESSES

A. The Random-Pulse Process Revisited

Let us look back at the stochastic process $\mathbf{x}(t)$ consisting of random pulses with random amplitudes and treated in Sec. 5-3E. Calculating its spectral density by taking the Fourier transform of (5.72), we find

$$\begin{aligned} S_{\mathbf{x}}(\omega) &= \lambda E(\mathbf{A}^2)\int_{-\infty}^{\infty}\int_{-\infty}^{\infty} h(u)h(\tau + u)e^{-j\omega\tau}\,du\,d\tau \\ &= \lambda E(\mathbf{A}^2)\int_{-\infty}^{\infty}\int_{-\infty}^{\infty} h(u)e^{j\omega u}h(\tau + u)e^{-j\omega(\tau+u)}\,du\,d\tau \\ &= \lambda E(\mathbf{A}^2)\int_{-\infty}^{\infty} h(u)e^{j\omega u}\,du\int_{-\infty}^{\infty} h(v)e^{-j\omega v}\,dv \\ &= \lambda E(\mathbf{A}^2)\,|\,H(j\omega)\,|^2, \end{aligned} \tag{5.148}$$

where

$$H(j\omega) = \int_{-\infty}^{\infty} h(u)e^{-j\omega u}\,du \tag{5.149}$$

is the spectrum of the shape $h(t)$ of an individual pulse. Comparing this with (5.127) we see that the spectral density of the random-pulse process is what would result if it were the output of a linear filter with transfer function $H(j\omega)$ and if the input to the filter were a stochastic process $\mathbf{z}(t)$ with uniform spectral density

$$S_{\mathbf{z}}(\omega) = \lambda E(\mathbf{A}^2), \tag{5.150}$$

at least for all frequencies ω where the filter gain $|H(j\omega)|$ is significantly different from zero. The output of the filter, in accordance with the definition of $\mathbf{x}(t)$ in (5.69), is a succession of copies of the impulse response

$$h(u) = \int_{-\infty}^{\infty} H(j\omega)e^{j\omega u}\,\frac{d\omega}{2\pi} \tag{5.151}$$

occurring at random times $\boldsymbol{\tau}_k$ and having independently random amplitudes $\mathbf{A}_k$. The input, therefore, must be a succession of delta functions,

$$\mathbf{z}(t) = \sum_{k=-\infty}^{\infty} \mathbf{A}_k\delta(t - \boldsymbol{\tau}_k) \tag{5.152}$$

with random "amplitudes" $\mathbf{A}_k$ and occurring at times $\boldsymbol{\tau}_k$ that constitute a Poisson point process. The spectral density $S_{\mathbf{z}}(\omega)$ of this process $\mathbf{z}(t)$, by (5.150), equals $\lambda E(\mathbf{A}^2)$, where λ is the rate of the point process and $E(\mathbf{A}^2)$ the mean-square amplitude; and its autocovariance function is

$$C_{\mathbf{z}}(\tau) = \lambda E(\mathbf{A}^2)\delta(\tau). \tag{5.153}$$

True delta functions cannot, of course, occur in nature; $\delta(t - \boldsymbol{\tau}_k)$ in (5.152) denotes a pulse signal that has a unit area, but is so narrow that our instruments cannot resolve it. When it enters the filter whose transfer function is $H(j\omega)$, the output is a copy $h(t - \boldsymbol{\tau}_k)$ of the impulse response of the filter. The process $\mathbf{z}(t)$ is called the *Poisson impulse process*.

B. The Gaussian Process

Now let us take $E(\mathbf{A}) = 0$ and imagine the rate λ increasing to infinity and the mean-square amplitude $E(\mathbf{A}^2)$ diminishing to zero in such a way that the spectral density $S_{\mathbf{z}}(\omega) = \lambda E(\mathbf{A}^2)$ of the Poisson impulse process remains fixed. The delta functions $\mathbf{A}_k\delta(t - \boldsymbol{\tau}_k)$, some positive, some negative, are now coming thick and fast, but their areas $|\mathbf{A}_k|$ are vanishingly small. A great many of these impulses enter the filter during any interval whose duration is of the order of magnitude of the response time of the filter. Out of the filter emerges a dense crowd of small, overlapping copies

$$\mathbf{A}_k h(t - \boldsymbol{\tau}_k)$$

of its impulse response. By virtue of the central limit theorem, the output $\mathbf{x}(t)$ of the filter,

$$\mathbf{x}(t) = \int_{-\infty}^{\infty} h(u)\mathbf{z}(t - u)\, du = \sum_{k=-\infty}^{\infty} \mathbf{A}_k h(t - \boldsymbol{\tau}_k), \tag{5.154}$$

sampled at an arbitrary time t, will be a Gaussian random variable, consisting as it does of the sum of a great many small, random integrants. Any probability density function $f_{\mathbf{A}}(A)$ of the amplitudes $\mathbf{A}_k$ with $E(|\mathbf{A}|^m)$ finite, $m \geq 3$, will do.

The joint probability density function

$$f_{\mathbf{x}}(x_1, t_1; x_2, t_2; \ldots ; x_n, t_n)$$

of n samples $\mathbf{x}_1 = \mathbf{x}(t_1)$, $\mathbf{x}_2 = \mathbf{x}(t_2)$, . . . , $\mathbf{x}_n = \mathbf{x}(t_n)$ of the output of the filter will also have the Gaussian form in (4.134), as we can see by the following argument. Form any linear combination

$$\boldsymbol{\xi} = \omega_1\mathbf{x}_1 + \omega_2\mathbf{x}_2 + \cdots + \omega_n\mathbf{x}_n \tag{5.155}$$

of these samples, where $\omega_1, \omega_2, \ldots, \omega_n$ are arbitrary constant weights. Then

$$\begin{aligned} \boldsymbol{\xi} &= \sum_{m=1}^{n} \omega_m \mathbf{x}(t_m) = \sum_{m=1}^{n} \omega_m \sum_{k=-\infty}^{\infty} \mathbf{A}_k h(t_m - \boldsymbol{\tau}_k) \\ &= \sum_{k=-\infty}^{\infty} \mathbf{A}_k \sum_{m=1}^{n} \omega_m h(t_m - \boldsymbol{\tau}_k) = \sum_{k=-\infty}^{\infty} \mathbf{A}_k \mathbf{X}_k, \end{aligned} \tag{5.156}$$

is itself the sum of a great many independently random components $\mathbf{A}_k\mathbf{X}_k$, where

$$\mathbf{X}_k = \sum_{m=1}^{n} \omega_m h(t_m - \boldsymbol{\tau}_k). \tag{5.157}$$

The number of terms $\mathbf{A}_k\mathbf{X}_k$ that effectively contribute to the sum increases with the increasing rate λ, and by the central limit theorem, the random variable $\boldsymbol{\xi}$ will also have a Gaussian probability density function in the limit $\lambda \to \infty$, $E(\mathbf{A}^2) \to 0$ that we are considering. This means that its characteristic function has the Gaussian form

$$\Phi_{\boldsymbol{\xi}}(u) = E(e^{j\boldsymbol{\xi}u}) = \exp\,[juE(\boldsymbol{\xi}) - \tfrac{1}{2}u^2 \operatorname{Var} \boldsymbol{\xi}]. \tag{5.158}$$

Substituting from (5.155), evaluating $E(\boldsymbol{\xi})$ and Var $\boldsymbol{\xi}$, and putting $u = 1$, one sees immediately that the joint characteristic function

$$\Phi_{\mathbf{x}}(\omega_1, \omega_2, \ldots, \omega_n) = E\{\exp\,[j(\omega_1\mathbf{x}_1 + \omega_2\mathbf{x}_2 + \cdots + \omega_n\mathbf{x}_n)]\} \tag{5.159}$$

has the Gaussian form in (4.140). Hence the random variables $\mathbf{x}_1 = \mathbf{x}(t_1)$, $\mathbf{x}_2 = \mathbf{x}(t_2)$, . . . , $\mathbf{x}_n = \mathbf{x}(t_n)$ are jointly Gaussian random variables.

A random process $\mathbf{x}(t)$, such as the output of our filter, the joint density functions of all samples $\mathbf{x}_k = \mathbf{x}(t_k)$ of which have the Gaussian form of (4.134), is called a *Gaussian stochastic process*. Those multivariate probability density functions are completely determined by the expected values

$$m_k = m(t_k) = E(\mathbf{x}_k) = E[\mathbf{x}(t_k)] \tag{5.160}$$

of the samples and by their covariances

$$\begin{aligned} c_{jk} &= \text{Cov}\,\{\mathbf{x}_j, \mathbf{x}_k\} = E[(\mathbf{x}_j - m_j)(\mathbf{x}_k - m_k)] \\ &= E\{[\mathbf{x}(t_j) - m(t_j)][\mathbf{x}(t_k) - m(t_k)]\} \\ &= C_{\mathbf{x}}(t_j, t_k). \end{aligned} \tag{5.161}$$

The matrix C of the quadratic form in (4.134) and (4.136) is the inverse of the $n \times n$ matrix c whose elements are these covariances c_{jk}. Not all Gaussian stochastic processes arise by filtering dense Poisson impulse processes. Often they serve as models for stochastic processes of which all that is known is the set of means and covariances in (5.160) and (5.161).

The result of any linear filtering of a Gaussian stochastic process is also a Gaussian stochastic process, for as we said at the end of Sec. 5-4, the integral in (5.88) is a weighted sum of samples, taken infinitely densely, of the process $\mathbf{x}(t)$, and as these are Gaussian random variables, their weighted sum is also Gaussian. The same argument as led to (5.159) convinces us that the joint probability density function of samples $\mathbf{y}(t_k)$ of the output of the linear filter at arbitrary times t_k will also have the Gaussian form.

A Gaussian stochastic process $\mathbf{x}(t)$ with expected value $E[\mathbf{x}(t)] = 0$ is often called *Gaussian noise*. We shall see in Sec. 5-7 that the thermal noise in a linear network of resistors, inductors, and capacitors, or at the terminals of an antenna, is a Gaussian stochastic process of this kind.

C. White Noise

The random process $\mathbf{z}(t)$ at the input to the filter of (5.154) is called *white noise* because its spectral density $\lambda E(\mathbf{A}^2)$ is uniform as a function of frequency—in false analogy to white light, whose spectral density, though broad, is not uniform in frequency. It is customary to designate the uniform spectral density of white noise by $N/2$,

$$S_{\mathbf{z}}(\omega) = \frac{N}{2}, \qquad -\infty < \omega < \infty. \tag{5.162}$$

Referring to (5.140) we see that if this white noise passes through the narrow-band pass filter of Fig. 5.14, the variance of the output is equal to NW, where W is the bandwidth of the filter. The quantity N is therefore called the *unilateral* spectral density of the white noise, NW representing the average total power passing through a filter of pass frequency Ω and bandwidth W, as though all the power were at positive frequencies. The effective autocovariance function of this white noise, the Fourier transform of (5.162), is

$$C_{\mathbf{z}}(\tau) = \frac{N}{2}\,\delta(\tau). \tag{5.163}$$

At extremely high frequencies the spectral density $S_{\mathbf{z}}(\omega)$ drops off to zero, however, and the average total power $C_{\mathbf{z}}(0)$ is not really infinite.

As we have seen, white noise can be thought of as a rapid succession of sharp, delta-function-like pulses of independently random amplitudes. The time during which each pulse rises and falls is too short for our instruments to measure, and in particular it is much shorter than the response time of any filter we can construct. Furthermore, a huge number of such random pulses occur in any interval whose duration is of the order of that response time. Although it is often customary to call such white noise "Gaussian," the central limit theorem suffices to determine that the output of any linear filter to which this noise is applied will be a Gaussian stochastic process, and the joint density functions of samples of the white noise itself—if it were indeed meaningful to speak of samples of such a process—are never needed.

D. The Brownian Process

The shot-noise process of Sec. 5-3D can be considered the output of a linear filter whose impulse response is $h(t)$ and whose input is a Poisson impulse process $\mathbf{z}(t)$ of unit amplitude, $E(\mathbf{A}) = E(\mathbf{A}^2) = 1$. When this special Poisson impulse process is integrated, beginning at time $t = 0$, the result,

$$\mathbf{n}(t) = \int_0^t \mathbf{z}(u)\,du, \tag{5.164}$$

is the Poisson counting process treated in Sec. 5-3A and B. The reader should show that the expected value and autocovariance function of the Poisson counting process can be obtained from (5.152), (5.153), and (5.164). The integral

$$\mathbf{x}(t) = \int_0^t \mathbf{z}(u)\,du \tag{5.165}$$

of white noise $\mathbf{z}(t)$ is likewise of fundamental importance. It is called the *Wiener–Lévy process* by mathematicians. Physicists call it the *Brownian process* because—as we shall see in Sec. 5-7—it represents in one dimension the position of a particle undergoing Brownian motion. During any perceptible interval $(0, t)$ so many impulses contribute to the integral in (5.165) that $\mathbf{x}(t)$ is, again by the central limit theorem, a Gaussian random process. A typical sample function consists of many infinitesimal random, positive or negative jumps.

The expected value of white noise $\mathbf{z}(t)$ being zero,

$$E[\mathbf{x}(t)] = 0.$$

The variance of the Brownian process is, by (5.153),

$$\begin{aligned} \operatorname{Var} \mathbf{x}(t) = E[\mathbf{x}(t)]^2 &= E\int_0^t\int_0^t \mathbf{z}(t_1)\mathbf{z}(t_2)\,dt_1\,dt_2 \\ &= D\int_0^t\int_0^t \delta(t_1 - t_2)\,dt_1\,dt_2 = D\int_0^t dt_1 = Dt, \end{aligned} \tag{5.166}$$

where $D = \lambda E(\mathbf{A}^2)$ is the bilateral spectral density of the white noise $\mathbf{z}(t)$. Hence the first-order probability density function of $\mathbf{x}(t)$ is

$$f_{\mathbf{x}}(x, t) = (2\pi Dt)^{-1/2} \exp\left(-\frac{x^2}{2Dt}\right). \tag{5.167}$$

In this context D is called the diffusion constant. The width $[\text{Var } \mathbf{x}(t)]^{1/2}$ of this density function is proportional to the square root $t^{1/2}$ of the time.

The increment

$$\mathbf{x}(t_2) - \mathbf{x}(t_1) = \int_{t_1}^{t_2} \mathbf{z}(t)\, dt \tag{5.168}$$

is independent of

$$\mathbf{x}(t_1) = \int_0^{t_1} \mathbf{z}(t)\, dt$$

for $t_2 > t_1 > 0$ because the delta functions coming in during the interval (t_1, t_2) occur independently of those coming in during $(0, t_1)$; they arrive at times $\boldsymbol{\tau}_k$ that make up an infinitely dense Poisson point process. Hence the conditional expected value of $\mathbf{x}(t_2)$, given that we observed the process at time t_1 to take on the value x_1, is

$$\begin{aligned} E[\mathbf{x}(t_2) \mid \mathbf{x}(t_1) = x_1] &= x_1 + E[\mathbf{x}(t_2) - \mathbf{x}(t_1) \mid \mathbf{x}(t_1) = x_1] \\ &= x_1, \qquad t_2 > t_1, \end{aligned}$$

and by (5.25) the autocovariance of the Brownian process is

$$\begin{aligned} C_{\mathbf{x}}(t_1, t_2) &= E_{\mathbf{x}_1}\{\mathbf{x}_1 E[\mathbf{x}(t_2) \mid \mathbf{x}(t_1) = \mathbf{x}_1]\} \\ &= E(\mathbf{x}_1^2) = \text{Var } \mathbf{x}(t_1) = Dt_1, \qquad t_2 > t_1. \end{aligned}$$

In general, therefore,

$$C_{\mathbf{x}}(t_1, t_2) = D \min (t_1, t_2). \tag{5.169}$$

This autocovariance function has the same form as that for the Poisson counting process given in (5.43), although the sample functions of the Brownian process look completely different.

When it is known that $\mathbf{x}(t_1) = x_1$, the process $\mathbf{x}(t)$ for $t > t_1$ can by (5.168) be treated as a Brownian process starting at x_1 at time t_1. Its expected value is x_1 and its variance is

$$\text{Var } [\mathbf{x}(t_2) \mid \mathbf{x}(t_1) = x_1] = D(t_2 - t_1), \qquad t_2 > t_1. \tag{5.170}$$

Hence the conditional probability density function of $\mathbf{x}(t_2) = \mathbf{x}_2$ is

$$f_{\mathbf{x}_2}(x_2, t_2 \mid \mathbf{x}(t_1) = x_1) = [2\pi D(t_2 - t_1)]^{-1/2} \exp \left[-\frac{(x_2 - x_1)^2}{2D(t_2 - t_1)} \right] \tag{5.171}$$

as in (5.167). The joint probability density function of two successive samples $\mathbf{x}_1 = \mathbf{x}(t_1)$, $\mathbf{x}_2 = \mathbf{x}(t_2)$ of the Brownian process is therefore

$$\begin{aligned} f_{\mathbf{x}}(x_1, t_1; x_2, t_2) &= (2\pi D t_1)^{-1/2}[2\pi D(t_2 - t_1)]^{-1/2} \\ &\quad \times \exp \left[-\frac{x_1^2}{2Dt_1} - \frac{(x_2 - x_1)^2}{2D(t_2 - t_1)} \right], \qquad t_2 > t_1. \end{aligned} \tag{5.172}$$

The reader should demonstrate that this bivariate Gaussian density function reduces to the form in (4.136) with the covariance matrix

$$\mathsf{c} = \mathsf{C}^{-1} = \begin{bmatrix} Dt_1 & Dt_1 \\ Dt_1 & Dt_2 \end{bmatrix}, \qquad t_1 < t_2. \tag{5.173}$$

The Brownian process is a Markov process; its behavior for times $t > t_2$ depends only on $\mathbf{x}(t_2)$ and not on how the process reached that value, for the impulses making up $\mathbf{z}(t')$, $t_2 < t' < t$, are independent of those in $(0, t_2)$. Hence for $t_3 > t_2 > t_1$,

$$\begin{aligned} & f_{\mathbf{x}_3}(x_3, t_3 \mid \mathbf{x}(t_1) = x_1, \mathbf{x}(t_2) = x_2) \\ &= f_{\mathbf{x}_3}(x_3, t_3 \mid \mathbf{x}(t_2) = x_2) \\ &= [2\pi D(t_3 - t_2)]^{-1/2} \exp\left[-\frac{(x_3 - x_2)^2}{2D(t_3 - t_2)}\right], \end{aligned} \tag{5.174}$$

from which the third-order joint probability density function is

$$\begin{aligned} & f_{\mathbf{x}}(x_1, t_1; x_2, t_2; x_3, t_3) \\ &= (2\pi D t_1)^{-1/2}[2\pi D(t_2 - t_1)]^{-1/2}[2\pi D(t_3 - t_2)]^{-1/2} \\ &\times \exp\left[-\frac{x_1^2}{2Dt_1} - \frac{(x_2 - x_1)^2}{2D(t_2 - t_1)} - \frac{(x_3 - x_2)^2}{2D(t_3 - t_2)}\right], \\ & 0 < t_1 < t_2 < t_3. \end{aligned} \tag{5.175}$$

Again it is an instructive exercise to reduce this to the form in (4.136) and verify the covariance matrix of $\mathbf{x}_1 = \mathbf{x}(t_1)$, $\mathbf{x}_2 = \mathbf{x}(t_2)$, $\mathbf{x}_3 = \mathbf{x}(t_3)$ involved.

The Brownian process is often treated as a limiting case of the following mechanism. Every τ seconds a coin is tossed, and successive tosses are statistically independent. A particle, at the origin at time $t = 0$, is moved s units to the right if the coin falls "heads" and s units to the left if the coin falls "tails." At time $n\tau$ the probability that the particle has progressed a total of ks units to the right and $(n - k)s$ units to the left, that is, the probability that it is to be found at $\mathbf{x} = (n - 2k)s$, is given by the binomial distribution,

$$\Pr[\mathbf{x}(n\tau) = (n - 2k)s] = \Pr(\mathbf{k} = k) = \binom{n}{k} p^k q^{n-k}, \tag{5.176}$$

where p and $q = 1 - p$ are the probabilities of heads and tails, respectively. The expected value of its position at time $n\tau$ is

$$E[\mathbf{x}(n\tau)] = sE(n - 2\mathbf{k}) = ns(1 - 2p) = ns(q - p), \tag{5.177}$$

where $\mathbf{k}$ is the binomial random variable, and the variance of its position is

$$\operatorname{Var} \mathbf{x}(n\tau) = 4s^2 \operatorname{Var} \mathbf{k} = 4ns^2pq \tag{5.178}$$

by (2.101). If $q > p$, the particle "drifts" to the right; if $q < p$, to the left. Its position is randomly dispersed about $E[\mathbf{x}(n\tau)]$ with a variance that is proportional to the number n of steps.

Now imagine the intervals τ between tosses to be very brief, and the length s of the steps to be very small. Let the probabilities q and p differ only slightly from $\frac{1}{2}$,

$$p = \tfrac{1}{2}(1 - \varepsilon), \qquad q = \tfrac{1}{2}(1 + \varepsilon), \qquad \varepsilon << 1,$$

and put

$$t = n\tau, \qquad D = \frac{s^2}{\tau}, \qquad \bar{v} = \frac{\varepsilon s}{\tau}.$$

At any finite time t, the number n will be very large, and by either the central limit theorem or the de Moivre–Laplace approximation (4.188), the random variable $\mathbf{x} = \mathbf{x}(t)$ will have a Gaussian distribution in the limit $\tau \to 0$, with s and ε decreasing in proportion to $\tau^{1/2}$. Then

$$E[\mathbf{x}(t)] = ns\varepsilon \to \bar{v}t, \qquad \text{Var } \mathbf{x}(t) = Dt,$$

and the probability density function of $\mathbf{x}(t)$ will be

$$f_{\mathbf{x}}(x, t) = (2\pi Dt)^{-1/2} \exp\left[-\frac{(x - \bar{v}t)^2}{2Dt}\right]. \tag{5.179}$$

The resulting random process is called "Brownian motion with drift," $\bar{v}$ corresponding to the drift velocity and D, as before, to the diffusion constant.

Again the increment $\mathbf{x}(t_2) - \mathbf{x}(t_1)$ of this process during an interval (t_1, t_2) subsequent to time t_1 is statistically independent of $\mathbf{x}(t_1)$, for the tosses of the coin in that interval are independent of those that went on during $(0, t_1)$. On this basis one can work out the joint density functions for this process as for the ordinary Brownian process. Drifting affects the expected value, but not the dispersion of this process.

5-7 THERMAL NOISE

A. The Brownian Motion

In 1827 the English botanist Robert Brown looked through a microscope at colloidal particles suspended in a liquid and discovered them dancing about in a lively, yet mysterious manner. Near the end of the nineteenth century particles of smoke were seen to undergo that same saccadic motion, which was then attributed to their being battered about by the invisible molecules of the air. The molecules strike a smoke particle extremely rapidly and at all angles, exerting a fluctuating force on it that can be considered a mechanical form of white noise. We shall analyze the resultant motion of the particle as a prelude to treating thermal noise in electrical circuits.

Consider first a particle of mass M suspended in a viscous fluid or gas and acted upon by a macroscopic force $F(t)$ in the x-direction. We confine our analysis to its position along the x-axis and to its velocity v in that direction. The velocity v acquired by the particle is assumed so small that the force by which the fluid resists acceleration of the particle is proportional to only the first power of v. Then the net force on the particle is, by Newton's law,

$$M \frac{dv}{dt} = F(t) - bv, \tag{5.180}$$

in which the constant b is proportional to the viscosity η of the fluid. For a spherical particle of radius a, $b = 6\pi\eta a$. When the force F is constant, the particle ultimately reaches a constant velocity equal to F/b. Under the influence of gravity, the particle drifts downward at constant velocity $(M - M_d)g/b$, where g is the acceleration of gravity and M_d is the mass of fluid the particle displaces. We shall analyze only its motion in the horizontal (x-) direction.

In the absence of a macroscopic applied force $F(t)$, the particle moves back and forth under the action of a random force $\mathbf{F}(t)$ caused by its numerous collisions with the molecules of the medium. Each collision lasts for so short a time that the force the impinging molecule exerts on the particle can be represented by a delta function with a random amplitude, $\mathbf{A}_k\delta(t - \boldsymbol{\tau}_k)$, occurring at a random time $\boldsymbol{\tau}_k$. When the medium and the suspended particle are in thermal equilibrium, the rate λ of collisions is constant, and the impinging molecules are so tiny and so numerous that the numbers of collisions in disjoint intervals of time can be considered statistically independent. The instants $\boldsymbol{\tau}_k$ of the collisions can therefore be considered a dense Poisson point process.

The velocity of the particle in the x-direction is now a stochastic process governed by a Langevin equation

$$M \frac{d\mathbf{v}}{dt} + b\mathbf{v} = \mathbf{F}(t), \tag{5.181}$$

in which the random force $\mathbf{F}(t)$ has the form in (5.152) and can be treated as a mechanical form of white noise with a uniform spectral density $\mathbf{S_F}(\omega) \equiv \mathbf{S_F}$. We take the position of the particle at time $t = 0$ as the origin of the x-coordinate, $\mathbf{x}(0) = 0$.

The impulse response $h(\tau)$ for the linear differential equation (5.181) is the inverse Laplace transform of

$$H(p) = (b + Mp)^{-1} = \frac{M^{-1}}{p + bM^{-1}} \tag{5.182}$$

and is therefore

$$h(\tau) = M^{-1} \exp\left(-\frac{b\tau}{M}\right) U(\tau). \tag{5.183}$$

Hence at time $t > 0$ the velocity of the particle is

$$\mathbf{v}(t) = M^{-1}\int_0^t e^{-b\tau/M}\mathbf{F}(t - \tau)\, d\tau. \tag{5.184}$$

After such a long time $t >> M/b$ that the initial transient has died away, the velocity $\mathbf{v}(t)$ can be considered as a stationary random process, and its spectral density is, by (5.127),

$$S_{\mathbf{v}}(\omega) = S_{\mathbf{F}}|H(j\omega)|^2 = \frac{S_{\mathbf{F}}}{b^2 + M^2\omega^2}. \tag{5.185}$$

The variance Var **v**, or the *mean-square velocity,* is thereupon

$$\text{Var } \mathbf{v} = \overline{v^2} = S_{\mathbf{F}}\int_{-\infty}^{\infty} \frac{d\omega/2\pi}{b^2 + M^2\omega^2} = \frac{S_{\mathbf{F}}}{2bM}, \tag{5.186}$$

and the mean kinetic energy of the particle in the x-direction is

$$\tfrac{1}{2}M\overline{v^2} = \frac{S_{\mathbf{F}}}{4b}. \tag{5.187}$$

The particle is now in thermal equilibrium with the surrounding medium. Like all the molecules of the medium, it possesses three degrees of freedom, corresponding to motion in the x-, y-, and z-directions. Statistical mechanics teaches that when a system is in thermal equilibrium at absolute temperature T(K) the average energy associated with each degree of freedom equals $\frac{1}{2}kT$, where $k = 1.38 \times 10^{-23}$ joule/deg is Boltzmann's constant; this is the *law of equipartition of energy*. Hence from (5.187)

$$\tfrac{1}{2}M\overline{v^2} = \frac{S_{\mathbf{F}}}{4b} = \tfrac{1}{2}kT,$$

and we find that the spectral density of the randomly fluctuating force $\mathbf{F}(t)$ must be

$$S_{\mathbf{F}} = 2kTb. \tag{5.188}$$

It is proportional to the constant b, which measures the resistance of the medium to acceleration of the particle by an applied force. Equivalently, b measures the rate at which coherent energy or *work* associated with the action of an applied force is converted into heat by friction between the particle and the fluid: under a constant force F the steady velocity is $v = F/b$, and the power converted to heat equals $Fv = F^2/b = bv^2$.

We find here one instance of a general law of nature, that whenever a mechanism exists by which coherent energy is converted into heat, that same mechanism manifests a randomly fluctuating force when the system is in thermal equilibrium, and the spectral density of that force is proportional to the same constant as determines the rate of conversion of work into heat. This law is given quantitative expression in the *fluctuation-dissipation theorem*.

The position $\mathbf{x}(t)$ of the particle after the fluctuating force $\mathbf{F}(t)$ has acted on it for a time t is, by (5.184),

$$\mathbf{x}(t) = \int_0^t \mathbf{v}(s)\,ds = b^{-1}\int_0^t (1 - e^{-b(t-u)/M})\mathbf{F}(u)\,du, \tag{5.189}$$

for the impulse response connecting the force $\mathbf{F}(t)$ and the displacement $\mathbf{x}(t)$ is the integral of that in (5.183), that is,

$$\int_0^\tau M^{-1}e^{-bs/M}\,ds = b^{-1}(1 - e^{-b\tau/M})U(\tau).$$

The autocovariance function of the force $\mathbf{F}(t)$ is

$$C_{\mathbf{F}}(u, v) = E[\mathbf{F}(u)\,\mathbf{F}(v)] = S_{\mathbf{F}}\delta(u - v),$$

and applying this we find for the variance of the displacement

$$\begin{aligned} \operatorname{Var} \mathbf{x}(t) &= b^{-2}\int_0^t\int_0^t (1 - e^{-b(t-u)/M})(1 - e^{-b(t-v)/M})\, E[\mathbf{F}(u)\mathbf{F}(v)]\, du\, dv \\ &= S_{\mathbf{F}} b^{-2}\int_0^t (1 - e^{-b(t-u)/M})^2\, du. \end{aligned} \tag{5.190}$$

After a long time $t >> M/b$,

$$\operatorname{Var} \mathbf{x}(t) = S_{\mathbf{F}} b^{-2} t = Dt,$$

where D is the diffusion constant of the Brownian process introduced in Sec. 5-6D. Hence by (5.188) we find that the diffusion constant is proportional to the absolute temperature of the gas,

$$D = S_{\mathbf{F}} b^{-2} = \frac{2kT}{b}. \tag{5.191}$$

This equation was derived by Einstein in his fundamental paper on the Brownian motion, which appeared along with two other epoch-making papers of his in Volume 17 (1905) of *Annalen der Physik*.* The subjects of the others were the special theory of relativity and the photoelectric effect. A few years later this relationship was verified experimentally by the French physicist Jean Perrin.† At the beginning of this century a few eminent physicists remained unconvinced of the molecular theory of the constitution of matter. The direct demonstration of the action of molecular forces in Brownian motion and the quantitative agreement between Einstein's formula and the experimental results vanquished all serious doubts.

Comparison of this treatment with that in Sec. 5-6D shows that the Brownian or Wiener–Lévy process discussed there is an accurate model of Brownian motion only for time intervals that are much longer than the response time M/b of the particle in interaction with the ambient medium. The autocovariance function in (5.169) and the second-order probability density function in (5.172) must be modified if times t_1 and t_2 are separated by an interval of the order of M/b or less. We leave these modifications as an exercise for the reader. The stochastic process $\mathbf{x}(t)$, $t > t_1$, is no longer a Markov process, for the future evolution of $\mathbf{x}(t)$ depends not only on its initial value at $t = t_1$, but also on that of its derivative $\mathbf{v}(t_1)$.

*A. Einstein, *Investigations on the Theory of the Brownian Motion* (R. Fürth, ed., and A. D. Cowper, trans.). New York: Dover, 1956. This book contains translations of Einstein's papers on this subject.

†J. B. Perrin, *Brownian Movement and Molecular Reality* (F. Soddy, trans.). London: Taylor & Francis, 1910.

B. Thermal Noise in a Resistor

A resistor contains a huge number of ions in a jumbled crystal lattice, and among them move the electrons. The whole structure is agitated by thermal motions, and these cause a multitude of tiny electric dipoles continually to form and vanish throughout the resistor as positive and negative charges briefly separate and reunite. These create a rapidly fluctuating voltage $\mathbf{z}(t)$ across the terminals of the resistor, and that voltage is the source of thermal noise in an electrical circuit. It consists of a dense succession of brief positive and negative pulses of random amplitude, as in (5.152), and its spectral density $S_{\mathbf{z}}(\omega)$ is uniform over a range of frequencies far broader than the frequency response of any instrument set to measure the voltage $\mathbf{z}(t)$.

Suppose that the resistor is connected to a capacitor as in Fig. 5.17. The thermal noise voltage $\mathbf{z}(t)$ can be considered as produced by a generator in series with the resistor, and it causes a fluctuating charge $\mathbf{q}(t)$ to appear on the capacitor. The differential equation for this random charge $\mathbf{q}(t)$ is

$$R\frac{d\mathbf{q}}{dt} + \frac{\mathbf{q}}{C} = \mathbf{z}(t). \tag{5.192}$$

By exactly the same procedure as we used to analyze (5.181), we find that after the circuit has been connected for a time long enough for the transients to have died away, $t >> RC$, so that the system is in a steadily fluctuating state, the spectral density of the charge on the capacitor is

$$S_{\mathbf{q}}(\omega) = \frac{C^2 S_{\mathbf{z}}}{1 + R^2C^2\omega^2}, \tag{5.193}$$

and the mean-square charge on the capacitor is

$$\text{Var}\,\mathbf{q}(t) = \overline{q^2} = \frac{CS_{\mathbf{z}}}{2R}. \tag{5.194}$$

The average energy stored in the electric field of the capacitor is

$$\frac{\overline{q^2}}{2C} = \frac{S_{\mathbf{z}}}{4R}. \tag{5.195}$$

The circuit in Fig. 5.17 possesses a huge number of degrees of freedom associated with the vibrations of the crystal lattices and the motions of the

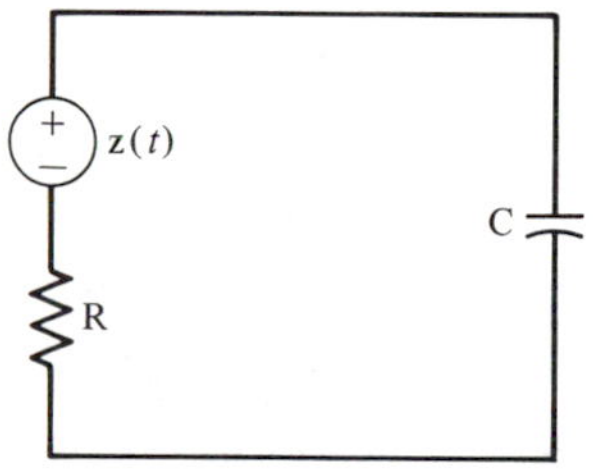

FIGURE 5.17 RC circuit.

electrons in the resistor. It possesses an additional degree of freedom associated with the charge on the capacitor, which determines the electric field between the plates. By the law of equipartition of energy, when the circuit is in thermal equilibrium at absolute temperature T, each degree of freedom accounts for an average energy of $\frac{1}{2}kT$,

$$\frac{\overline{q^2}}{2C} = \frac{S_{\mathbf{z}}}{4R} = \tfrac{1}{2}kT,$$

and hence the spectral density of the fluctuating voltage $\mathbf{z}(t)$ across the resistor is

$$S_{\mathbf{z}}(\omega) = 2kTR. \tag{5.196}$$

This result is called *Nyquist's law.*

For example, the root-mean-square voltage across the terminals of a 1 nF (10^{-3} μF) capacitor connected to a resistor and in thermal equilibrium at a temperature of 300K = 27°C amounts to $(kT/C)^{1/2} = (1.38 \times 10^{-23} \times 300/10^{-9})^{1/2}$ V $\cong 2$ μV.

The reader should verify that if the resistor is instead connected to an inductor L as in Fig. 5.18, and if one writes down the equation for the fluctuating current $\mathbf{i}(t)$ in the circuit, calculates the average energy $\frac{1}{2}L$ Var $\mathbf{i}$ stored in the magnetic field of the inductor, and equates it to $\frac{1}{2}kT$, one obtains the same result as in (5.196) for the spectral density of the voltage $\mathbf{z}(t)$. A lossless network of inductors and capacitors behaves at any frequency like either an inductor or a capacitor, and an analysis of the same kind would yield the same result for the spectral density of the voltage across a resistor connected to it. Indeed, since the thermal noise voltage $\mathbf{z}(t)$ arises from random atomic and electronic processes within the resistor itself, we conclude that whatever linear network the resistor is connected to, it will behave in thermal equilibrium as though there were in series with it an ideal generator of a stochastic voltage $\mathbf{z}(t)$ whose spectral density equals $2kTR$.

Again thermal fluctuations are associated with the dissipation of work into heat: the resistance R determines the rate at which the coherent energy of, say, an ideal voltage source of constant voltage E is changed into heat when it is connected to a resistor, the power dissipated there being $E^2/R = I^2R$, where I is the resulting electric current. The conversion to heat occurs because the energy is divided among a huge number of chaotically moving ions and electrons. Those

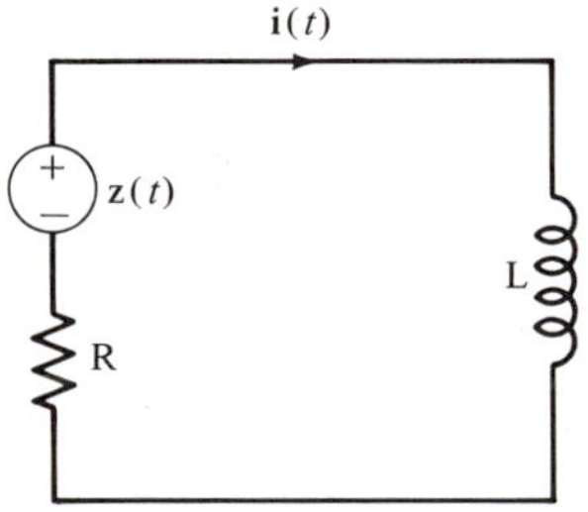

FIGURE 5.18 RL circuit.

chaotic motions cause a randomly fluctuating voltage $\mathbf{z}(t)$ to appear across the terminals of the resistor, and its spectral density is proportional to that same dissipation parameter R.

C. Thermal Noise in Linear Networks

If a linear network consists of a number of resistors connected to inductors and capacitors, the voltage across each resistor fluctuates independently. The charges on the capacitors and the fluxes in the inductors fluctuate as a result of these random voltages, and because the network is linear, the contributions to these fluctuating charges and fluxes from the several resistors can be calculated separately and superimposed. When the linear network is in thermal equilibrium and all its resistors are at the same temperature, a simpler approach to this problem is based on what is called the generalized Nyquist law, which states that the fluctuating voltage $\mathbf{v}(t)$ across any pair of terminals of the network has a spectral density given by

$$S_{\mathbf{v}}(\omega) = 2kT \operatorname{Re}[Z_e(j\omega)], \tag{5.197}$$

where $Z_e(j\omega)$ is the equivalent impedance measured across that pair of terminals, "Re" standing for "real part"; Re $[Z_e(j\omega)]$ is the equivalent resistance of the network at angular frequency ω. Again, k is Boltzmann's constant and T the absolute temperature.

EXAMPLE 5.14. RC Circuit

In the circuit of Fig. 5.17 the equivalent impedance at the terminals of the capacitor is

$$Z_e(j\omega) = \left(\frac{1}{R} + j\omega C\right)^{-1} = \frac{R}{1 + j\omega RC}.$$

Hence the spectral density of the fluctuating voltage $\mathbf{v}(t) = \mathbf{q}(t)/C$ across the capacitor is, by (5.197),

$$S_{\mathbf{v}}(\omega) = 2kT \operatorname{Re}\left(\frac{R}{1 + j\omega RC}\right) = \frac{2kTR}{1 + \omega^2R^2C^2},$$

in agreement with (5.193) and (5.196) since $S_{\mathbf{q}}(\omega) = C^2S_{\mathbf{v}}(\omega)$. ■

An equivalent form of the generalized Nyquist law states that when the network is in thermal equilibrium at absolute temperature T, the spectral density of the randomly fluctuating current $\mathbf{i}(t)$ in any branch is

$$S_{\mathbf{i}}(\omega) = 2kT \operatorname{Re}[Y_e(j\omega)], \tag{5.198}$$

where $Y_e(j\omega)$ is the equivalent admittance across two terminals formed by opening that branch of the circuit.

EXAMPLE 5.15. RL Circuit

Opening the circuit of Fig. 5.18 yields a series connection of a resistor and an inductor, whose total impedance is $R + j\omega L$ and whose admittance is $(R + j\omega L)^{-1}$. Hence the spectral density of the current $\mathbf{i}(t)$ in the RL circuit is

$$S_{\mathbf{i}}(\omega) = 2kT \text{ Re } [(R + j\omega L)^{-1}] = \frac{2kTR}{R^2 + \omega^2 L^2}. \tag{5.199}$$

■

From (5.197) we can derive the autocovariance function $C_{\mathbf{v}}(\tau)$ of the voltage $\mathbf{v}(t)$ across the terminals. It is the Fourier integral of the spectral density,

$$\begin{aligned} C_{\mathbf{v}}(\tau) &= 2kT \int_{-\infty}^{\infty} \text{Re } [Z_e(j\omega)] e^{j\omega\tau} \frac{d\omega}{2\pi} \\ &= kT \int_{-\infty}^{\infty} [Z_e(j\omega) + Z_e^*(j\omega)] e^{j\omega\tau} \frac{d\omega}{2\pi} \\ &= kT[z_e(\tau) + z_e(-\tau)], \end{aligned}$$

where

$$z_e(\tau) = \int_{-\infty}^{\infty} Z_e(j\omega) e^{j\omega\tau} \frac{d\omega}{2\pi} \tag{5.200}$$

is the impulse response associated with the equivalent impedance $Z_e(j\omega)$, and since impulse responses must be real,

$$z_e(-\tau) = \left(\int_{-\infty}^{\infty} Z_e(j\omega) e^{-j\omega\tau} \frac{d\omega}{2\pi} \right)^* = \int_{-\infty}^{\infty} Z_e^*(j\omega) e^{j\omega\tau} \frac{d\omega}{2\pi}.$$

The impulse response $z_e(\tau)$, however, equals zero for $\tau < 0$, and $z_e(-\tau) \equiv 0$ for $\tau > 0$. Hence for any value of the delay variable τ, the autocovariance function of the voltage is

$$C_{\mathbf{v}}(\tau) = kTz_e(|\tau|). \tag{5.201}$$

By the same method one can show that the autocovariance function of the fluctuating current $\mathbf{i}(t)$ in a branch of the network is

$$C_{\mathbf{i}}(\tau) = kTy_e(|\tau|), \tag{5.202}$$

where

$$y_e(\tau) = \int_{-\infty}^{\infty} Y_e(j\omega) e^{j\omega\tau} \frac{d\omega}{2\pi} \tag{5.203}$$

is the impulse response associated with the equivalent admittance across a pair of terminals formed by opening that branch.

EXAMPLE 5.14. RC Circuit (continued)

The impulse response associated with the impedance $Z_e(j\omega)$ across the capacitor in Fig. 5.17 is

$$z_e(\tau) = C^{-1} e^{-\tau/RC} U(\tau),$$

and hence the autocovariance function of the voltage $v(t)$ across the capacitor is

$$C_{\mathbf{v}}(\tau) = \frac{kT}{C} e^{-|\tau|/RC}.$$

In particular,

$$\overline{v^2} = \text{Var}\,\mathbf{v}(t) = C_{\mathbf{v}}(0) = \frac{kT}{C},$$

and the mean energy stored in the electric field of the capacitor is as before $\frac{1}{2}C\overline{v^2} = \frac{1}{2}kT$. ■

EXAMPLE 5.15. RL Circuit (continued)

The impulse response associated with the admittance

$$Y_e(j\omega) = \frac{1}{R + j\omega L}$$

across a pair of terminals formed by opening the RL circuit of Fig. 5.18 is

$$y_e(\tau) = L^{-1}e^{-R\tau/L}U(\tau),$$

and by (5.202) the autocovariance function of the random current $\mathbf{i}(t)$ in that circuit is

$$C_{\mathbf{i}}(\tau) = \frac{kT}{L}e^{-R|\tau|/L},$$

from which one can deduce that the mean energy stored in the magnetic field of the inductor is also $\frac{1}{2}kT$. ■

If all we want is the variance of a fluctuating voltage or current, it is unnecessary even to determine the impulse response $z_e(\tau)$ or $y_e(\tau)$. From (5.201),

$$\text{Var}\,\mathbf{v}(t) = kTz_e(0^+),$$

but by the initial-value theorem of Laplace transforms,

$$z_e(0^+) = \lim_{p\to\infty} pZ_e(p).$$

Hence for the voltage $\mathbf{v}(t)$ across two terminals,

$$\text{Var}\,\mathbf{v}(t) = kT \lim_{p\to\infty} pZ_e(p), \tag{5.204}$$

and similarly for the current $\mathbf{i}(t)$ in a branch,

$$\text{Var}\,\mathbf{i}(t) = kT \lim_{p\to\infty} pY_e(p). \tag{5.205}$$

EXAMPLE 5.16. RLC Circuit

In the RLC circuit of Fig. 5.19, we seek the mean-square voltage on the capacitor. The equivalent impedance across its terminals is, in the Laplace domain,

$$Z_e(p) = \frac{R + pL}{pC(R + pL + 1/pC)},$$

whence

$$\lim_{p\to\infty} pZ_e(p) = \frac{1}{C},$$

and by (5.204), Var $\mathbf{v}(t) = \overline{v^2} = kT/C$, which of course also follows from the fact with which we began our analysis—that the mean energy stored in the capacitor in thermal equilibrium is $\frac{1}{2}kT$.

To find the mean-square current in the circuit of Fig. 5.19, we break it to form a

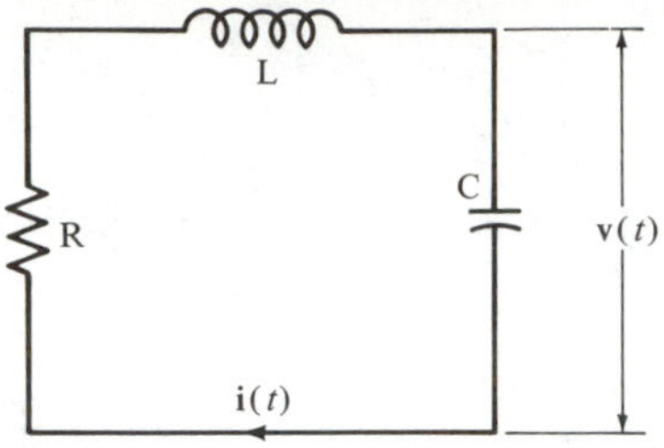

FIGURE 5.19 RLC circuit.

series connection of R, L, and C, whose total impedance is $R + pL + (pC)^{-1}$. The admittance across the terminals so formed is

$$Y_e(p) = [R + pL + (pC)^{-1}]^{-1} = \frac{pC}{1 + pC(R + pL)}$$

and $\lim_{p\to\infty} pY_e(p) = L^{-1}$, whence by (5.205),

$$\text{Var } \mathbf{i}(t) = \overline{i^2} = \frac{kT}{L},$$

in agreement with the fact that the mean energy stored in the magnetic field of the inductor is $\frac{1}{2}kT$. ■

D. Power Transfer Between Two Networks

The generalized Nyquist relation (5.197) can be derived by applying the laws of circuit theory to evaluating the expression

$$S_{\mathbf{v}}(\omega) = 2kT \sum_k R_k \mid H_k(j\omega) \mid^2,$$

where R_k is the resistance of the kth resistor in the network and $H_k(j\omega)$ is the transfer function between a voltage in series with that resistor and the voltage across the terminals where $\mathbf{v}(t)$ is observed. The summation includes each resistor in the network, all of them at the same absolute temperature T. We shall instead take a thermodynamical approach.

First we consider the transfer of noise power between two networks whose open-circuit impedances are, respectively, $Z_1(j\omega)$ and $Z_2(j\omega)$, which are connected as shown in Fig. 5.20. In series with the first is an ideal generator of a random voltage $\mathbf{z}_1(t)$, which can be considered as the Thévenin equivalent open-circuit voltage due to all the independent sources of thermal noise in the left-hand network, one for each resistor therein. In series with the second is an ideal generator of a random voltage $\mathbf{z}_2(t)$, the Thévenin open-circuit voltage accounting for all the thermal sources in the right-hand network. The two stochastic processes $\mathbf{z}_1(t)$ and $\mathbf{z}_2(t)$ are statistically independent.

Let $\mathbf{v}_2(t)$ be the random voltage across impedance $Z_2(j\omega)$ due to the generator of $\mathbf{z}_1(t)$, and let $\mathbf{i}(t)$ be the random current flowing around the loop. Then the average power dissipated in the right-hand network is $E[\mathbf{v}_2(t)\, \mathbf{i}(t)]$. Let

$$H_{1\to i}(j\omega) = (Z_1 + Z_2)^{-1} = \mathcal{F}h_{1\to i}(\tau) \tag{5.206}$$

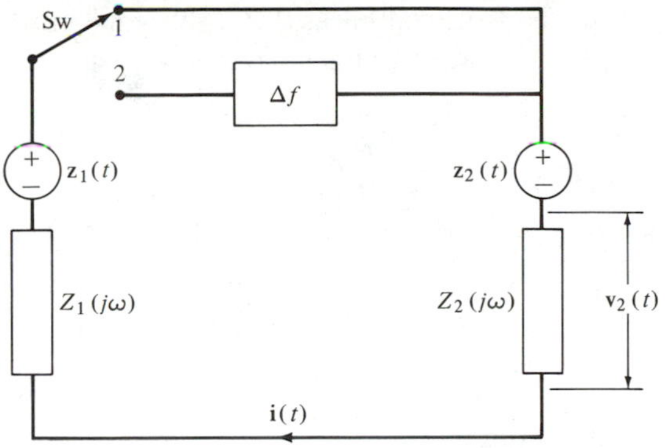

FIGURE 5.20 Series connection of equivalent impedances.

be the transfer function connecting the current in the loop with the voltage $\mathbf{z}_1(t)$; $h_{1\to i}(\tau)$ is the associated impulse response, $\mathcal{F}$ standing for "Fourier transform." Then

$$\mathbf{i}(t) = \int_0^\infty h_{1\to i}(\tau)\mathbf{z}_1(t-\tau)\,d\tau. \tag{5.207}$$

Similarly, the transfer function connecting the voltage $\mathbf{v}_2(t)$ with the voltage $\mathbf{z}_1(t)$ is

$$H_{1\to 2}(j\omega) = \frac{Z_2}{Z_1 + Z_2} = \mathcal{F}h_{1\to 2}(\tau), \tag{5.208}$$

and that voltage across the second network due to the first is

$$\mathbf{v}_2(t) = \int_0^\infty h_{1\to 2}(\tau)\mathbf{z}_1(t-\tau)\,d\tau. \tag{5.209}$$

Hence the average power dissipated in Z_2 and due to $\mathbf{z}_1(t)$ is

$$\begin{aligned} P_{12} &= E[\mathbf{v}_2(t)\mathbf{i}(t)] \\ &= E\int_0^\infty\int_0^\infty h_{1\to 2}(s_1)h_{1\to i}(s_2)\mathbf{z}_1(t-s_1)\mathbf{z}_1(t-s_2)\,ds_1\,ds_2 \\ &= \int_0^\infty\int_0^\infty h_{1\to 2}(s_1)h_{1\to i}(s_2)C_1(s_1-s_2)\,ds_1\,ds_2 \\ &= \int_{-\infty}^\infty H^*_{1\to 2}(j\omega)H_{1\to i}(j\omega)S_1(\omega)\,\frac{d\omega}{2\pi} \end{aligned} \tag{5.210}$$

in terms of the autocovariance function $C_1(\tau)$ and the spectral density $S_1(\omega)$ of the voltage $\mathbf{z}_1(t)$. The last step involved the convolution theorem. Using (5.206) and (5.208) we can write this transferred power as

$$P_{12} = \int_{-\infty}^\infty \frac{Z_2^*(j\omega)S_1(\omega)}{|Z_1+Z_2|^2}\,\frac{d\omega}{2\pi}.$$

Because the power P_{12} must be real, and because $Z_2^*(j\omega)$ is the only factor in the integrand that is complex, this average transferred power equals

$$P_{12} = \int_{-\infty}^{\infty} \frac{\mathrm{Re}[Z_2(j\omega)]S_1(\omega)}{|Z_1 + Z_2|^2} \frac{d\omega}{2\pi}. \tag{5.211}$$

By the same kind of analysis, the power transferred from network 2 to network 1 is

$$P_{21} = \int_{-\infty}^{\infty} \frac{\mathrm{Re}[Z_1(j\omega)]S_2(\omega)}{|Z_1 + Z_2|^2} \frac{d\omega}{2\pi}, \tag{5.212}$$

where $S_2(\omega)$ is the spectral density of $\mathbf{z}_2(t)$.

At thermal equilibrium, $P_{12} = P_{21}$. If, furthermore, by moving the switch "Sw" from position 1 to position 2, we inserted between network 1 and network 2 a narrowband filter—shown in Fig. 5.20 as "Δf"—that passes only frequencies in the neighborhood of an arbitrary frequency ω, the average power passing from network 1 to network 2 must still be the same as that passing from network 2 to network 1; that is, the integrands of (5.211) and (5.212) must be equal at all frequencies ω. Hence

$$\frac{S_1(\omega)}{\mathrm{Re}[Z_1(j\omega)]} = \frac{S_2(\omega)}{\mathrm{Re}[Z_2(j\omega)]}. \tag{5.213}$$

Now suppose that network 2 is a resistor R. We know from (5.196) that $S_2(\omega)/\mathrm{Re}[Z_2(j\omega)] = 2kT$, and hence, whatever network 1 may contain, the spectral density of the voltage $\mathbf{z}_1(t)$ across its terminals must be

$$S_1(\omega) = 2kT\,\mathrm{Re}[Z_1(j\omega)],$$

whence the generalized Nyquist law (5.197).

A similar analysis, carried through with the Norton equivalent of each network, yields (5.198). Alternatively, we can derive (5.198) from (5.197) as follows. At any frequency ω, a sinusoidal generator of phasor voltage E in series with an impedance $Z_e(j\omega)$ is equivalent to an ideal current generator producing a phasor current $I = E/Z_e(j\omega)$; the current generator is in parallel with the impedance. Figure 5.21 shows these equivalent circuits; $Y_e(j\omega) = 1/Z_e(j\omega)$. Hence the spectral density of the fluctuating current $\mathbf{i}(t)$ must be, by (5.127),

$$S_{\mathrm{i}}(\omega) = \frac{S_{\mathrm{v}}(\omega)}{|Z_e(j\omega)|^2} = 2kT\frac{\mathrm{Re}[Z_e(j\omega)]}{|Z_e(j\omega)|^2} = 2kT\,\mathrm{Re}\left[\frac{1}{Z_e(j\omega)}\right]$$
$$= 2kT\,\mathrm{Re}[Y_e(j\omega)],$$

as in (5.198).

E. Maximum Power Transfer

The power transferred from network 1 to network 2 in Fig. 5.20 through a filter passing only frequencies in a band of width $\Delta f = \Delta\omega/2\pi$ hertz at frequency ω is, from (5.211) and (5.197),

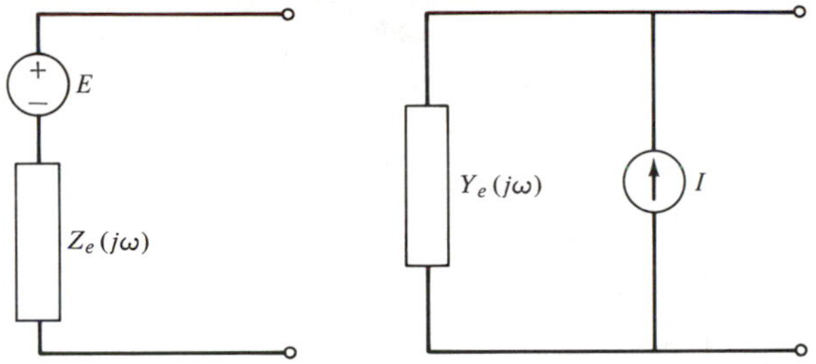

FIGURE 5.21 Thévenin and Norton equivalents.

$$\Delta P_{12} = 2\,\frac{S_1(\omega)\,\mathrm{Re}[Z_2(j\omega)]}{|Z_1 + Z_2|^2}\,\Delta f = 4kT\,\frac{\mathrm{Re}[Z_1(j\omega)]\mathrm{Re}[Z_2(j\omega)]}{|Z_1 + Z_2|^2}\,\Delta f. \tag{5.214}$$

Considering network 2 as a load on network 1, we ask for what load this transferred power ΔP_{12} is greatest. The denominator in (5.214) is

$$|\,\mathrm{Re}\,(Z_1 + Z_2)\,|^2 + |\,\mathrm{Im}\,(Z_1 + Z_2)\,|^2 \geq |\,\mathrm{Re}\,(Z_1 + Z_2)\,|^2$$

and is smallest when $\mathrm{Im}\,Z_2 = -\,\mathrm{Im}\,Z_1$. Writing $\mathrm{Re}\,Z_1 = R_1$, $\mathrm{Re}\,Z_2 = R_2$, we find

$$P_{12} \leq 4kT\,\frac{R_1R_2}{(R_1 + R_2)^2}\,\Delta f \leq kT\,\Delta f \tag{5.215}$$

with equality attained when $R_1 = R_2$, for

$$(R_1 + R_2)^2 \geq (R_1 + R_2)^2 - (R_1 - R_2)^2 = 4R_1R_2.$$

Hence the maximum power is drawn from network 1 when network 2 is *matched* to it in the sense that

$$Z_2(j\omega) = Z_1^*(j\omega).$$

The quantity $kT\,\Delta f$ is the maximum thermal power that can be drawn from the network on the left in a band of frequencies of width Δf hertz.

If the combination of networks 1 and 2 is in thermal equilibrium, an equal power will flow from network 2 to network 1. Thermal equilibrium is not necessary, however. If network 1 is kept at temperature T by a surrounding heat bath, and network 2 is at absolute zero, an average power equal to $kT\,\Delta f$ watts, supplied by the heat bath, flows from network 1 to network 2 through the pass filter of width Δf hertz, but no power flows back. It is in this sense that $kT\,\Delta f$ can be termed the maximum available power in that frequency band.

The system on the left, provided that it is linear and maintained at temperature T, need not be a network of resistors, inductors, and capacitors. It might be a transmission line terminated at the far end by some impedance Z_T, as in Fig. 5.22, or it might be an antenna in a large enclosure, as in Fig. 5.23. In either case there will appear across the terminals a fluctuating voltage $\mathbf{z}(t)$ whose spectral density equals

$$S_{\mathbf{z}}(\omega) = 2kT\,\mathrm{Re}[Z_e(j\omega)],$$

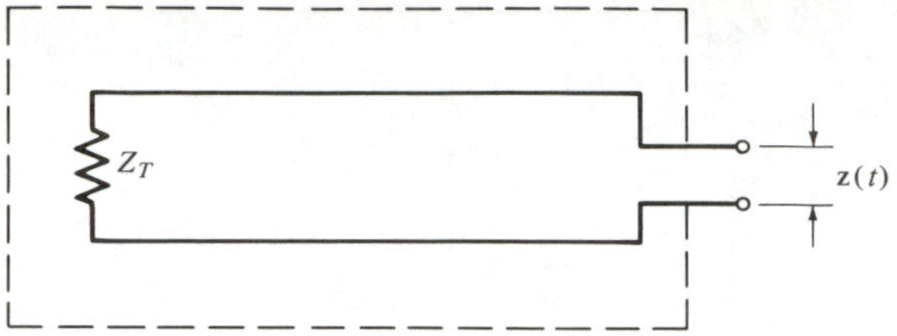

FIGURE 5.22 Transmission line.

where T is the absolute temperature of the surroundings and $Z_e(j\omega)$ is the impedance as measured across the terminals, "looking into" the transmission line or antenna. Throughout the enclosure containing the antenna (Fig. 5.23) are the fluctuating electromagnetic fields of the blackbody radiation that fills it, and the voltage $\mathbf{z}(t)$ arises from the excitation of fluctuating currents in the antenna by that ambient radiation. The main component of the equivalent impedance $Z_e(j\omega)$ of the antenna is its radiation resistance $R_a(\omega)$, and the spectral density of the voltage $\mathbf{z}(t)$ across the terminals of the antenna is approximately $2kTR_a$. It is the radiation resistance R_a that measures the rate at which electromagnetic energy would be radiated out into space and eventually absorbed as heat, were a sinusoidal generator of frequency ω connected to the terminals. Here is yet another instance of the relation between dissipation and thermal fluctuation.

The universe, fortunately, is not in thermal equilibrium. Nevertheless, an antenna whose beam is directed into space picks up randomly fluctuating electromagnetic fields of all frequencies. It is customary to measure the strength of that incoming radiation by an effective temperature $T_e(\omega)$ K, which is so defined that the spectral density of the fluctuating voltage $\mathbf{z}(t)$ at the terminals is given by $2kT_e(\omega)R_a(\omega)$, where $R_a(\omega)$ is the radiation resistance of the antenna at frequency ω. Equivalently, the maximum power that a matched receiver can draw in a frequency band of width Δf hertz at frequency ω is $kT_e(\omega)\,\Delta f$. The effective temperature will depend on the direction in which the antenna is pointing;

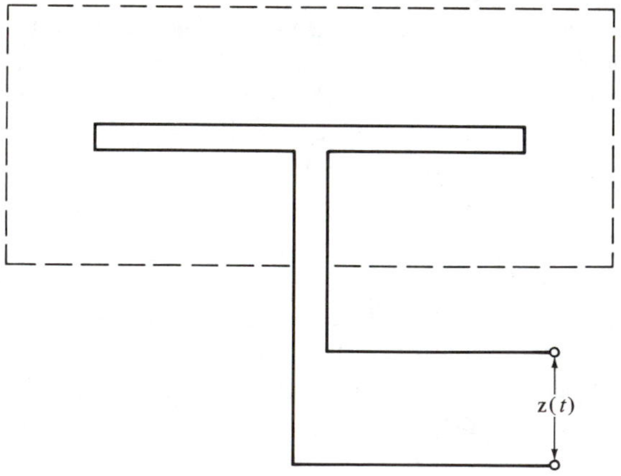

FIGURE 5.23 Antenna.

it may, for instance, be looking at a radio star. Included in $T_e(\omega)$ is the celebrated 3K background radiation that is said to be left from the initial "big bang" that started the universe on its erratic evolution.

The total power available from a linear system in thermal equilibrium at temperature T is not infinite, as the constant spectral density kT would seem to indicate. Thermal fluctuations are subject to the laws of quantum mechanics, which modifies the generalized Nyquist law, replacing the factor kT everywhere by the *Planck factor*

$$\hbar\omega(e^{\hbar\omega/kT} - 1)^{-1}$$

where $h = 2\pi\hbar = 6.626 \times 10^{-34}$ joule-sec is Planck's constant. This factor reduces to kT when $\hbar\omega << kT$, but must be taken into account when $\hbar\omega \gtrsim kT$, that is, for frequencies $\omega/2\pi \gtrsim kT/h$. For $T = 300\text{K} = 27°\text{C}$, $kT/h = 6.25 \times 10^{12}$ Hz, which corresponds to light of wavelength $hc/kT = 48\ \mu\text{m}$, a wavelength in the far infrared. For $\hbar\omega >> kT$, the Planck factor decreases exponentially to zero, extinguishing all thermal spectral densities at frequencies beyond the optical range.

F. Amplifier Noise

An antenna can be modeled as an ideal voltage source in series with an impedance $Z_a(j\omega)$. The output $v(t)$ of this ideal source includes both the signal incident on the antenna and the thermal noise it picks up from its surroundings. An amplifier connected across its terminals will amplify both the signal and the noise, but the amplifier generates additional noise because of the random electronic processes going on inside it. In a transistor, for instance, the random arrival of electrons or holes at the collector creates a form of shot noise, to which is added a fluctuation due to the random formation and recombination of electrons and holes in the semiconducting material. If the amplifier contributed no such excess noise, the maximum average noise power that could be drawn from it by a matched load would equal $GkT\,\Delta f$ in a band of frequencies of width Δf hertz, where T is the effective temperature of the thermal noise and G is the power gain of the amplifier. If the unilateral spectral density of the actual noise output of the amplifier equals N at the frequencies being passed, the ratio

$$F = \frac{N\,\Delta f}{GkT\,\Delta f} = \frac{N}{GkT}$$

of the maximum average power actually available in that band of frequencies to the ideal value $GkT\,\Delta f$ is called the *noise figure* of the amplifier. Necessarily $F \geq 1$. In measuring noise figures of amplifiers, it is customary to set the effective temperature T of the noise at the input equal to $290\text{K} = 17°\text{C}$. Techniques for carrying out these measurements and for designing amplifiers with low values of $F - 1$ are discussed in books on electronics.

The effective noise figure of a number of amplifiers connected in cascade can be derived by the following argument. Let the first of two cascaded amplifiers have power gain G_1 and noise figure F_1; let the second have gain G_2 and noise figure F_2. The unilateral spectral density of the noise entering the second amplifier is

$$N_1 = G_1F_1kT,$$

and when amplified by the second amplifier it contributes $G_2G_1F_1kT$ to the output noise spectral density. The additional noise supplied by the second amplifier is $(F_2 - 1)G_2kT$ by the definition of the noise figure F_2, and the total spectral density at its output is therefore

$$G_2(G_1F_1 + F_2 - 1)kT = G_1G_2F_ckT$$

in terms of the effective noise figure F_c of the combination, whose gain is G_1G_2. Hence we find that

$$F_c = F_1 + \frac{F_2 - 1}{G_1}.$$

For three amplifiers in cascade, by analogous reasoning,

$$F_c = F_1 + \frac{F_2 - 1}{G_1} + \frac{F_3 - 1}{G_1G_2},$$

and so on. The overall noise figure F_c is dominated by that of the first stage of amplification when its gain G_1 is large.

5-8 MEASUREMENT OF STOCHASTIC PROCESSES

Parameters such as the mean and the variance characterize the first-order probability density function $f_{\mathbf{x}_1}(x_1, t_1)$ of a stochastic process $\mathbf{x}(t)$, and the autocovariance function $C_{\mathbf{x}}(t_1, t_2)$ partially characterizes its second-order density function. These parameters can be estimated at arbitrary instants t_1, t_2 of time by the method described in Sec. 4-6. A large number of trials of the chance experiment defining the stochastic process are conducted, samples of the realization $\mathbf{x}(t)$ arising in each trial are taken at times t_1, t_2, and the averages required by (4.168), (4.169), and (4.180) are calculated. The labor of estimating the mean $E[\mathbf{x}(t_1)]$, the variance Var $\mathbf{x}(t_1)$, and the autocovariance function $C_{\mathbf{x}}(t_1, t_2)$ in this manner for any comprehensive selection of instants of time would be enormous.

A. Estimating the Mean of a Stationary Process

If the stochastic process $\mathbf{x}(t)$ is known or deemed to be stationary, at least over a long period of time, one would hope to be able to estimate these parameters from a single realization of the process. Indeed, in many situations more than a single realization may be unavailable. A meteorologist, for instance, may wish to model the temperature or barometric pressure at a particular station as a stochastic process and needs to evaluate its mean and its autocovariance function, but only one record of temperature and pressure exists. In studying the noise in a semiconductor amplifier under various conditions of applied voltage, ambient temperature, and the like, it may not be feasible to set up more than one device under a particular set of conditions, let alone the great number that would be required for the averages called for by the methods of Sec. 4-6.

A natural candidate for an estimate of the mean value m of a stationary process $\mathbf{x}(t)$ is the output, at an arbitrary time t, of a linear filter that in effect averages over some stretch of the past of the process, as did the integrator studied in Example 5.11. We therefore take as an estimate $\hat{\mathbf{m}}$ of the mean value $E[\mathbf{x}(t)]$ the output

$$\hat{\mathbf{m}} = \mathbf{y}(t) = \int_0^\infty h(s)\mathbf{x}(t - s)\, ds \tag{5.216}$$

at time t of a filter whose impulse response $h(s)$ is normalized so that

$$\int_0^\infty h(s)\, ds = 1. \tag{5.217}$$

Then the expected value of the estimate is

$$\begin{aligned} E(\hat{\mathbf{m}}) = E[\mathbf{y}(t)] &= \int_0^\infty h(s)E[\mathbf{x}(t - s)]\, ds \\ &= m\int_0^\infty h(s)\, ds = m, \end{aligned} \tag{5.218}$$

and the estimate $\hat{\mathbf{m}}$ is unbiased. Its variance or mean-square error is, as in (5.94),

$$\text{Var}\, \hat{\mathbf{m}} = \text{Var}\, \mathbf{y}(t) = \int_0^\infty \int_0^\infty h(s_1)h(s_2)C_\mathbf{x}(s_1 - s_2)\, ds_1\, ds_2, \tag{5.219}$$

in deriving which we used the fact that the process $\mathbf{x}(t)$ is stationary; $C_\mathbf{x}(\tau)$ is its autocovariance function.

If this method is to be useful, the mean-square error Var $\hat{\mathbf{m}}$ should be as small as possible, and the only hope of reducing it is to integrate over ever longer segments of the input. That would be of no use, however, for the constant process introduced in Example 5.1, whose autocovariance function, as shown in Example 5.5, is constant and equal to Var $\mathbf{x}$. For this process

$$\text{Var}\, \hat{\mathbf{m}} = \text{Var}\, \mathbf{x} \left[\int_0^\infty h(s)\, ds\right]^2 = \text{Var}\, \mathbf{x}$$

is unaffected by lengthening the integration time T_{int} of the filter. The reason is not hard to see: the estimate $\hat{\mathbf{m}}$ yielded by the constant process $\mathbf{x}(t) = \mathbf{x}$ equals the value of $\mathbf{x}$ that happens to arise in the chance experiment generating $\mathbf{x}(t)$, no matter for how long a time the filter integrates. Nothing new ever occurs in this process, and observing a particular sample function tells us nothing about the expected value of $\mathbf{x}(t)$ with respect to the ensemble from which it was drawn by our chance experiment.

If Var $\hat{\mathbf{m}}$ defined in (5.219) is to be the smaller, the longer the integration time T_{int}, the autocovariance function $C_\mathbf{x}(\tau)$ of the process $\mathbf{x}(t)$ must drop off to zero as the delay $|\tau|$ increases beyond all bounds. Samples of the process

$\mathbf{x}(t)$ taken at times t_1 and t_2 must be the less correlated, the longer the interval $|t_2 - t_1|$ between them. Then the spectral density $S_{\mathbf{x}}(\omega)$ can be defined as the Fourier transform of $C_{\mathbf{x}}(\tau)$, and by (5.138)

$$\text{Var } \hat{\mathbf{m}} = \text{Var } \mathbf{y}(t) = \int_{-\infty}^{\infty} | H(j\omega) |^2 S_{\mathbf{x}}(\omega) \frac{d\omega}{2\pi} \tag{5.220}$$

in terms of the transfer function $H(j\omega)$ of the averaging filter. The variance of its output, relative to that of its input, is given by

$$\frac{\text{Var } \mathbf{y}(t)}{\text{Var } \mathbf{x}(t)} = \int_{-\infty}^{\infty} \left| \frac{H(j\omega)}{H(0)} \right|^2 S_{\mathbf{x}}(\omega) \frac{d\omega}{2\pi} \Bigg/ \int_{-\infty}^{\infty} S_{\mathbf{x}}(\omega) \frac{d\omega}{2\pi}, \tag{5.221}$$

for by (5.217) $H(0) = 1$. The integration time T_{int} is roughly the reciprocal $1/W$ of the bandwidth of the filter, which can be defined for such a low-pass filter as

$$W = T_{\text{int}}^{-1} = \int_{-\infty}^{\infty} \left| \frac{H(j\omega)}{H(0)} \right|^2 \frac{d\omega}{2\pi}$$
$$= \frac{\displaystyle\int_0^{\infty} [h(s)]^2 \, ds}{\left[\displaystyle\int_0^{\infty} h(s) \, ds \right]^2}. \tag{5.222}$$

The correlation time τ_c of the stochastic process $\mathbf{x}(t)$ (i.e., the width of its autocovariance function) is roughly the reciprocal of the bandwidth of its spectral density,

$$\tau_c \sim S_{\mathbf{x}}(0) \Bigg/ \int_{-\infty}^{\infty} S_{\mathbf{x}}(\omega) \frac{d\omega}{2\pi} = \int_{-\infty}^{\infty} \frac{C_{\mathbf{x}}(\tau)}{C_{\mathbf{x}}(0)} \, d\tau. \tag{5.223}$$

Now when $T_{\text{int}} >> \tau_c$, $W\tau_c << 1$, the factor $|H(j\omega)/H(0)|^2$ in the integrand of the numerator of (5.221), which is peaked at $\omega = 0$, is much narrower than the factor $S_{\mathbf{x}}(\omega)$, and (5.221) becomes approximately

$$\frac{\text{Var } \hat{\mathbf{m}}}{\text{Var } \mathbf{x}} = \frac{\text{Var } \mathbf{y}(t)}{\text{Var } \mathbf{x}(t)} \simeq \frac{S_{\mathbf{x}}(0) \displaystyle\int_{-\infty}^{\infty} | H(j\omega)/H(0) |^2 \, d\omega/2\pi}{\displaystyle\int_{-\infty}^{\infty} S_{\mathbf{x}}(\omega) \, d\omega/2\pi}$$
$$= W\tau_c = \frac{\tau_c}{T_{\text{int}}}. \tag{5.224}$$

In order for the averaging filter to yield a good estimate of the mean $\hat{\mathbf{m}}$ of the process by (5.216), we want Var $\hat{\mathbf{m}}$ << Var $\mathbf{x}(t)$, and therefore $T_{\text{int}} >> \tau_c$: the filter must integrate over a time much longer than the correlation time τ_c of the process.

We have assumed here that $S_{\mathbf{x}}(0) \neq 0$; otherwise, (5.223) would be meaningless. The mean m of $\mathbf{x}(t)$ corresponds to a dc component of the process, and if $S_{\mathbf{x}}(0)$ equaled zero, the "noise" $\mathbf{x}(t) - m$ would possess negligible components with frequencies in the neighborhood of dc. The decrease of the ratio Var $\mathbf{y}(t)$: Var $\mathbf{x}(t)$ with increasing integration time T_{int} would then be much more rapid than as indicated by (5.224).

B. Estimating the Correlation Function of a Stationary Process

The correlation function $R_{\mathbf{x}}(t_1, t_2)$ of a stationary stochastic process, like the autocovariance function $C_{\mathbf{x}}(t_1, t_2)$, is a function only of the duration $|t_2 - t_1|$ of the interval between the times t_1 and t_2 at which it is sampled. It can be written in terms of $\tau = t_2 - t_1$ as

$$R_{\mathbf{x}}(\tau) = E[\mathbf{x}(t)\mathbf{x}(t - \tau)], \tag{5.225}$$

and it is an even function of $\tau : R_{\mathbf{x}}(-\tau) = R_{\mathbf{x}}(\tau)$. The value $R_{\mathbf{x}}(0)$ for $\tau = 0$ is the mean-square value $E[\mathbf{x}(t)]^2$ of the stochastic process; it is sometimes loosely called the "average power" of the process.

A natural candidate for an estimator of the correlation function is the output $\mathbf{y}(t)$ of an averaging filter whose input is the product $\mathbf{x}(t)\mathbf{x}(t - \tau)$, as shown in Fig. 5.24. We suppose the impulse response $h(s)$ of the averaging filter normalized as in (5.217). The output of the filter is

$$\mathbf{y}(t) = \int_0^\infty h(s)\mathbf{x}(t - s)\mathbf{x}(t - \tau - s)\, ds, \tag{5.226}$$

and its expected value is

$$\begin{aligned} E[\mathbf{y}(t)] &= \int_0^\infty h(s)E[\mathbf{x}(t - s)\mathbf{x}(t - \tau - s)]\, ds \\ &= \int_0^\infty h(s)R_{\mathbf{x}}(\tau)\, ds = R_{\mathbf{x}}(\tau). \end{aligned} \tag{5.227}$$

The output $\mathbf{y}(t)$ of this "correlator" observed at any time t is therefore an unbiased estimator of the correlation function $R_{\mathbf{x}}(\tau)$. The delay τ must be capable of variation over a comprehensive range of values. If the process $\mathbf{x}(t)$ is available as an electrical signal, this can be done by passing it through a delay line with a number of closely spaced taps, among which the observer can switch. If $\mathbf{x}(t)$ is recorded on magnetic tape, the tape can be passed through two transducers or readout heads whose distance apart can be varied and whose outputs are multiplied and integrated as indicated by Fig. 5.24.

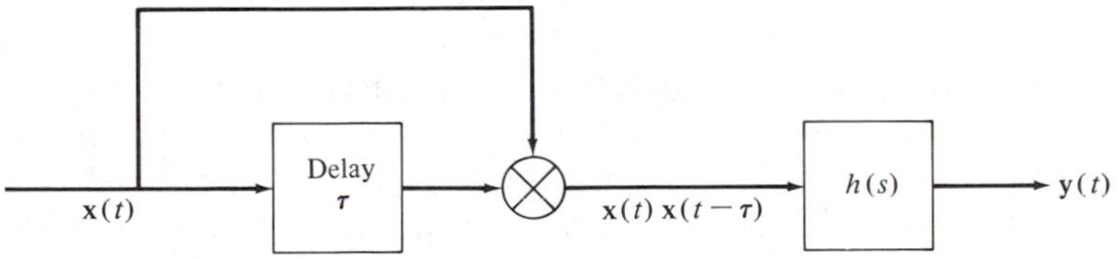

FIGURE 5.24 Correlator.

The product

$$\mathbf{z}(t) = \mathbf{x}(t)\mathbf{x}(t - \tau)$$

fluctuates roughly twice as rapidly as the process $\mathbf{x}(t)$ itself, and the purpose of the integrating filter $h(s)$ is to smooth those fluctuations. If we assume for simplicity that the input $\mathbf{x}(t)$ is a Gaussian stochastic process with mean value zero, the spectral density $S_{\mathbf{z}}(\omega)$ of the product $\mathbf{z}(t)$ is

$$S_{\mathbf{z}}(\omega) = \int_{-\infty}^{\infty} S_{\mathbf{x}}(u - \tfrac{1}{2}\omega)S_{\mathbf{x}}(u + \tfrac{1}{2}\omega)(1 + \cos 2\tau u)\,\frac{du}{2\pi}, \tag{5.228}$$

whose derivation we leave as an exercise for the reader. Because $S_{\mathbf{x}}(u - \tfrac{1}{2}\omega) = S_{\mathbf{x}}(\tfrac{1}{2}\omega - u)$, the spectral density $S_{\mathbf{z}}(\omega)$ of the product process $\mathbf{z}(t)$,

$$\begin{aligned} S_{\mathbf{z}}(\omega) &= \int_{-\infty}^{\infty} S_{\mathbf{x}}(\tfrac{1}{2}\omega - u)S_{\mathbf{x}}(\tfrac{1}{2}\omega + u)(1 + \cos 2\tau u)\,\frac{du}{2\pi} \\ &= \int_{-\infty}^{\infty} S_{\mathbf{x}}(v)S_{\mathbf{x}}(\omega - v)[1 + \cos 2\tau(v - \tfrac{1}{2}\omega)]\,\frac{dv}{2\pi}, \end{aligned}$$

closely resembles the self-convolution of the spectral density $S_{\mathbf{x}}(\omega)$ of the input, except for a slight modification due to the factor $(1 + \cos 2\tau u)$ in the integrand of (5.228). The spectral density $S_{\mathbf{z}}(\omega)$ is therefore approximately twice as broad as $S_{\mathbf{x}}(\omega)$, and the correlation time of $\mathbf{z}(t)$ will be roughly $\tfrac{1}{2}\tau_c$, half that of the input $\mathbf{x}(t)$. In order to reduce the fluctuations of $\mathbf{z}(t) = \mathbf{x}(t)\mathbf{x}(t - \tau)$ the filter $h(s)$ should integrate over a time T_{int} much longer than $\tfrac{1}{2}\tau_c$. The variance of the output $\mathbf{y}(t)$ of the correlator is

$$\operatorname{Var} \mathbf{y}(t) = \int_{-\infty}^{\infty} |H(j\omega)|^2 S_{\mathbf{z}}(\omega)\,\frac{d\omega}{2\pi}, \tag{5.229}$$

where $H(j\omega)$ is the transfer function of the integrating filter, and it can be evaluated from (5.228) for special assumptions about the spectral density $S_{\mathbf{x}}(\omega)$ of the input.

The autocovariance function $C_{\mathbf{x}}(\tau)$ of a stationary process $\mathbf{x}(t)$ is estimated by subtracting from the estimate of the correlation function $R_{\mathbf{x}}(\tau)$ the square of the estimate $\hat{\mathbf{m}}$ of the mean value $m = E[\mathbf{x}(t)]$ as determined by the method of part A.

C. Ergodicity

The methods of parts A and B for estimating the mean and the correlation function of a stationary random process utilize a single realization or sample function of the process. In effect, they assume that temporal averages over that realization will, except for residual fluctuations due to the finite effective length T_{int} of the sample, be equivalent to averages over the ensemble of processes with respect to which the mean and the correlation function are defined as in (5.16) and (5.18). The longer the averaging time T_{int}, the smaller the residual fluctuations. It is thus expected that in terms of ordinary temporal averages of a single realization of $\mathbf{x}(t)$,

$$E[\mathbf{x}(t)] = m = \lim_{T\to\infty} \frac{1}{T} \int_{-T/2}^{T/2} \mathbf{x}(t)\, dt \tag{5.230}$$

and

$$E[\mathbf{x}(t)\mathbf{x}(t - \tau)] = R_{\mathbf{x}}(\tau) = \lim_{T\to\infty} \frac{1}{T} \int_{-T/2}^{T/2} \mathbf{x}(t)\mathbf{x}(t - \tau)\, dt. \tag{5.231}$$

Stochastic processes for which ensemble averages can be replaced by temporal averages over a single realization are said to be *ergodic*.

We have seen in part A that the constant process $\mathbf{x}(t) \equiv \mathbf{x}$, $-\infty < t < \infty$, is not ergodic; the temporal average in (5.230) will equal whatever constant value $\mathbf{x}$ the sample function bears, and depending on its probability density function $f_{\mathbf{x}}(x)$, that value may differ widely from the ensemble average

$$m = E[\mathbf{x}(t)] = \int_{-\infty}^{\infty} x f_{\mathbf{x}}(x, t)\, dx.$$

For the sinusoidal process in Examples 5.3 and 5.7 the temporal average in (5.231) is

$$\begin{aligned}
&\lim_{T\to\infty} \frac{1}{T} \int_{-T/2}^{T/2} \mathbf{A}^2 \cos(\omega t + \psi) \cos(\omega t - \omega\tau + \psi)\, dt \\
&= \lim_{T\to\infty} \frac{\mathbf{A}^2}{2T} \int_{-T/2}^{T/2} [\cos \omega\tau + \cos(2\omega t - \omega\tau + 2\psi)]\, dt \\
&= \mathbf{A}^2 \cos \omega\tau,
\end{aligned}$$

and it depends on the amplitude $\mathbf{A}$ of the particular sample function utilized; the correlation function—provided that $\mathbf{A}$ and $\boldsymbol{\psi}$ are independent and $\boldsymbol{\psi}$ is uniformly distributed over $(0, 2\pi)$—is the ensemble average

$$R_{\mathbf{x}}(\tau) = E(\mathbf{A}^2) \cos \omega\tau$$

and may differ considerably from this temporal average. The sinusoidal process $\mathbf{A} \cos(\omega t + \boldsymbol{\psi})$ is also not ergodic. Processes like these that depend on only a finite number of random parameters are usually not ergodic.

If a stochastic process is to be ergodic, a single realization must in the course of time take on configurations closely resembling the entire ensemble of processes. Something new and unpredictable must at least potentially happen at every instant, perhaps only a change in some derivative of $\mathbf{x}(t)$ of high order. Stationary filtered white noise and time-invariant nonlinear transformations thereof can for the most part be taken to be ergodic. Indeed, the assumption that a stationary physical process is ergodic is essential to almost any mathematical model fitting it into the framework of the probabilistic theory of stochastic processes.

D. Estimating the Spectral Density of a Stationary Process

The spectral density of a stationary stochastic process can of course be estimated by first estimating the autocovariance function $C_{\mathbf{x}}(\tau)$ as in part B at a sufficient number of delays τ and then taking the Fourier transform of the result numerically. Building variable delay lines may be more difficult than building narrowband pass filters, however, and an approach based on the interpretation in (5.138)–(5.140) is often more feasible.

In the spectrum analyzer depicted in Fig. 5.25 the process $\mathbf{x}(t)$ passes through a narrowband pass filter (NBF) tuned to a pass frequency Ω, whose output $\mathbf{y}(t)$ is squared by a rectifier and integrated by a low-pass filter of impulse response $h(s)$. The transfer function of the narrowband pass filter is, as shown in Fig. 5.26(a),

$$K_\Omega(j\omega) = G(\omega - \Omega) + G^*(-\omega - \Omega), \tag{5.232}$$

in which the function $G(\omega - \Omega)$ is peaked at $\omega = \Omega$ and has a width W in frequency much smaller than $\Omega/2\pi$. The impulse response of this filter is

$$\begin{aligned} k_\Omega(s) &= \int_{-\infty}^{\infty} K_\Omega(j\omega)\, e^{j\omega s} \frac{d\omega}{2\pi} \\ &= \int_{-\infty}^{\infty} G(\omega) e^{j(\omega+\Omega)s} \frac{d\omega}{2\pi} + \int_{-\infty}^{\infty} G^*(\omega) e^{-j(\omega+\Omega)s} \frac{d\omega}{2\pi} \\ &= 2 \operatorname{Re}\,[g(s)e^{j\Omega s}], \end{aligned} \tag{5.233}$$

where

$$g(s) = \int_{-\infty}^{\infty} G(\omega) e^{j\omega s} \frac{d\omega}{2\pi}, \tag{5.234}$$

generally complex, is the Fourier transform of the "low-frequency" function $G(\omega)$, which is peaked at $\omega = 0$. The duration of the *envelope* $|g(s)|$ is long enough to contain many cycles of $\cos \Omega s$ and $\sin \Omega s$, and the impulse response $k_\Omega(s)$ looks somewhat like the sketch in Fig. 5.26(b).

At any time t the output of the spectrum analyzer is

$$\mathbf{z}(t) = \int_0^{\infty} h(s)[\mathbf{y}(t - s)]^2\, ds, \tag{5.235}$$

where $h(s)$, normalized as in (5.217), is the impulse response of the integrating filter and

$$\mathbf{y}(t) = \int_0^{\infty} k_\Omega(s)\mathbf{x}(t - s)\, ds$$

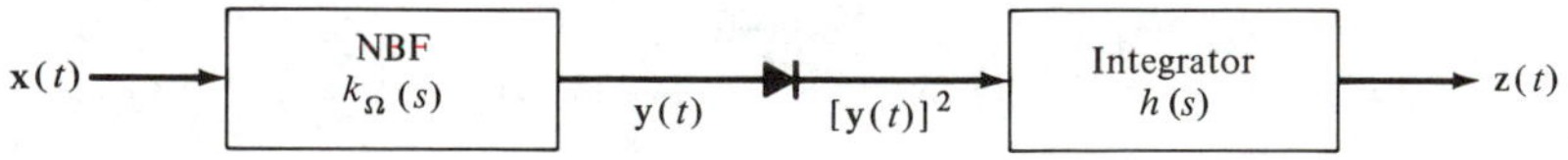

FIGURE 5.25 Spectrum analyzer.

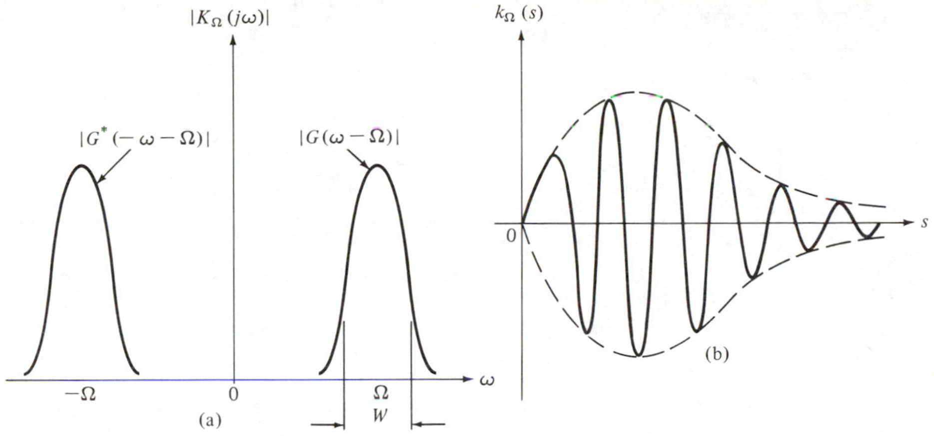

FIGURE 5.26 The narrowband pass filter (NBF)

is the output of the narrowband pass filter. Again the function of the integrating filter is to smooth the inevitable fluctuations of $[\mathbf{y}(t)]^2$.

The expected value of the output $\mathbf{z}(t)$ of the spectrum analyzer is

$$
\begin{aligned}
E[\mathbf{z}(t)] &= E[\mathbf{y}(t)]^2 = \int_{-\infty}^{\infty} |K_\Omega(j\omega)|^2 S_{\mathbf{x}}(\omega)\,\frac{d\omega}{2\pi} \\
&= \int_{-\infty}^{\infty} |G(\omega - \Omega) + G^*(-\omega - \Omega)|^2 S_{\mathbf{x}}(\omega)\,\frac{d\omega}{2\pi} \qquad (5.236) \\
&\simeq 2\int_{-\infty}^{\infty} |G(u)|^2 S_{\mathbf{x}}(u + \Omega)\,\frac{du}{2\pi}
\end{aligned}
$$

if we assume that $K_\Omega(0) = 0$ and neglect the cross-product

$$G(\omega - \Omega)G(-\omega - \Omega)$$

and its complex conjugate appearing when we form the absolute square of $K_\Omega(j\omega)$. The expected value $E[\mathbf{z}(t)]$ of the output is therefore proportional to a weighted average of the spectral density $S_{\mathbf{x}}(\omega)$, which is, so to speak, "smeared" over a neighborhood of the frequency Ω. The narrower the pass filter $K_\Omega(j\omega)$ of Fig. 5.26(a), the more nearly the expected value $E[\mathbf{z}(t)]$ is proportional to the value of the spectral density of the input $\mathbf{x}(t)$ at the pass frequency Ω;

$$E[\mathbf{z}(t)] \simeq 2S_{\mathbf{x}}(\Omega)\int_{-\infty}^{\infty} |G(u)|^2\,\frac{du}{2\pi}, \qquad (5.237)$$

provided that the spectral density $S_{\mathbf{x}}(u + \Omega)$ in (5.237) is nearly constant over the range of values of u in which $|G(u)|^2$ differs significantly from zero. By applying the output $\mathbf{z}(t)$ to a meter and tuning the pass filter over a range of frequencies Ω, the estimate $\mathbf{z}(t) = c\hat{\mathbf{S}}_{\mathbf{x}}(\Omega)$, c an instrumental constant, can be plotted versus the pass frequency Ω.

The arrangement shown in Fig. 5.27 is more convenient when the spectral density being measured is concentrated in the neighborhood of some high an-

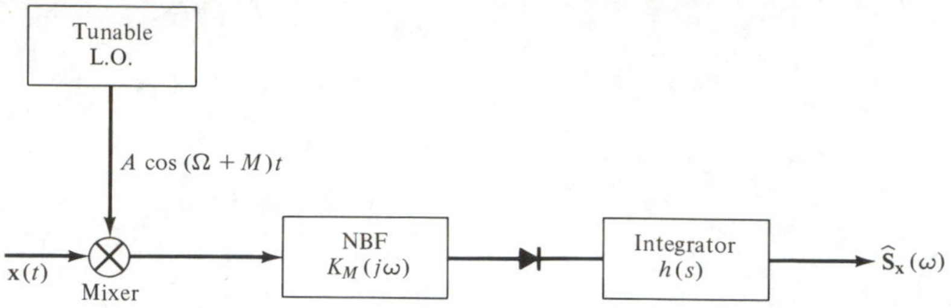

FIGURE 5.27 Heterodyne spectrum analyzer.

gular frequency U. The output $A \cos (\Omega + M)t$ of a tunable local oscillator is multiplied by the input $\mathbf{x}(t)$ in a mixer. A component of frequency ω in the input generates components of frequencies $\Omega + M + \omega$ and $\Omega + M - \omega$ in the output of the mixer, which is fed to a narrowband pass filter of fixed pass frequency M and bandwidth W. The local oscillator is tunable over a band of frequencies about $U + M$ whose width equals or exceeds that of the spectral density $S_{\mathbf{x}}(\Omega)$ under measurement. The components of frequency $\Omega + M + \omega$ from the mixer fall outside the passband of the narrowband filter. The components of frequency $\Omega + M - \omega$ that get through are rectified and integrated to yield an estimate $\hat{\mathbf{S}}_{\mathbf{x}}(\Omega)$ of the spectral density of the input at angular frequency Ω.

This spectrum analyzer much resembles a heterodyne AM radio receiver. Indeed, if you disconnected the speaker of an AM receiver and fed the output current instead to a sluggish ammeter, the meter reading as you tuned across the dial would trace out an estimate of the spectral density of the radio waves being picked up by the antenna over a range of frequencies from about 540 kHz to about 1600 kHz, but the estimate would be smeared, as in (5.236), over a range of frequencies approximately 10 kHz in width. The performance of the

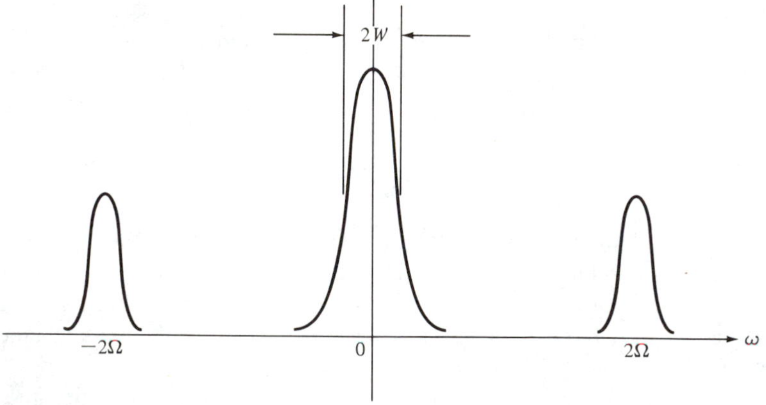

FIGURE 5.28 Spectral density of rectifier output.

spectrum analyzer in Fig. 5.27 is much like that of the one in Fig. 5.25, but for simplicity we return to the one in Fig. 5.25 for further analysis.

The spectral density of the output $\mathbf{y}(t)$ of the narrowband pass filter in Fig. 5.25 is, for $\omega > 0$,

$$S_{\mathbf{y}}(\omega) = |G(\omega - \Omega) + G^*(-\omega - \Omega)|^2 S_{\mathbf{x}}(\omega)$$
$$\simeq |G(\omega - \Omega)|^2 S_{\mathbf{x}}(\Omega)$$

if we again neglect the cross-product terms and assume that the bandwidth of the filter is so small that the spectral density $S_{\mathbf{x}}(\omega)$ is nearly constant over the region of frequencies where $|G(\omega - \Omega)|^2$ is significant. Similarly

$$S_{\mathbf{y}}(\omega) \simeq |G(-\omega - \Omega)|^2 S_{\mathbf{x}}(\Omega), \qquad \omega < 0.$$

Again assuming that the input $\mathbf{x}(t)$ is a Gaussian stochastic process, we can use (5.228), with $\tau = 0$, and with $S_{\mathbf{y}}(\omega)$ replacing $S_{\mathbf{x}}(\omega)$ there, to show that the spectral density $S_{\mathbf{r}}(\omega)$ of the output $\mathbf{r}(t) = [\mathbf{y}(t)]^2$ of the rectifier will be

$$\begin{aligned} S_{\mathbf{r}}(\omega) &= 2\int_{-\infty}^{\infty} S_{\mathbf{y}}(u - \tfrac{1}{2}\omega) S_{\mathbf{y}}(u + \tfrac{1}{2}\omega)\,\frac{du}{2\pi} \\ &\simeq 2[S_{\mathbf{x}}(\Omega)]^2 \int_{-\infty}^{\infty} [|G(u - \tfrac{1}{2}\omega - \Omega)|^2 |G(u + \tfrac{1}{2}\omega - \Omega)|^2 \\ &\quad + |G(-u + \tfrac{1}{2}\omega - \Omega)|^2 |G(-u - \tfrac{1}{2}\omega - \Omega)|^2]\,\frac{du}{2\pi} \\ &= 4|S_{\mathbf{x}}(\Omega)|^2 \int_{-\infty}^{\infty} |G(v)|^2 |G(v + \omega)|^2\,\frac{dv}{2\pi} \end{aligned} \qquad (5.238)$$

for frequencies ω within the passband of the low-pass integrating filter $h(s)$, whose bandwidth we assume much smaller than the frequency $\Omega/2\pi$ at which the spectral density $S_{\mathbf{x}}(\omega)$ is being estimated. The output of the rectifier will also contain components of frequencies in the neighborhood of $\pm 2\Omega$, but these are extinguished by the integrating filter. Figure 5.28 shows the entire spectral density $S_{\mathbf{r}}(\omega)$. Only the central peak is of concern here.

The central peak of the spectral density $S_{\mathbf{r}}(\omega)$ of the output of the rectifier is seen from (5.238) to be proportional to the self-convolution of the function $|G(v)|^2$, and its width will therefore be roughly twice the bandwidth W of the narrowband pass filter $k_{\Omega}(s)$ in Fig. 5.25. The fluctuations in $\mathbf{r}(t) = [\mathbf{y}(t)]^2$ thus have a bandwidth of the order of $2W$ and a correlation time τ_c of the order of $(2W)^{-1}$. In order to reduce them to a small fraction of their root-mean-square value, as we saw in Example 5.11 and in part A of this section, the integration time T_{int} of the low-pass filter $h(s)$ in Fig. 5.25 must be much longer than $\tau_c \sim (2W)^{-1}$, that is,

$$2WT_{\text{int}} \gg 1.$$

Now W is the width over which the spectral density $S_{\mathbf{x}}(\omega)$ is smeared in (5.236), which gives the expected value of the output of the spectrum analyzer. The more precise an estimate we desire of $S_{\mathbf{x}}(\omega)$, the smaller we must take the bandwidth W of the narrowband pass filter $k_\Omega(s)$ and—a fortiori—the longer must be the effective integration time T_{int} of the filter $h(s)$ that smoothes the fluctuations of the output $[\mathbf{y}(t)]^2$ of the rectifier. If we need to resolve fine structure such as a local peak in the spectral density $S_\mathbf{x}(\omega)$ of the input, and if that peak has a frequency width of the order of W_0, we must take

$$T_{\text{int}}^{-1} \ll W \ll W_0.$$

Unless $T_{\text{int}} >> W^{-1} >> W_0^{-1}$, the fluctuations in the output $\mathbf{z}(t)$ of the spectrum analyzer due to the random nature of the input $\mathbf{x}(t)$ may mask the fine structure in its spectral density $S_\mathbf{x}(\omega)$ that we are trying to measure.

Indeed, from (5.238) and (5.237), the ratio of the mean-square fluctuation of the output to the square of its expected value is

$$\frac{\operatorname{Var} \mathbf{z}(t)}{\{E[\mathbf{z}(t)]\}^2} \simeq \frac{\displaystyle\int_{-\infty}^{\infty}\int_{-\infty}^{\infty} \left|\frac{H(j\omega)}{H(0)}\right|^2 |G(v)|^2|G(v+\omega)|^2 \frac{d\omega\, dv}{(2\pi)^2}}{\left[\displaystyle\int_{-\infty}^{\infty} |G(v)|^2 \frac{dv}{2\pi}\right]^2} \tag{5.239}$$

and when $T_{\text{int}} >> W^{-1}$ the transfer function $H(j\omega)$ of the integrating filter is so narrowly peaked about $\omega = 0$ that this can be written approximately as

$$\begin{aligned}\frac{\operatorname{Var} \mathbf{z}(t)}{\{E[\mathbf{z}(t)]\}^2} &\simeq \int_{-\infty}^{\infty} \left|\frac{H(j\omega)}{H(0)}\right|^2 \frac{d\omega}{2\pi} \int_{-\infty}^{\infty} |G(v)|^4 \frac{dv}{2\pi} \Bigg/ \left[\int_{-\infty}^{\infty} |G(v)|^2 \frac{dv}{2\pi}\right]^2 \\ &= (WT_{\text{int}})^{-1}\end{aligned}$$

when the integration time T_{int} is defined as in (5.222) and the bandwidth W of the narrowband pass filter is defined by

$$W = \left[\int_{-\infty}^{\infty} |G(v)|^2 \frac{dv}{2\pi}\right]^2 \Bigg/ \int_{-\infty}^{\infty} |G(v)|^4 \frac{dv}{2\pi}. \tag{5.240}$$

The root-mean-square relative fluctuation at the output of the spectrum analyzer is therefore roughly proportional to $(WT_{\text{int}})^{-1/2}$ when $WT_{\text{int}} >> 1$.

E. Digital Spectral Estimation

The spectral density of a stationary process $\mathbf{x}(t)$ can be estimated digitally if the process varies slowly enough so that it can be sampled at regular intervals and so that the samples can be converted from analog to digital form and fed to a general- or special-purpose digital computer. The method is based on the interpretation (5.142) of the spectral density. A sample function $\mathbf{x}(t)$ of duration T is written as a Fourier series over $(0, T)$,

$$\mathbf{x}(t) = \sum_{m=-\infty}^{\infty} \mathbf{c}_m \exp(j\omega_m t), \qquad \omega_m = \frac{2\pi m}{T}, \quad 0 < t < T, \tag{5.241}$$

in which the real and imaginary parts of the coefficients

$$\mathbf{c}_m = \mathbf{c}_{mx} + j\mathbf{c}_{my} = \frac{1}{T}\int_0^T \mathbf{x}(t)\exp(-j\omega_m t)\,dt \tag{5.242}$$

are random variables subject to the condition

$$\mathbf{c}_m^* = \mathbf{c}_m, \qquad \mathbf{c}_{mx} = \mathbf{c}_{-m,x}, \qquad \mathbf{c}_{my} = -\mathbf{c}_{-m,y}$$

imposed by the fact that the process $\mathbf{x}(t)$ is real.

According to (5.142),

$$TE|\mathbf{c}_m|^2 = TE(\mathbf{c}_{mx}^2 + \mathbf{c}_{my}^2) \simeq S_{\mathbf{x}}(\omega_m)$$

when the duration T of the sample is much greater than the correlation time τ_c of the input $\mathbf{x}(t)$. We can therefore take

$$\hat{\mathbf{S}}_m = T|\mathbf{c}_m|^2 = T^{-1}\left|\int_0^T \mathbf{x}(t)\exp(-j\omega_m t)\,dt\right|^2 \tag{5.243}$$

as an estimate of the spectral density $S_{\mathbf{x}}(\omega)$ at angular frequency $\omega = \omega_m = 2\pi m/T$. This is called the *periodogram estimate*.

By an analysis like that in part D it can be shown that the variance of this estimate $\hat{\mathbf{S}}_m$ is of the order of $[S_{\mathbf{x}}(\omega_m)]^2$ when $\mathbf{x}(t)$ is a Gaussian random process. Furthermore,

$$\lim_{T\to\infty} TE(\mathbf{c}_n\mathbf{c}_m^*) = 0, \tag{5.244}$$

so that when a very long segment $(0, T)$ of the process $\mathbf{z}(t)$ is utilized, $T >> \tau_c$, the Fourier coefficients $\mathbf{c}_m$ are nearly uncorrelated. As a result, when one plots the periodogram estimate $\hat{\mathbf{S}}_m$ as a function of $\omega_m = 2\pi m/T$, it is found to exhibit a most erratic appearance, with dense, random oscillations whose root-mean-square magnitude in the neighborhood of any frequency ω_m is of the order of the expected value $E(\hat{\mathbf{S}}_m) \simeq S_{\mathbf{x}}(\omega_m)$. Some way of smoothing these fluctuations must be found.

One technique is to calculate the estimate $\hat{\mathbf{S}}_m$ in (5.243) for a large number N of independent realizations or sample functions of the process and to average the results at each frequency $\omega_m = 2\pi m/T$. The variance of the estimate will then be reduced by a factor N^{-1}, but one has utilized a sample of $\mathbf{x}(t)$ of total length NT.

Alternatively, one can smooth the fluctuations in the periodogram estimate by averaging over neighboring values, forming a weighted sum such as

$$\hat{\mathbf{S}}_m' = \sum_{k=-\infty}^{\infty} f_k \hat{\mathbf{S}}_{m-k}. \tag{5.245}$$

The sequence of coefficients f_k that are convolved with the periodogram values $\hat{\mathbf{S}}_m$ makes up what is called a *window function*. A typical one might look like that sketched in Fig. 5.29. Let its width in the angular frequency domain be $2\pi W_s$, so that $|f_k| << f_0$ for $k >> W_sT$. The expected value of the smoothed periodogram estimate at frequency ω_m is then

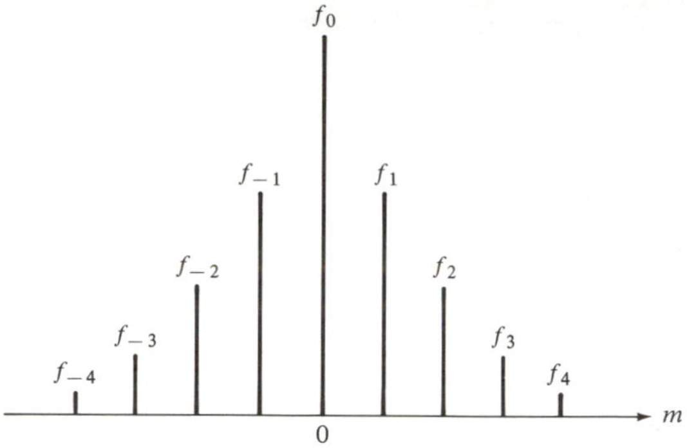

FIGURE 5.29 Window function.

$$E(\hat{\mathbf{S}}_m') \simeq \sum_{k=-\infty}^{\infty} f_k S_{\mathbf{x}}(\omega_{m-k}), \tag{5.246}$$

and it is a smeared version of the true spectral density $\mathbf{S}_x(\omega)$ over a neighborhood of ω_m whose width is of the order of $2\pi W_s$. One has accepted reduced resolution in exchange for a reduced error variance. The mean-square fluctuations can be shown to have been reduced to a relative magnitude of the order of

$$\frac{\text{Var}\ \hat{\mathbf{S}}_m'}{[S_{\mathbf{x}}(\omega_m)]^2} \sim \frac{1}{W_s T}. \tag{5.247}$$

What is happening in this procedure is much like what happens in the spectrum analyzer of part D. The length T of the sample in (5.241) corresponds to the integration time T_{int} of the averaging filter $h(s)$ in Fig. 5.25, for that integration time determines the duration of the input $\mathbf{x}(t)$ that effectively contributes to the output $\mathbf{z}(t)$ observed at a particular time t. The window sequence f_k corresponds to the function $G(\omega)$ appearing in the transfer function $K_\Omega(\omega)$ of the narrowband pass filter (NBF) in Fig. 5.25, which determines over what range of frequencies about Ω the spectrum analyzer, as indicated by (5.236), effectively smoothes the empirical spectral density. As in part D, for good resolution of fine details of the spectral density $S_{\mathbf{x}}(\omega)$ the bandwidth W_s of the window function must be rather smaller than the expected width W_0 of those details; and because in order to reduce the fluctuations in the estimate, the number $W_s T$ must be much greater than 1, a sample function of $\mathbf{x}(t)$ of total duration T much greater than W_0^{-1} must be utilized,

$$T \gg W_s^{-1} \gg W_0^{-1}. \tag{5.248}$$

The Fourier integral for $\mathbf{c}_m$ appearing in (5.242) is in practice evaluated by sampling the input $\mathbf{x}(t)$ at intervals of length δ and applying the fast Fourier transform routine to evaluating the sums

$$\mathbf{c}_m = \frac{1}{T}\int_0^T \mathbf{x}(t)e^{-j\omega_m t}\,dt = M^{-1}\sum_{k=0}^{M-1} \mathbf{x}(k\delta)\exp\left(-\frac{2\pi jkm}{M}\right), \tag{5.249}$$

where $M = T/\delta$ is the number of samples $\mathbf{x}(k\delta)$. Thus the periodogram estimate is replaced by the digital estimate

$$\hat{\mathbf{S}}_m^{(d)} = TM^{-2} \left| \sum_{k=0}^{M-1} \mathbf{x}(k\delta) \exp\left(-\frac{2\pi jkm}{M} \right) \right|^2 .$$

This is then transformed back to the "time" domain by the inverse discrete Fourier transform,

$$\hat{\mathbf{C}}_k = \frac{1}{T} \sum_{m=0}^{M-1} \hat{\mathbf{S}}_m^{(d)} \exp\left(\frac{2\pi jmk}{M} \right) \tag{5.250}$$

to yield estimates of values of the autocovariance function $C_{\mathbf{x}}(\tau)$ at discrete values $k\delta$ of the delay variable τ.

In this time domain the discrete convolution in (5.245) becomes a simple multiplication of the samples $\hat{\mathbf{C}}_k$ by the discrete Fourier transform ϕ_k of the window sequence $\{f_k\}$. Typical such window sequences in the time domain are the triangle

$$\phi_m = \begin{cases} 1 - \dfrac{|m|}{M'}, & -M' < m < M', \\ 0 & |m| > M', \end{cases} \tag{5.251}$$

with $M' = M/W_sT$, and the "raised cosine"

$$\phi_m = \begin{cases} \dfrac{1}{2}\left[1 + \cos\left(\dfrac{\pi m}{M'} \right) \right] = \cos^2\left(\dfrac{\pi m}{2M'} \right), & |m| < M', \\ 0, & |m| > M'. \end{cases} \tag{5.252}$$

One then transforms the product sequence $\phi_m \hat{\mathbf{C}}_m$ back to the frequency domain by yet another fast Fourier transform in order to obtain the digital equivalent of the smoothed periodogram estimate (5.245).

The resulting estimate of the spectral density suffers from *aliasing*: It is a periodic function of the frequency $\omega_m = 2\pi m/T$ with period $2\pi M/T = 2\pi/\delta$. In order for the periodically repeated copies of the estimated spectral density to be cleanly separated, the sampling interval δ must be so short that $2\pi/\delta$ exceeds the total width of the spectral density $S_{\mathbf{x}}(\omega)$ being estimated. This criterion in conjunction with (5.248) determines the number M of samples that must be processed by the digital computer. Numerical analysts and workers in the field of digital signal processing have developed various rules of thumb for selecting good window sequences and numerous artifices for conducting these computations efficiently. Details are to be found in books on this subject.*

*R. B. Blackman and J. W. Tukey, *The Measurement of Power Spectra*. New York: Dover, 1958. G. M. Jenkins and D. G. Watts, *Spectral Analysis and Its Applications*. San Francisco: Holden-Day, 1968. A. V. Oppenheim and R. W. Schafer, *Digital Signal Processing*. Englewood Cliffs, N.J.: Prentice-Hall, 1975 (see Chap. 11, pp. 532–574).

PROBLEMS

5-1 The random process $\mathbf{x}(t) = e^{-\boldsymbol{\mu} t}$, $t > 0$, depends on a random parameter $\boldsymbol{\mu}$ whose probability density function is $f_{\boldsymbol{\mu}}(\mu)$, in terms of which calculate the first- and second-order probability density functions $f_{\mathbf{x}}(x, t)$ and $f_{\mathbf{x}}(x_1, t_1; x_2, t_2)$ for this process.

5-2 Suppose that the random variables $\mathbf{A}$ and $\boldsymbol{\psi}$ in Example 5.3 are statistically independent, that the amplitude $\mathbf{A}$ has a Rayleigh distribution,

$$f_{\mathbf{A}}(A) = \left(\frac{A}{s^2}\right) \exp\left(-\frac{A^2}{2s^2}\right) U(A),$$

and that the phase $\boldsymbol{\psi}$ is uniformly distributed over $(0, 2\pi)$. Calculate the first-, second-, and third-order probability density functions of the stochastic process

$$\mathbf{x}(t) = \mathbf{A}\cos(\omega t + \boldsymbol{\psi}).$$

[*Hint:* Introduce new random variables

$$\mathbf{u} = \mathbf{A}\cos\boldsymbol{\psi}, \qquad \mathbf{v} = \mathbf{A}\sin\boldsymbol{\psi},$$

and determine their joint density function and the transformation from $(\mathbf{u}, \mathbf{v})$ to $(\mathbf{x}_1, \mathbf{x}_2)$.]

5-3 For the random process of Problem 5-1, calculate the mean value $E[\mathbf{x}(t)]$ and the autocovariance function $C_{\mathbf{x}}(t_1, t_2)$, assuming that the random variable $\boldsymbol{\mu}$ is uniformly distributed over $(0, M)$.

5-4 The independent random variables $\boldsymbol{\xi}$ and $\boldsymbol{\eta}$ have expected value 0 and the variance σ^2. Find the autocovariance function of the random process

$$\mathbf{x}(t) = \boldsymbol{\xi}\cos\omega t + \boldsymbol{\eta}\sin\omega t$$

for fixed ω. Note that $\boldsymbol{\xi}$ and $\boldsymbol{\eta}$, once chosen in a chance experiment, are fixed for all time.

5-5 Calculate the autocovariance function $C_{\mathbf{x}}(t_1, t_2)$ of the stochastic process $\mathbf{x}(t)$ depicted in Fig. 5.2 and described in Example 5.4, in terms of the functions $f_0(t)$ and $f_1(t)$. Define the new stochastic process $\mathbf{y}(t) = \mathbf{x}(t - \mathbf{d})$, where $\mathbf{d}$ is uniformly distributed over $(0, T)$. Calculate its autocovariance function $C_{\mathbf{y}}(t_1, t_2)$.

5-6 A stochastic process "with independent increments" is a process $\mathbf{x}(t)$ such that for all times $t_1 < t_2 < t_3$ the change $\mathbf{x}(t_3) - \mathbf{x}(t_2)$ is independent of the change $\mathbf{x}(t_2) - \mathbf{x}(t_1)$. Determine its autocovariance function $C_{\mathbf{x}}(t_1, t_2)$ in terms of Var $\mathbf{x}(t_1)$ for $t_2 > t_1$, assuming that $\mathbf{x}(0) = 0$; $0 < t_1 < t_2$.

5-7 The process $\mathbf{x}(t)$ consists of jumps of independently random amplitudes $\mathbf{A}_1, \mathbf{A}_2, \ldots, \mathbf{A}_k, \ldots$ at times $\boldsymbol{\tau}_1, \boldsymbol{\tau}_2, \ldots, \boldsymbol{\tau}_k, \ldots$ that form a Poisson point process with rate λ. The process $\mathbf{x}(t)$ starts at 0 at $t = 0$. A typical sample function is shown in Fig. 5P-7.

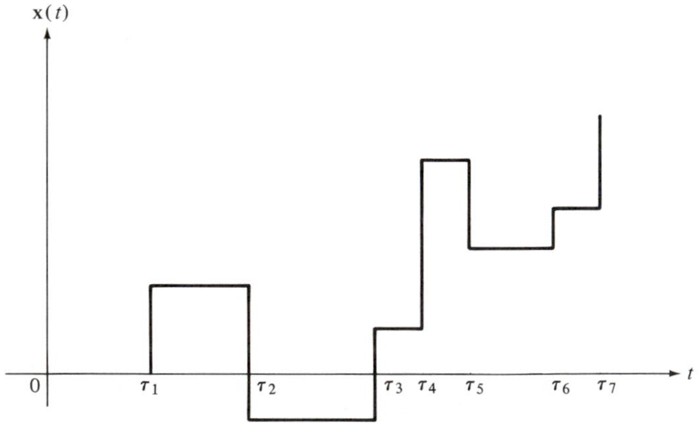

FIGURE 5P-7 Process with random jumps.

In terms of the probability density function $f_{\mathbf{A}}(A)$ of the amplitudes of the jumps, calculate the expected value $E[\mathbf{x}(t)]$ and the covariance function $C_{\mathbf{x}}(t_1, t_2)$ of this process. Under what conditions does this process become a Brownian process? (*Hint:* Use Problem 5-6 and Sec. 4-8.)

5-8 Find the probability density function of the kth "switching time" $\boldsymbol{\tau}_k$ in a Poisson point process with constant rate as described in Sec. 5-3. This is known as the Erlang distribution. Find the expected value and the variance of this kth switching time $\boldsymbol{\tau}_k$. [*Hint:* The event $(t < \boldsymbol{\tau}_k < t + dt)$ is the same as the event "$k - 1$ points $\boldsymbol{\tau}_i$ occurred in $(0, t)$ and one point occurred in $(t, t + dt)$."]

5-9 The input $\mathbf{x}(t)$ to a rectifier is a stationary Gaussian stochastic process with mean zero and autocovariance function $C_{\mathbf{x}}(\tau)$. The output of the rectifier, which is memoryless, is

$$\mathbf{y}(t) = A\{\exp [\mu \mathbf{x}(t)] - 1\}.$$

Find the autocovariance function of the output. (*Hint:* Note the resemblance between the integral you must evaluate and that for the joint characteristic function of a pair of Gaussian random variables. Most of the work on this problem has already been done.)

5-10 The spectral density of a stationary random process $\mathbf{x}(t)$ with mean zero is given by

$$S_{\mathbf{x}}(\omega) = \begin{cases} a^2 - \omega^2, & |\omega| < a, \\ 0, & |\omega| > a. \end{cases}$$

Find Var $\mathbf{x}(t)$.

5-11 Show that the first-order characteristic function of the stochastic process treated in Sec. 5-3E is given by

$$\Phi_{\mathbf{x}}(\omega) = \exp \left\{ \lambda \int_{-\infty}^{\infty} [\Phi_{\mathbf{A}}(\omega h(u)) - 1]\, du \right\}$$

in terms of the characteristic function $\Phi_{\mathbf{A}}(\omega) = E(e^{j\omega \mathbf{A}})$ of the independently random amplitudes $\mathbf{A}$ of the random pulses. [*Hint:* In the exponents in (5.64) replace $\boldsymbol{\varepsilon}_m$ by $\boldsymbol{\varepsilon}_m \mathbf{A}_m$. In what corresponds to the first line of (5.65), average over the distribution of $\mathbf{A}_m$, introducing $\Phi_{\mathbf{A}}(\cdot)$ at this point.]

5-12 The random process $\mathbf{x}(t)$ consists of a signal $\cos \boldsymbol{\omega} t$ with a random angular frequency $\boldsymbol{\omega}$ that is uniformly distributed over $\Omega - \frac{1}{2}\Delta < \boldsymbol{\omega} < \Omega + \frac{1}{2}\Delta$. Find the expected value and the autocovariance function of the process $\mathbf{x}(t)$.

5-13 The input $\mathbf{x}(t)$ to the perfect integrator of Example 5.11 is a stationary random process with expected value zero and autocovariance function

$$C_{\mathbf{x}}(\tau) = \begin{cases} 1 - \dfrac{|\tau|}{S}, & -S < \tau < S, \\ 0, & |\tau| > S. \end{cases}$$

Find the variance Var $\mathbf{y}(t)$ of the output, assuming that $T < S$.

5-14 The stochastic process $\mathbf{x}(t)$ is stationary and Gaussian with mean value zero, autocovariance function $C_{\mathbf{x}}(\tau)$, and spectral density $S_{\mathbf{x}}(\omega)$. In terms of $C_{\mathbf{x}}(\tau)$ find the autocovariance function $C_{\mathbf{y}}(\tau)$ of the output $\mathbf{y}(t) = [\mathbf{x}(t)]^2$ of a quadratic rectifier to which $\mathbf{x}(t)$ is applied. In terms of $S_{\mathbf{x}}(\omega)$ find the spectral density $S_{\mathbf{y}}(\omega)$ of $\mathbf{y}(t)$. [*Hint:* Use (4.147).]

5-15 The spectral density function of a stationary random process $\mathbf{x}(t)$ is

$$S_{\mathbf{x}}(\omega) = \begin{cases} 1 - \dfrac{|\omega|}{W}, & -W < \omega < W, \\ 0, & |\omega| > W. \end{cases}$$

Find the autocovariance function of this process.

5-16 A linear system with input $\mathbf{x}(t)$ and output $\mathbf{y}(t)$ obeys the linear differential equation

$$\frac{d^2\mathbf{y}}{dt^2} + 3\frac{d\mathbf{y}}{dt} + 2\mathbf{y} = \mathbf{x}(t), \qquad t \geq 0.$$

The input $\mathbf{x}(t)$ is white noise with spectral density $S_{\mathbf{x}}(\omega) \equiv 1$ and mean zero, which is turned on at time $t = 0$, at which time

$$\mathbf{y} = 2, \qquad \frac{d\mathbf{y}}{dt} = 1.$$

Find the expected value $E[\mathbf{y}(t)]$ and the variance Var $\mathbf{y}(t)$ of the output for $t > 0$.

5-17 The Gaussian random process $\mathbf{x}(t)$ that is the solution of the linear differential equation

$$\frac{d\mathbf{x}}{dt} + \mathbf{x}(t) = \mathbf{z}(t),$$

where $\mathbf{z}(t)$ is white noise with spectral density $S_{\mathbf{z}}(\omega) \equiv 1$, is called the Ornstein–Uhlenbeck process. Taking $\mathbf{x}(0) = x_0$, find its first- and second-order probability density functions $f_{\mathbf{x}}(x, t)$ and $f_{\mathbf{x}}(x_1, t_1; x_2, t_2)$ for $t, t_1, t_2 > 0$. Find the conditional density function $f_{\mathbf{x}_2}(x_2, t_2 \mid \mathbf{x}(t_1) = x_1)$ for $t_2 > t_1$.

5-18 A Brownian process $\mathbf{x}(t)$ that is known to start at $\mathbf{x}(0) = 0$ and to pass through 0 at time $T > 0$, [i.e., $\mathbf{x}(T) = 0$] is called the "Brownian bridge." Find the conditional probability density functions

$$f_{\mathbf{x}}(x, t \mid \mathbf{x}(0) = \mathbf{x}(T) = 0), \qquad 0 < t < T,$$

and

$$f_{\mathbf{x}}(x_1, t_1; x_2, t_2 \mid \mathbf{x}(0) = \mathbf{x}(T) = 0), \qquad 0 < t_1 < t_2 < T,$$

from the joint density functions of the original Brownian process given in Sec. 5-6D, assuming zero drift velocity. Calculate the conditional covariance

$$\text{Cov}\,\{\mathbf{x}(t_1), \mathbf{x}(t_2) \mid \mathbf{x}(0) = \mathbf{x}(T) = 0\}.$$

5-19 The Brownian process of Sec. 5-6D, instead of starting at $\mathbf{x}(0) = 0$, starts at a random point $\mathbf{x}(0) = \mathbf{x}_0$ that has a Gaussian probability density function with mean 0 and variance σ^2. For $t_2 > t_1 > 0$, calculate the first- and second-order probability density functions $f_{\mathbf{x}}(x_1, t_1)$ and $f_{\mathbf{x}}(x_1, t_1; x_2, t_2)$. Assume zero drift velocity.

5-20 The rectified Brownian process is $\mathbf{y}(t) = |\mathbf{x}(t)|$, where $\mathbf{x}(t)$ is the Brownian process of Sec. 5-6D, starting at $\mathbf{x}(0) = 0$ and having zero drift velocity. Find the first- and second-order joint density functions of $\mathbf{y}(t)$.

5-21 Write down the covariance matrix of $\mathbf{x}_1 = \mathbf{x}(t_1)$, $\mathbf{x}_2 = \mathbf{x}(t_2)$, $\mathbf{x}_3 = \mathbf{x}(t_3)$ for the Brownian process of Sec. 5-6D (zero drift velocity), form the inverse of this matrix, and show that when that is used in (4.136), the result agrees with the third-order density function in (5.175).

5-22 The random process $\mathbf{x}(t)$ of Problem 5-15 is applied as a voltage to the *RLC* network of Fig. 5P-22. Neglecting the thermal noise of the resistor, find the spectral density $S_{\mathbf{y}}(\omega)$ of the voltage $\mathbf{y}(t)$ across the inductor.

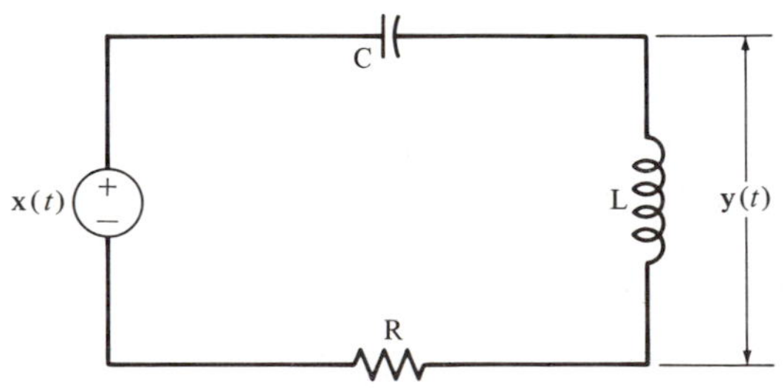

FIGURE 5P-22 RLC circuit.

5-23 In the *RLC* circuit of Fig. 5.19 determine the spectral density and the autocovariance function of the voltage $\mathbf{v}(t)$ across the capacitor. Do the same for the current $\mathbf{i}(t)$ through the inductor. Assume that the circuit is in thermal equilibrium.

5-24 For the physical Brownian motion treated in Sec. 5-7A, assume that the particle starts at the origin with zero velocity, $\mathbf{x}(0) = \mathbf{v}(0) = 0$. For $t_2 > t_1 > 0$ calculate the joint probability density function $f_{\mathbf{x}}(x_1, t_1; x_2, t_2)$ of $\mathbf{x}_1 = \mathbf{x}(t_1)$ and $\mathbf{x}_2 = \mathbf{x}(t_2)$, and the conditional density function $f_{\mathbf{x}_2}(x_2, t_2 \mid \mathbf{x}(t_1) = x_1)$. Express the latter in terms of $E[\mathbf{x}(t_2) \mid \mathbf{x}(t_1) = x_1]$ and $\text{Var}\,[\mathbf{x}(t_2) \mid \mathbf{x}(t_1) = x_1]$, and calculate these.

5-25 The circuit of Fig. 5P-25 is in thermal equilibrium at absolute temperature T. Find the spectral densities of the voltage $\mathbf{v}_c(t)$ across the capacitor and the current $\mathbf{i}(t)$ through the inductor. Indicate how you would calculate the autocovariance functions of these stochastic processes.

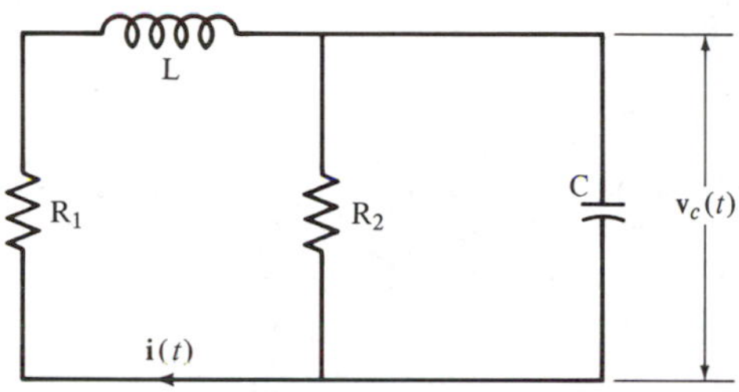

FIGURE 5P-25 Circuit for Problem 5-25.

5-26 The *RLC* circuit in Fig. 5.19 is in thermal equilibrium at absolute temperature T. Find the correlation $E[\mathbf{v}_C(t)\mathbf{v}_L(t)]$ between the voltage $\mathbf{v}_C(t)$ on the capacitor and the voltage $\mathbf{v}_L(t)$ across the inductor.

5-27 **(a)** Consider differentiation as a linear operation whose equivalent transfer function is $H(j\omega) = j\omega$. Find the spectral density of the first derivative

$$\mathbf{y}(t) = \mathbf{x}'(t) = \frac{d\mathbf{x}}{dt}$$

of a stationary random process $\mathbf{x}(t)$ in terms of the spectral density $S_{\mathbf{x}}(\omega)$ of $\mathbf{x}(t)$. From that determine the autocovariance function $C_{\mathbf{y}}(\tau)$ in terms of $C_{\mathbf{x}}(\tau)$, the autocovariance function of $\mathbf{x}(t)$. If $\mathbf{x}(t)$ is a stationary Gaussian process, what is the first-order probability density function of $\mathbf{y}(t)$?

(b) Find the spectral density of the kth derivative

$$\mathbf{x}^{(k)}(t) = \frac{d^k\mathbf{x}(t)}{dt^k}$$

of the stationary process $\mathbf{x}(t)$ in terms of $S_{\mathbf{x}}(\omega)$. This kth derivative is said to "exist" only if $\text{Var}\,\mathbf{x}^{(k)}(t) < \infty$. What condition does the existence of $\mathbf{x}^{(k)}(t)$ place on the behavior of the spectral density $S_{\mathbf{x}}(\omega)$ as $|\omega| \to \infty$?

(c) Show that $\mathbf{x} = \mathbf{x}(t)$ and $\mathbf{y} = \mathbf{x}'(t)$ for the same times t are uncorrelated random variables.

(d) Find the joint probability density function of $\mathbf{x} = \mathbf{x}(t)$ and $\mathbf{y} = \mathbf{x}'(t)$ when $\mathbf{x}(t)$ is a Gaussian stochastic process.

5-28 The stationary random process $\mathbf{x}(t)$ has the autocovariance function

$$C_{\mathbf{x}}(\tau) = 2e^{-|\tau|} - e^{-2|\tau|}.$$

What is the greatest integer k such that the kth derivative $\mathbf{x}^{(k)}(t) = d^k\mathbf{x}/dt^k$ exists in the sense of Problem 5-27(b)? Prove your answer.

5-29 In estimating the mean of a stationary process $\mathbf{x}(t)$, the integrating filter has an impulse response

$$h(s) = \mu e^{-\mu s} U(s),$$

and the autocovariance function of the input is

$$C_{\mathbf{x}}(\tau) = \sigma^2 e^{-\alpha|\tau|}.$$

Calculate the bandwidth W of the integrating filter by (5.222) and the correlation time τ_c of the input by (5.223).

5-30 In the correlator of Fig. 5.24 assume that $h(s)$ is a perfect integrator,

$$h(s) = \begin{cases} T^{-1}, & 0 < s < T, \\ 0, & s < 0, \quad s > T. \end{cases}$$

Show that the variance of the output $\mathbf{y}(t)$ is bounded by

$$\operatorname{Var} \mathbf{y}(t) \le T^{-1} \int_{-\infty}^{\infty} \left| [C_{\mathbf{x}}(u)]^2 + C_{\mathbf{x}}(\tau + u) C_{\mathbf{x}}(\tau - u) \right| du$$

when the input process $\mathbf{x}(t)$ is Gaussian with mean zero and autocovariance function $C_{\mathbf{x}}(\tau)$. Evaluate this bound for the input autocovariance function in Problem 5-29.

5-31 Show that the spectral density of the process $\mathbf{z}(t) = \mathbf{x}(t)\mathbf{x}(t - \tau)$ is as given in (5.228) when the process $\mathbf{x}(t)$ is Gaussian with mean zero. [*Hint:* Use (4.147).]

5-32 For an integrating filter with a Gaussian frequency response

$$|H(j\omega)| = \exp\left(-\tfrac{1}{4}T^2\omega^2\right)$$

and a narrowband pass filter with a transfer function as in (5.232) with

$$G(\omega) = \exp\left(-\frac{\omega^2}{4V^2}\right),$$

use (5.239) to evaluate the ratio $\operatorname{Var} \mathbf{z}(t)/\{E[\mathbf{z}(t)]^2\}$ in terms of V and T for the output of the spectrum analyzer, assuming that $S_{\mathbf{x}}(\omega)$ is constant over the region where $K_{\Omega}(\omega)$ is significant.

5-33 A high-Q simply resonant circuit has an impulse response of the approximate form

$$k_{\Omega}(s) = e^{-\mu s} \cos \Omega s \; U(s),$$

with $\mu \ll \Omega$. Assuming that the narrowband pass filter in the spectrum analyzer in Fig. 5.25 is of this form, calculate the bandwidth W defined by eq. (5.240).

5-34 Assuming that the stationary process $\mathbf{x}(t)$ is Gaussian with mean zero, evaluate Var $\hat{\mathbf{S}}_m$ for the periodogram estimate $\hat{\mathbf{S}}_m$ in (5.243) in terms of the autocovariance function $C_{\mathbf{x}}(\tau)$ of $\mathbf{x}(t)$. Then use (5.123) to express the result in terms of $S_{\mathbf{x}}(\omega)$. [*Hint:* Use (4.147).]

5-35 Define the "quasiderivative" of the Poisson counting process as

$$\mathbf{w}_T(t) = \frac{\mathbf{n}(t + T) - \mathbf{n}(t)}{T}$$

for $T > 0$. Calculate the autocovariance function $C_{\mathbf{w}}(t_1, t_2)$ of $\mathbf{w}_T(t)$ and show how it goes into that of the derivative process (Poisson impulse process) $\mathbf{z}(t) = d\mathbf{n}/dt$ when $T \to 0$. [*Hint:* Represent $\mathbf{w}_T(t)$ as a shot-noise process and use the results of Sec. 5-3C, or use (5.43).]

5-36 The random process $\mathbf{z}(t)$ is white noise of unilateral spectral density N as described in Sec. 5-6C. The functions $f_k(t)$, $k = 1, 2, \ldots$, are orthonormal over the interval $(0, T)$,

$$\int_0^T f_k(t) f_m(t)\, dt = \delta_{km} = \begin{cases} 1, & k = m, \\ 0, & k \ne m. \end{cases}$$

Calculate the joint probability density function of the n Gaussian random variables

$$\mathbf{x}_k = \int_0^T f_k(t)\,\mathbf{z}(t)\,dt,$$

$k = 1, 2, \ldots, n$.

5-37 A signal $s(t)$, $0 < t < T$, is to be detected in white noise $\mathbf{z}(t)$ of unilateral spectral density N by observing the input $\mathbf{v}(t)$ to a receiver during $(0, T)$. The receiver decides between two hypotheses about the input,

$$H_0{:}\quad \mathbf{v}(t) = \mathbf{z}(t),$$
$$H_1{:}\quad \mathbf{v}(t) = \mathbf{z}(t) + s(t).$$

The prior probabilities of the hypotheses are p_0 and p_1,

$$p_0 + p_1 = 1.$$

The receiver bases its decisions on the Gaussian random variables

$$\mathbf{v}_k = \int_0^T f_k(t)\mathbf{v}(t)\,dt, \qquad k = 1, 2, \ldots, n,$$

where the $f_k(t)$ are a set of orthonormal functions on $(0, T)$ as in Problem 5-36.

Calculate the conditional probabilities of hypotheses H_0 and H_1, given that particular values $\mathbf{v}_k = v_k$, $1 \le k \le n$, have been observed. Assume that the receiver uses Bayes's rule, as discussed in Sec. 1-11, to make its decisions. For what region R_1 of the n-dimensional space of the data $\mathbf{v}_1, \mathbf{v}_2, \ldots, \mathbf{v}_n$ does the receiver choose hypothesis H_1, that is, decide that the signal is present? On what linear combination

$$\mathbf{g} = \sum_{k=1}^{n} a_k \mathbf{v}_k$$

of the data does this receiver equivalently base its decisions, choosing H_1 when $\mathbf{g} > g_0$ and H_0 when $\mathbf{g} < g_0$ for some "decision level" g_0? Calculate the probability density functions of the random variable $\mathbf{g}$ under each hypothesis, and from them determine the probability P_e of error incurred by this receiver. Express this in terms of the energy-to-noise ratio

$$d^2 = \frac{2}{N}\int_0^T [s(t)]^2\,dt$$

in the limit $n \to \infty$. [*Hint:* Expand $s(t)$ in a series of the same orthonormal functions,

$$s(t) = \sum_{k=0}^{\infty} s_k f_k(t), \qquad 0 < t < T.$$

How are the coefficients s_k determined? Express d^2 in terms of them.]

5-38 The input to a linear filter is white noise $\mathbf{z}(t)$ of unilateral spectral density N. The impulse response of the filter is $h(\tau)$. In terms of these calculate the cross-covariance function of the input and the output of the filter, as defined in (5.34).

5-39 Assume that the spectral density at the input to the correlator of Fig. 5.24 is

$$S_{\mathbf{x}}(\omega) = S_0 \exp\left(-\frac{\omega^2}{2A^2}\right).$$

Evaluate the spectral density $S_{\mathbf{z}}(\omega)$ of $\mathbf{z}(t) = \mathbf{x}(t)\,\mathbf{x}(t-\tau)$ by (5.228). Let the integrating filter at the output of the correlator have the gain function

$$|H(j\omega)| = \exp\left(-\frac{\omega^2}{4B^2}\right).$$

Calculate the ratio $\operatorname{Var}\mathbf{y}(t)/\{E[\mathbf{y}(t)]\}^2$ for the output $\mathbf{y}(t)$ of the correlator in terms of A and B with no assumptions about their relative magnitudes.

INDEX